Lecture Notes in Physics

Edited by H. Araki, Kyoto, J. Ehlers, München, K. Hepp, Zürich
R. L. Jaffe, Cambridge, MA, R. Kippenhahn, München, D. Ruelle, Bures-sur-Yvette
H. A. Weidenmüller, Heidelberg, J. Wess, Karlsruhe and J. Zittartz, Köln
Managing Editor: W. Beiglböck

381

J. Casas-Vázquez D. Jou (Eds.)

Rheological Modelling: Thermodynamical and Statistical Approaches

Proceedings of the Meeting Held at the
Bellaterra School of Thermodynamics
Autonomous University of Barcelona
Sant Feliu de Guíxols, Catalonia, Spain
24–28 September 1990

Springer-Verlag
Berlin Heidelberg GmbH

Editors
José Casas-Vázquez
David Jou
Departament de Física, Universitat Autònoma de Barcelona
E-08193 Bellaterra, Catalonia, Spain

ISBN 978-3-662-13868-7 ISBN 978-3-540-46569-0 (eBook)
DOI 10.1007/978-3-540-46569-0

Originally published by Springer-Verlag Berlin Heidelberg New York in 1991
Softcover reprint of the hardcover 1st edition 1991

2153/3140-543210 – Printed on acid-free paper

Preface

The analysis of rheological systems provides stimulating challenges to the development of statistical mechanics and nonequilibrium thermodynamics. A sound understanding of the observed behaviour of such systems in terms of microscopic models and the use of nonequilibrium thermodynamics to restrict the possible forms of constitutive equations are goals which have been a source of inspiration in the respective fields. In nonequilibrium thermodynamics, rheological systems do not admit the simple local-equilibrium hypothesis: a number of internal variables must be considered, so the form of the entropy cannot be assumed a priori; but it can be studied by prescribing a functional dependence on a set of these internal variables. In statistical mechanics, the geometrical complexity of the macromolecules requires new techniques for the analysis of transport phenomena. Furthermore, even those fluids composed of the simplest monatomic particles exhibit several kinds of rheological behaviour when they are subjected to high frequencies or to intense thermodynamic forces, so that rheological analysis becomes an important ingredient in the study of fast and nonlinear phenomena.

These were the motivations for devoting the fifth session of the Bellaterra School of Thermodynamics (BeST) to the relation between rheological modelling, statistical mechanics and nonequilibrium thermodynamics. The aim of BeST is to gather together leading scientists in these areas, graduate students, and other researchers interested in the current developments in these fields. The school is addressed to fundamental questions, both theoretical and experimental, rather than to direct applications, To join the pedagogical features of a school with the stimulus of fresh information on recent advances is the goal pursued at each meeting of the Bellaterra School.

The lecturers invited to the School are well known for their contributions to different aspects of rheology, statistical mechanics or nonequilibrium thermodynamics. Bird provides a clear introduction to the relation between kinetic models, macroscopic rheological equations and rheometric measurements. Joseph presents an account of the mathematical and physical problems associated with hyperbolicity arising in the context of viscoelasticity. Hess studies the microscopic consequences of shear flows on the structure of liquids and compares the results obtained with kinetic theory and nonequilibrium molecular dynamics simulations. Rubí considers the role that time dependent hydrodynamic interactions play in the dynamics of polymer solutions. Grmela studies the interrelations between several levels of description of rheological systems, from the macroscopic thermodynamics to the statistical descriptions based on distribution functions, and the conditions of compatibility between such different descriptions. Carreau and Grmela discuss conformation tensor rheological models, in which distribution functions of molecular state variables are used as internal variables in the theory.

Quemada analyses the complex phenomenology concerning the viscous behaviour of concentrated suspensions, with emphasis on biological fluids. Wolf describes the changes that a shear flow produces on the phase diagram of polymer solutions. Jongschaap contributes a broad account of the problems of rheological modelling using a thermodynamic approach which may be applied to many types of models in a unified and systematic way. Finally, de Gennes deals with new approaches to the problem of adhesion, in particular various forms of the constitutive law relating the applied stress and the rate of opening of the junction width in a thin ribbon.

The reader will also find the text of some of the seminars delivered at the School, which complement the insights into rheology that may be gained from the perspectives of statistical mechanics and nonequilibrium thermodynamics. Thus, the book includes such topics as hard sphere suspensions, an EIT approach to rheology, and the formalism of fractional derivatives.

We gratefully acknowledge the financial support of the DGICyT of the Spanish Ministry of Education and Science, of the DGU and the CIRIT of the Government of Catalonia, of the Autonomous University of Barcelona, of the savings bank La Caixa, and, last but not least, of the City council of Sant Feliu de Guíxols and Murlà Park Hotel, in the Costa Brava, where this meeting was held.

J. Casas-Vázquez
D. Jou

Contents

Lectures

Viscoelastic Behavior of Polymeric Liquids 1
R.B. Bird

Experimental Evidence for Non-Newtonian Behavior of Polymeric Liquids
Rheometry and Material Functions
Continuum Mechanical Ideas and Empirical Constitutive Equations
Kinetic Theory Ideas and Dumbbell Models
Kinetic Theory Ideas and Chain Models

Problems Associated with the Elasticity of Liquids 22
D.D. Joseph

Physical Phenomena Associated with Hyperbolicity and Change of Type
Conceptual Ideas
Mathematical Theory

Rheology and Shear Induced Structure of Fluids 51
S. Hess

Introduction
Pressure Tensor, Viscosity Coefficients
The Structure of Streaming Fluids
Concluding Remarks

On the Dynamics of Polymers in Solution 74
J.M. Rubí, J. Bonet Avalos and D. Bedeaux

Introduction
Equation of Motion of a Polymer
Polymer Dynamics
Conclusions

Mesoscopic Dynamics and Thermodynamics: Applications to Polymeric Fluids 99
M. Grmela

Introduction
Hierarchy of Descriptions
Thermodynamics - Geometry of the State Space
Dynamics - Physical Foundation of Thermodynamics
Rheological Modelling
Thermodynamics and Dynamics of Driven Systems
Concluding Remarks

Conformation Tensor Rheological Models 126
P.J. Carreau and M. Grmela

Introduction
Governing Equations
Predictions and Comparison with Experiments
Concluding Remarks

Biofluids as Structured Media: Rheology and Flow Properties of Blood 158
D. Quemada

Introduction
Main Characteristics of Blood Flows
Blood as a Concentrated Dispersion
Blood as a Shear-Thinning Fluid
Blood as a Thixotropic and Viscoelastic Fluid
Modelling of Blood Microcirculation
Concluding Remarks

Phase Separation of Flowing Polymer Solutions 194
B.A. Wolf

Introduction
Procedures and Observations
Calculation of Phase Diagrams
Discussion

Towards a Unified Formulation of Microrheological Models 215
R.J.J. Jongschaap

Introduction
Theory
Applications
Discussion

Adhesion and Rheology (Abstract) 248
P.G. de Gennes

Seminars

Rheology of Hard Sphere Suspensions 250
B.U. Felderhof

Extended Irreversible Thermodynamics Versus Rheology 257
G. Lebon, D. Jou and J. Casas-Vázquez

Objectivity and the Extended Thermodynamic Description of Rheology 278
P.C. Dauby

Convection in Viscoelastic Fluids 292
C.Pérez-García, J. Martínez-Mardones and J. Millán

Fractional Relaxation Equations for Viscoelasticity and Related Phenomena 309
T.F. Nonnenmacher

Relaxation Functions of Rheological Constitutive Equations with Fractional Derivatives: Thermodynamical Constraints 321
C. Friedrich

A Simple One Dimensional Model Showing Glasslike Dynamical Behavior 331
J.J. Brey and M.J. Ruiz-Montero

Statistical Conformation of a Polymer in a Nematic Medium Under a Shear Flow Using the Rouse Model 344
Y. Thiriet, R. Hocquart, F. Lequeux and J.F. Palierne

On the Modelling of Stationary Heat Transfer by the Use of Dissipative Networks 356
G. Brunk

Thermomechanics of Porous Media Filled with a Fluid 367
W. Derski

VISCOELASTIC BEHAVIOR OF
POLYMERIC LIQUIDS

R. Byron Bird
Chemical Engineering Department and Rheology Research Center
University of Wisconsin-Madison, Madison, WI 53706-1691, USA

A survey is given of the three main approaches to the study of the non-linear constitutive equations for polymeric liquids: rheometric measurements; continuum mechanics results for special flows and useful empiricisms; and kinetic theories, which make use of various mechanical models to represent the polymer molecules. For further discussions see [1,2].

1. Experimental evidence for non-Newtonian behavior of polymeric liquids [1, Chapter 2]

For incompressible fluids the laws of conservation of mass and momentum lead to the equations of continuity and motion [3, Chapter 3]:

Eq. of Continuity: $(\nabla \cdot \mathbf{v}) = 0,$ (1.1)

Eq. of Motion: $\rho \frac{D\mathbf{v}}{Dt} = -[\nabla \cdot \pi] + \rho \mathbf{g}.$ (1.2)

Here ρ is the fluid density, $\mathbf{v}(\mathbf{r},t)$ the velocity field, and $\mathbf{g}$ the gravitational acceleration. The stress tensor π may be written as the sum of two terms,

$$\pi = p\delta + \tau, \qquad (1.3)$$

in which p is a pressure (not uniquely determined for incompressible fluids) and τ is the "extra stress tensor", which is zero at equilibrium; δ is the unit tensor.

For liquids made up of small molecules, the mechanical behavior is well described by the incompressible Newtonian constitutive equation [4] (this is just the linear term in the ordered expansion in eq. (3.14)):

$$\tau = -\mu(\nabla \mathbf{v} + (\nabla \mathbf{v})^{\dagger}) \equiv -\mu \gamma_{(1)}, \qquad (1.4)$$

in which $(\nabla \mathbf{v})^\dagger$ indicates the transpose of $(\nabla \mathbf{v})$, and $\gamma_{(1)}$ is the rate-of-strain tensor. The Newtonian viscosity μ may be determined from any viscometer for which eqs. (1-4) can be solved (for steady-state or unsteady-state flow).

Polymer solutions and polymer melts cannot be described by eq. (1.4) and are therefore *non-Newtonian*. One of the challenges in polymer fluid dynamics is that of obtaining a constitutive equation for τ, containing a small number of physically interpretable parameters, that is capable of describing a wide range of fluid responses.

Many experiments indicate that polymeric liquids are non-Newtonian [1]:

(A) In a cone-and plate viscometer the torque needed to turn the cone is proportional to the angular velocity of the cone for Newtonian fluids, but not so for polymeric liquids. This means that for polymers the shear stress is not proportional to the velocity gradient. Consequently, the viscosity (defined as the proportionality factor between shear stress and velocity gradient) is not a constant. In fact, the viscosity may decrease by factors of ten to a thousand as the velocity gradient increases. Such a dramatic decrease in viscosity cannot be ignored in fluid dynamics calculations. Even in extremely dilute polymer solutions this shear thinning behavior is observed.

(B) A cylindrical container with inside radius R is filled with a Newtonian fluid. When a rotating disk of radius R is placed in contact with the fluid surface, it imparts a tangential motion to the liquid (the "primary flow"); at steady-state there is, in addition to the primary flow, a "secondary flow" outward along the rotating disk, caused by the centrifugal force, and then downward along the cylinder wall, inward along the bottom, and upward near the cylinder axis. For polymeric fluids, however, there is a reversal of the secondary flow in the disk-and-cylinder system: the fluid moves inward along the rotating disk, down along the axis and then upward at the cylinder wall!

(C) If a Newtonian fluid flows down an inclined semicylindrical trough, the liquid surface is flat (except near the edges where meniscus effects may be seen). Polymeric liquids, however, display a slightly convex surface in tilted-trough flow.

(D) When the pumping of a Newtonian fluid through a tube is suddenly stopped, the fluid comes to rest almost at once. But when one stops supplying a driving force for the flow of a polymeric liquid in a tube, the polymeric liquid begins to retreat and then gradually comes to rest. This recoil is symptomatic of a fluid which can "remember" its past kinematic experience; the fluid does not retreat all the way to its initial configuration, since as time proceeds it "forgets" its kinematic history. Polymeric fluids are often referred to as "fluids with fading memory."

(E) When a Newtonian liquid is being siphoned, the siphon ceases to function when the end of the tube is removed from the liquid. For a polymeric liquid, however, the tube may be withdrawn from the liquid and the siphoning continues; there may be a gap of several centimeters between the liquid surface and the end of the tube. This effect is called the tubeless siphon.

Many other experiments showing that polymeric liquids are qualitatively different from Newtonian fluids are described in [1,12], where the original literature references are given. The behavior in expt. A is obviously not explainable by eq. (1.4). Expts. B and C can be explained if in steady shear flow the normal stresses (τ_{xx}, τ_{yy}, τ_{zz}) are all different from each other; such "normal-stress phenomena" cannot be explained by eq. (1.4). To describe expt. D it is necessary that the stress at the present time t depend on the kinematic tensors at all past times t'. Finding a replacement for eq. (1.4) is an enormous challenge, requiring the collaboration of the experimental rheologist, the continuum mechanicist, and the kinetic theorist.

2. Rheometry and material functions [5]

Much of our present knowledge about the flow behavior of polymeric liquids has been obtained from measurements of stress components in carefully controlled flows. The two most studied flows are *shear flows* for which

$$v_x = \dot{\gamma}(t)y; \quad v_y = 0; \quad v_z = 0, \tag{2.1}$$

where $\dot{\gamma}(t)$ is the "shear rate," and *elongational flows* for which

$$v_x = -\frac{1}{2}\dot{\varepsilon}(t)x; \; v_y = -\frac{1}{2}\dot{\varepsilon}(t)y; \; v_z = \dot{\varepsilon}(t)z \tag{2.2}$$

where $\dot{\varepsilon}(t)$ is the "elongation rate." There is a general feeling among polymer fluid dynamicists that continuum or molecular theories that can describe the above flows will probably be successful for more complex flows.

2.1 Shear flows

For steady-state shear flow, with constant $\dot{\gamma}$, three material functions $\eta(\dot{\gamma})$, $\Psi_1(\dot{\gamma})$, and $\Psi_2(\dot{\gamma})$ are defined by

$$\tau_{yx} = -\eta\dot{\gamma}, \tag{2.3}$$

$$\tau_{xx} - \tau_{yy} = -\Psi_1\dot{\gamma}^2, \tag{2.4}$$

$$\tau_{yy} - \tau_{zz} = -\Psi_2\dot{\gamma}^2. \tag{2.5}$$

The viscosity η, and the first normal stress coefficient Ψ_1 are positive, and the second normal stress coefficient Ψ_2 is negative. The secondary flow in the disk-and-cylinder experiment is related to Ψ_1, and the negative Ψ_2 explains the convex surface in the tilted-trough flow. The ratio Ψ_2/Ψ_1 varies between about -0.1 and -0.4 for most fluids.

The most widely studied time-dependent shear flow is the homogeneous small-amplitude oscillatory shearing motion, with $\dot{\gamma}(t) = \mathrm{Re}\{\dot{\gamma}^0 e^{i\omega t}\}$, where $\dot{\gamma}^0$ may be complex. For this flow $\tau_{yx}(t) = \mathrm{Re}\{\tau^0_{yx}\, e^{i\omega t}\}$, where τ^0_{yx} is also complex. Then

$$\tau^0_{yx} = -\eta^*\, \dot{\gamma}^0, \tag{2.6}$$

where $\eta^*(\omega) = \eta'(\omega) - i\eta''(\omega)$ is the complex viscosity; $\eta'(\omega)$ describes the

Figure 1:

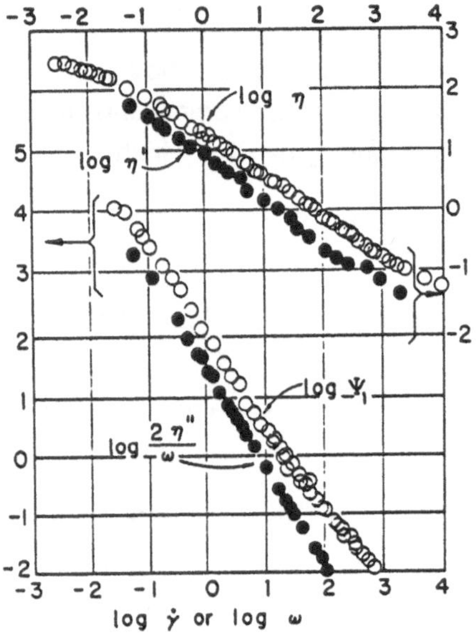

viscous response and $\eta''(\omega)$ describes the elastic response. Normal stresses oscillate with a frequency 2ω, but only a few measurements have been reported.

Fig. 1 shows data replotted from [6] on η, η', and η'' (in Pa·s) and Ψ_1 (in Pa·s^2), and with $\dot{\gamma}$ and ω in s^{-1}, for a 1.5% solution of a polyacrylamide in a water-glycerine mixture. The η and η' curves both begin at η_0 (the zero-shear-rate viscosity), and η' lies somewhat below η; similarly Ψ_1 and $2\eta''/\omega$ both begin at $\Psi_{1,0}$ (zero-shear-rate first normal stress coefficient) and $2\eta''/\omega$ lies below Ψ_1. This behavior is typical for most polymeric liquids.

Another much studied time-dependent experiment is the measurement of the development of the stresses when $\dot{\gamma}$ undergoes a step function change from $\dot{\gamma} = 0$ to $\dot{\gamma} = \dot{\gamma}_0$ at $t = 0$. For $t > 0$ both the time dependent $\eta^+(t) = \tau_{yx}/(-\dot{\gamma}_0)$ and $\Psi_1^+(t) = (\tau_{xx}-\tau_{yy})/(-\dot{\gamma}_0^2)$ go through maxima before decreasing to their steady-state values. For some polymer melts the time at which the maximum occurs in Ψ_1 is about twice that for the maximum in η^+, and the times for the maxima are inversely proportional to $\dot{\gamma}_0$.

Many other unsteady shear experiments have been performed, including stress relaxation after shear flow, stress relaxation after sudden shearing displacement, and superposition of steady flow and oscillatory motion.

Figure 2:

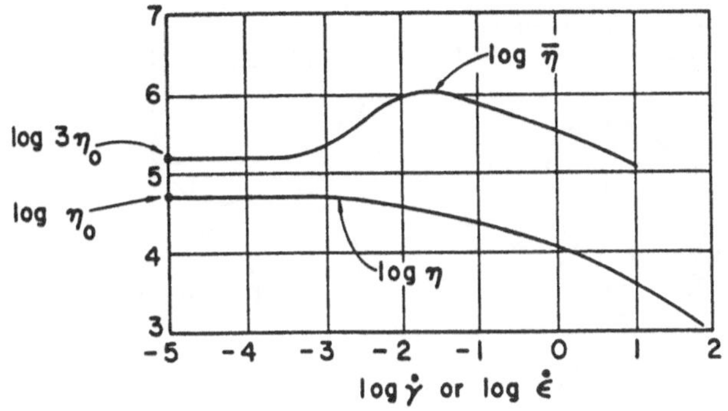

2.2 Elongational Flows [7]

For steady-state elongational flow the elongational viscosity $\bar{\eta}(\dot{\varepsilon})$ is defined by

$$\tau_{zz} - \tau_{xx} = -\bar{\eta}\dot{\varepsilon} \tag{2.7}$$

($\dot{\varepsilon} < 0$ corresponds to biaxial stretching). For dilute solutions, many molecular theories suggest that $\bar{\eta}$ is monotone increasing with $\bar{\eta}$ possibly going to rather high values [8], although convincing experimental proof of this is lacking. Recent melt data show $\bar{\eta}$ increasing with $\dot{\varepsilon}$, going through a maximum, and then decreasing; for melts $\bar{\eta}$ seems to vary less with $\dot{\varepsilon}$ than η varies with $\dot{\gamma}$. In Fig. 2 data for $\bar{\eta}(\dot{\varepsilon})$ for a LDPE melt are shown along with the data for $\eta(\dot{\gamma})$ [9,10].

When $\dot{\varepsilon}$ goes from 0 for $t < 0$ to a constant value $\dot{\varepsilon}_0$ for $t > 0$, the quantity $\bar{\eta}^+(t) = (\tau_{zz} - \tau_{xx})/(-\dot{\varepsilon}_0)$ can be measured. Most data show $\bar{\eta}^+$ to be a monotone increasing function of time, but overshoot effects have also been observed.

3. Continuum mechanical ideas and empirical constitutive equations [1,11,12]

To describe the kinematics of a moving fluid, it is useful to introduce the notion of a convected coordinate system [11], which is imprinted in the fluid and moves with it. This coordinate system may be so chosen that, at the present time t, it coincides with a space-fixed cartesian coordinate system. In Fig. 3 a typical element from the coordinate system is shown at some past time t' and at the present time t. For past times the coordinate system is in general nonorthogonal. At each fluid particle a set of convected base vectors $\hat{g}_i(\mathbf{r},t,t')$ can be defined; they are the base vectors at time t' associated with a fluid particle which is located at **r** at time t. At time t' = t the base vectors coincide

Figure 3:

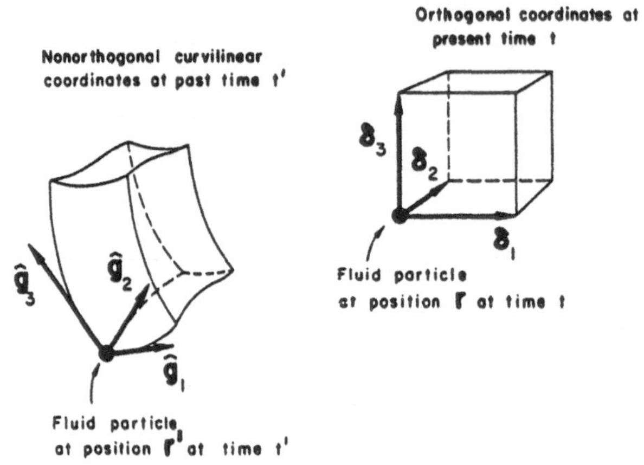

with the unit vectors δ_i. A set of reciprocal convected base vectors $\hat{g}^i$ is given by relations like $\hat{g}^1 = [\hat{g}_2 \times \hat{g}_3]$ for incompressible fluids. As the fluid particle moves along, the convected base vectors change with time as follows [1, p. 486]:

$$\frac{\partial}{\partial t'}\hat{g}_i(\mathbf{r},t,t') = [\hat{g}_i(\mathbf{r},t,t')\cdot\nabla\mathbf{v}(\mathbf{r},t,t')], \tag{3.1}$$

$$\frac{\partial}{\partial t'}\hat{g}^i(\mathbf{r},t,t') = -\,[\nabla\mathbf{v}(\mathbf{r},t,t')\cdot\hat{g}^i(\mathbf{r},t,t')]. \tag{3.2}$$

The convected base vectors may be described in terms of their cartesian components [1, p. 485]:

$$\hat{g}_i(\mathbf{r},t,t') = \sum_j \delta_j \Delta_{ji}(\mathbf{r},t,t'), \tag{3.3}$$

$$\hat{g}^i(\mathbf{r},t,t') = \sum_j E_{ij}(\mathbf{r},t,t')\delta_j. \tag{3.4}$$

Both Δ_{ij} and E_{ij} become the Kronecker delta δ_{ij} at time t' = t. From eqs. (3.1)-(3.4) it can be shown that the relations

$$\frac{\partial}{\partial t'}\Delta_{ij}(\mathbf{r},t,t') = \sum_n (\nabla\mathbf{v})^\dagger_{in}(\mathbf{r},t,t')\Delta_{nj}(\mathbf{r},t,t'), \tag{3.5}$$

$$\frac{\partial}{\partial t'}E_{ij}(\mathbf{r},t,t') = -\sum_n E_{in}(\mathbf{r},t,t')(\nabla\mathbf{v})^\dagger_{nj}(\mathbf{r},t,t'), \tag{3.6}$$

describe how Δ_{ij} and E_{ij} change with t' following a fluid particle, "**r**,t".

3.1 Stress and strain tensors and their derivatives

The convected base vectors are of interest because the dot products, $\hat{g}_{ij} = (\hat{g}_i \cdot \hat{g}_j)$ or $\hat{g}^{ij} = (\hat{g}^i \cdot \hat{g}^j)$, describe the deformation of a fluid element independently of its orientation in space. Of particular interest is the deformation at time t' relative to the reference configuration at time t, which can be written with the help of eqs. (3.3) and (3.4) as

$$\left[\hat{g}_{ij}(\mathbf{r},t,t') - \hat{g}_{ij}(\mathbf{r},t,t)\right] = \sum_k \Delta'_{ki}\Delta'_{kj} - \delta_{ij} \equiv B^{-1'}_{ij} - \delta_{ij} \equiv \gamma^{[0]'}_{ij}, \tag{3.7}$$

$$-[\hat{g}^{ij}(\mathbf{r},t,t') - \hat{g}^{ij}(\mathbf{r},t,t)] = \delta_{ij} - \sum_k E'_{ik}E'_{jk} \equiv \delta_{ij} - B'_{ij} \equiv \gamma'_{[0]ij}, \tag{3.8}$$

in which primes on the tensor components indicate functions of **r**, t and t'. Both relative finite strain tensors $\gamma^{[0]}(\mathbf{r},t,t')$ and $\gamma_{[0]}(\mathbf{r},t,t')$ are zero for t' = t, and both simplify to the infinitesimal strain tensor of linear viscoelasticity. The tensors $\mathbf{B}^{-1}$ and $\mathbf{B}$ are called the Cauchy and Finger strain tensors, respectively [1, p. 427].

Successive time derivatives $\left(\partial/\partial t', \partial^2/\partial t'^2, \cdots\right)$ may be taken of eqs. (3.7) and (3.8) to generate rate-of-strain tensors, second-rate-of-strain tensors, etc., $\gamma^{[n]}$, $\gamma_{[n]}$ at time t', by using eqs. (3.5) and (3.6). The nth rate of strain tensors at time t' = t, $\gamma^{(n)}$ and $\gamma_{(n)}$, may also be generated by "convected differentiation" [1, p. 492],

$$\gamma^{[n+1]}(\mathbf{r},t,t) = \gamma^{(n+1)}(\mathbf{r},t) = \frac{D}{Dt}\gamma^{(n)} + \{(\nabla\mathbf{v})\cdot\gamma^{(n)} + \gamma^{(n)}\cdot(\nabla\mathbf{v})^\dagger\}, \tag{3.9}$$

$$\gamma_{[n+1]}(\mathbf{r},t,t) = \gamma_{(n+1)}(\mathbf{r},t) = \frac{D}{Dt}\gamma_{(n)} - \{(\nabla\mathbf{v})^\dagger\cdot\gamma_{(n)} + \gamma_{(n)}\cdot(\nabla\mathbf{v})\}, \tag{3.10}$$

in which $\gamma^{(1)}$ and $\gamma_{(1)}$ are both defined to be $\nabla\mathbf{v} + (\nabla\mathbf{v})^\dagger$.

The stress tensor at the fluid particle "**r**,t" at time t' may be written in terms of covariant or contravariant convected components.

$$\tau(\mathbf{r},t,t') = \sum_i \sum_j \hat{\mathbf{g}}_i(\mathbf{r},t,t')\hat{\mathbf{g}}_j(\mathbf{r},t,t')\hat{\tau}^{ij}(\mathbf{r},t,t') \tag{3.11}$$

$$= \sum_i \sum_j \hat{\mathbf{g}}^i(\mathbf{r},t,t')\hat{\mathbf{g}}^j(\mathbf{r},t,t')\hat{\tau}_{ij}(\mathbf{r},t,t'). \tag{3.12}$$

It may also be written in terms of the fixed cartesian components

$$\tau(\mathbf{r},t,t') = \sum_i \sum_j \boldsymbol{\delta}_i\boldsymbol{\delta}_j\tau_{ij}(\mathbf{r},t,t'). \tag{3.13}$$

The convected and fixed components are related according to the usual rules of tensor transformations [1, p. 495]. The derivatives $(\partial/\partial t', \partial^2/\partial t'^2, \cdots)$ of $\hat{\tau}^{ij}$ and $\hat{\tau}_{ij}$ can then be formed; this gives sets of tensors $\tau^{[n]}$ and $\tau_{[n]}$ at time t'. At time t' = t the convected time derivatives, $\tau^{(n)}$ and $\tau_{(n)}$, are obtained exactly as in eqs. (3.9) and (3.10) with $\tau^{(0)}$ and $\tau_{(0)}$ both being equal to τ.

3.2 Constitutive equations

In a landmark publication Oldroyd [11] set down certain postulates regarding constitutive equations: (1) The relation between the stress and kinematic tensors for a fluid element should not depend on the kinematics of neighboring elements. (2) The constitutive equation should not depend on the motion of the element as a whole in space. Therefore the constitutive equation can involve any relation among the convected components of the stress tensor and the $\hat{g}_{ij}$ or $\hat{g}^{ij}$, involving differentiations or integrations of these quantities with respect to time; in terms of fixed components (which are more useful for solving the equations of continuity and motion), the constitutive equation can involve any of the time t' quantities $\gamma^{[n]}$, $\gamma_{[n]}$, $\tau^{[n]}$, $\tau_{[n]}$ (the indices in brackets indicate quantities evaluated at time t') or the same quantities at t' = t (indices in parentheses).

Some progress has been made in obtaining general results for restricted classes of flows based upon Oldroyd's postulates:

(a) Retarded-motion expansion [13]. For slow flows which are also slowly-varying in time the stress tensor can be expanded in a series:

$$\tau = -b_1\gamma_{(1)} + b_2\gamma_{(2)} - b_{11}\{\gamma_{(1)}\cdot\gamma_{(1)}\} + b_3\gamma_{(3)}$$

$$-b_{12}\{\gamma_{(1)}\cdot\gamma_{(2)} + \gamma_{(2)}\cdot\gamma_{(1)}\} - b_{1:11}(\gamma_{(1)}:\gamma_{(1)})\gamma_{(1)} + \cdots, \tag{3.14}$$

in which the b's are constants. The first term on the right side gives the Newtonian fluid. Retention of the first three terms (including all terms quadratic in velocity gradients) gives the "second-order fluid." Retention of all six terms shown in eq. (3.14) gives the "third-order fluid." Because the series in eq. (3.14) converges very slowly, it is of limited value in polymer fluid dynamics. Some kinetic theories give relations between the b's and molecular structure [1, p. 302; 2, p. 207]. Note that $b_1 = \eta_0$, $b_2 = \frac{1}{2}\Psi_{1,0}$, and $b_{11} = \Psi_{2,0}$.

(b) Criminale-Ericksen-Filbey equation [14]. For steady-state shear flows (e.g., axial flow in tubes, helical flow in annuli, Couette flow, cone-and-plate flow) the most general expression for the stress tensor consistent with Oldroyd's postulates is

$$\tau = -\eta\gamma_{(1)} + \frac{1}{2}\Psi_1\gamma_{(2)} - \Psi_2\{\gamma_{(1)}\cdot\gamma_{(1)}\}, \qquad (3.15)$$

in which η, Ψ_1, and Ψ_2 are the properties defined in eqs. (2.3)-(2.5); they are all functions of the shear rate $\dot{\gamma} = \sqrt{\frac{1}{2}(\gamma_{(1)}:\gamma_{(1)})}$. Eq. (3.15) has a structure similar to that of the second-order fluid, but the coefficients are functions rather than constants, and there is no restriction to flows with small velocity gradients.

(c) Memory integral expansion [15]. For flows with small deformations, but with arbitrary time responses, the stress tensor can be expanded as

$$\tau = \int_{-\infty}^{t} M_1(t-t')\gamma'_{[0]}dt' + \frac{1}{2}\int_{-\infty}^{t}\int_{-\infty}^{t} M_2(t-t',t-t'')\{\gamma'_{[0]}\cdot\gamma''_{[0]} + \gamma''_{[0]}\cdot\gamma'_{[0]}\}\, dt''dt' + \cdots \qquad (3.16)$$

The first term in this expansion gives the Lodge rubberlike liquid [12], which for infinitesimal deformations simplifies to the constitutive equation for linear viscoelasticity. Note that eq. (3.16) incorporates the idea of fading memory.

3.3 Widely used empiricisms

Because of the limited applicability of eqs. (3.14)-(3.16), dozens of empirical constitutive equations have been proposed [1, Chapters 7 and 8]. As an example of a differential constitutive equation we cite the widely used Oldroyd 8-constant model suggested in 1958 [11]:

$$\tau + \lambda_1\tau_{(1)} + \frac{1}{2}\lambda_3\{\gamma_{(1)}\cdot\tau + \tau\cdot\gamma_{(1)}\}$$

$$+ \frac{1}{2}\lambda_5(\mathrm{tr}\,\tau)\gamma_{(1)} + \frac{1}{2}\lambda_6(\tau:\gamma_{(1)})\delta$$

$$= -\underline{\eta_0[\gamma_{(1)} + \lambda_2\, \gamma_{(2)}} + \lambda_4\{\gamma_{(1)}\cdot\gamma_{(1)}\}$$

$$+ \frac{1}{2}\lambda_7(\gamma_{(1)}:\gamma_{(1)})\delta] \tag{3.17}$$

in which η_0 is the zero-shear-rate viscosity and the λ_j ($j = 1, 2, \cdots 7$) are time constants. This model includes many special cases, such as the "Oldroyd-B model" or "convected Jeffreys model" (dashed underlined terms) [1, p. 352].

As an illustration of an integral constitutive equation we cited the factorized K-BKZ model [16; 1, p. 436]

$$\tau(t) = \int_{-\infty}^{t} M(t-t')\left[\frac{\partial W}{\partial I_1}\gamma_{[0]} + \frac{\partial W}{\partial I_2}\gamma^{[0]}\right] dt' \tag{3.18}$$

in which $M(t-t')$ is the linear viscoelastic memory function (M_1 in eq. 3.16), and $W(I_1,I_2)$ the so-called "potential function," which depends on the strain invariants $I_1 = \mathrm{tr}\,\mathbf{B}$ and $I_2 = \mathrm{tr}\,\mathbf{B}^{-1}$, and which must satisfy the condition $(\partial W/\partial I_1)_{3,3} + (\partial W/\partial I_2)_{3,3} = 1$.

Deducing empirical constitutive equations from rheometric data is a difficult and frustrating exercise in "tensorial curve fitting." In recent years the emphasis has shifted to the use of kinetic theories to provide guidelines as to the form of constitutive equations.

4. Kinetic theory ideas and dumbbell models

To introduce the main ideas used in kinetic theories, we use some highly oversimplified mechanical models. In particular, we discuss the elastic dumbbell model (two "beads" joined by a "spring") and the rigid dumbbell (two "beads" joined by a rigid "rod"); for a discussion of molecular models and their origins, see [1, Chapter 11].

4.1 Dilute solutions of elastic dumbbell [2, Chapter 13]

We can write an equation of motion for each bead of the dumbbell indicating that the mass of the bead times its acceleration is equal to the sum of all forces acting on the bead. When we neglect the inertial terms containing the bead masses, we get a "force balance" among the various forces:

$$\mathbf{F}_\nu^{(h)} + \mathbf{F}_\nu^{(b)} + \mathbf{F}_\nu^{(\phi)} = 0 \qquad (\nu = 1,2) \tag{4.1}$$

in which

$$\mathbf{F}_\nu^{(h)} = -[\zeta_0\cdot([\dot{\mathbf{r}}_\nu] - (\mathbf{v}_\nu + \mathbf{v}'_\nu))] \tag{4.2}$$

$$\mathbf{F}_\nu^{(b)} = -\frac{1}{\Psi}\left[\frac{\partial}{\partial \mathbf{r}_\nu}\cdot([m(\dot{\mathbf{r}}_\nu - \mathbf{v})(\dot{\mathbf{r}}_\nu - \mathbf{v})]\Psi)\right] \tag{4.3}$$

$$\mathbf{F}_\nu^{(\phi)} = -\frac{\partial}{\partial \mathbf{r}_\nu}\phi \tag{4.4}$$

In these equations $\mathbf{r}_\nu$ is the location of the νth bead, m is the mass of a bead, ϕ is the potential energy associated with the spring, and $\Psi(\mathbf{r}_1,\mathbf{r}_2,t)$ is the configurational distribution function for the dumbbell. The double brackets [ ] indicate an average in the velocity space. We now discuss each of the contributions to the force balance.

Equation 4.2 describes the hydrodynamic force acting on bead ν. According to this expression the force is proportional to the difference between the bead velocity $\dot{\mathbf{r}}_\nu$ (appropriately averaged with respect to the velocity distribution) and the velocity $(\mathbf{v}_\nu + \mathbf{v}'_\nu)$ of the solution at bead ν. The velocity $\mathbf{v}_\nu = \mathbf{v}_0 + [\boldsymbol{\kappa}\cdot\mathbf{r}_\nu]$ is the imposed homogeneous flow field at bead ν (here $\boldsymbol{\kappa}(t)$ and $\mathbf{v}_0(t)$ are position independent), and $\mathbf{v}'_\nu$ is the perturbation of the flow field at bead ν resulting from the motion of the other bead ("hydrodynamic interaction"). According to eq. 4.2, the hydrodynamic drag force is not necessarily collinear with the velocity difference since the coefficient of proportionality is a tensor $\boldsymbol{\zeta}_0$, (the "friction tensor"). The beads actually execute very tortuous paths as they move about in the solvent, but by using the velocity-space average of $\dot{\mathbf{r}}_\nu$ we obtain a kind of "smoothed out" drag force.

Equation 4.3 is a smoothed-out Brownian motion force. The true Brownian motion force would be a rapidly and irregularly fluctuating function. Instead of the latter we use a statistically averaged force, the origin of which can be understood from a complete phase-space kinetic theory [1, Chapter 18]. It should be noted that the expression for the Brownian force has the form of the divergence of a momentum flux. In almost all kinetic theories published so far, equilibration in momentum space (Maxwellian velocity distribution) has been tacitly assumed; then the Brownian force becomes $\mathbf{F}_\nu^{(b)} = -kT(\partial \ln \Psi/\partial \mathbf{r}_\nu)$.

Equation 4.4 gives the force $\mathbf{F}_\nu^{(\phi)}$ on the νth bead resulting from the intramolecular potential energy, that is, just the force acting through the spring in the dumbbell. Since the forces on the two beads are equal and opposite, it is useful to define a "connector force" $\mathbf{F}^{(c)}$ by $\mathbf{F}^{(c)} = \mathbf{F}_1^{(\phi)} = -\mathbf{F}_2^{(\phi)}$.

We now assume that the friction tensor is a multiple of the unit tensor ($\boldsymbol{\zeta}_0 = \zeta\boldsymbol{\delta}$), that the Maxwellian velocity distribution is used in the Brownian motion term, and that hydrodynamic interaction is neglected; then the equations of motion become:

$$-\zeta([\dot{\mathbf{r}}_\nu] - \mathbf{v}_0 - [\boldsymbol{\kappa}\cdot\mathbf{r}_\nu]) - kT\frac{\partial}{\partial \mathbf{r}_\nu}\ln\Psi + \mathbf{F}_\nu^{\phi} = 0 \quad (\nu = 1,2) \tag{4.5}$$

When these two equations are added together and then divided by 2, we get the equation of motion for the center of mass $\mathbf{r}_c = (1/2)(\mathbf{r}_1 + \mathbf{r}_2)$; when they are subtracted, we get that for the connector vector $\mathbf{Q} = \mathbf{r}_2 - \mathbf{r}_1$:

$$[\dot{\mathbf{r}}_c] = \mathbf{v}_0 + [\boldsymbol{\kappa}\cdot\mathbf{r}_c] \tag{4.6}$$

$$[\dot{\mathbf{Q}}] = [\boldsymbol{\kappa}\cdot\mathbf{Q}] - \frac{2kT}{\zeta}\frac{\partial}{\partial \mathbf{Q}}\ln\psi - \frac{2}{\zeta}\mathbf{F}^{(c)} \tag{4.7}$$

Here we have introduced $\psi(\mathbf{Q},t)$ which is related to $\Psi(\mathbf{r}_1,\mathbf{r}_2,t)$ by $\Psi = n\psi$, where n is the number density of dumbbells. The first of the equations above shows that the center of mass of the dumbbell moves on the average with the solution velocity at the location of the center of mass. The second equation is used presently to obtain the diffusion equation for $\psi(\mathbf{Q},t)$.

The distribution function $\Psi(\mathbf{r}_1,\mathbf{r}_2,t)$ must satisfy a continuity equation:

$$\frac{\partial\Psi}{\partial t} = -\left(\frac{\partial}{\partial \mathbf{r}_1}\cdot[\dot{\mathbf{r}}_1]\Psi\right) - \left(\frac{\partial}{\partial \mathbf{r}_2}\cdot[\dot{\mathbf{r}}_2]\Psi\right) \tag{4.8}$$

which accounts for conservation of system points in the six-dimensional configuration space. By a change of variables this can be rewritten for $\psi(\mathbf{Q},t)$ as:

$$\frac{\partial\psi}{\partial t} = -\left(\frac{\partial}{\partial \mathbf{Q}}\cdot[\dot{\mathbf{Q}}]\psi\right) \tag{4.9}$$

This is the continuity equation in the three-dimensional internal configuration space.

Substitution of $[\dot{\mathbf{Q}}]$ from eq. 4.7 into eq. 4.9 gives the "diffusion equation":

$$\frac{\partial\psi}{\partial t} = -\left(\frac{\partial}{\partial \mathbf{Q}}\cdot\{[\boldsymbol{\kappa}\cdot\mathbf{Q}]\psi - \frac{2kT}{\zeta}\frac{\partial}{\partial \mathbf{Q}}\psi - \frac{2}{\zeta}\mathbf{F}^{(c)}\psi\}\right) \tag{4.10}$$

This describes the way the distribution of configurations changes with time when the time-dependent homogeneous velocity field is described by $\boldsymbol{\kappa}(t)$ and the dumbbell spring force is given as $\mathbf{F}^{(c)}$. This is the basic differential equation in the elementary elastic-dumbbell kinetic theory.

Next we turn to the expression for the stress tensor, which contains three contributions: the solvent contribution; a contribution from the tensions in the springs, $\mathbf{F}^{(c)}$; and a contribution from the bead momentum flux:

$$\boldsymbol{\pi} = \boldsymbol{\pi}_s - n\langle\mathbf{Q}\mathbf{F}^{(c)}\rangle + nm\sum_{\nu=1}^{2}\langle(\dot{\mathbf{r}}_\nu - \mathbf{v})(\dot{\mathbf{r}}_\nu - \mathbf{v})\rangle \tag{4.11}$$

Here $\langle\ \rangle$ stands for an average in the phase space of the dumbbell. The term containing $\mathbf{F}^{(c)}$ is usually the most important of the three terms. Equation 4.11 is valid whether or not hydrodynamic interaction is included and whether or not the friction tensor ζ is isotropic; also no assumption has been introduced regarding the velocity distribution.

For a velocity distribution Maxwellian about **v** the last term in eq. 4.11 becomes isotropic, and the expression for the extra stress tensor τ then becomes

$$\text{Kramers form} \qquad \tau = -\eta_s\gamma_{(1)} - n\langle QF^{(c)}\rangle + nkT\,\delta \tag{4.12}$$

$$\text{Giesekus form} \qquad \tau = -\eta_s\gamma_{(1)} + \frac{n}{4}\zeta\langle QQ\rangle_{(1)} \tag{4.13}$$

The Giesekus form [17] is obtained from the Kramers form [18] by using the second moment of eq. 4.10 (see 1, p. 70).

For Hookean dumbbells ($F^{(c)} = HQ$, where H is the spring constant), $\langle QQ\rangle$ can be eliminated between eqs. 4.12 and 4.13 to get the "convected Jeffreys model" (dashed-line terms in eq. 3.17) with:

$$\begin{aligned} &\text{Zero-shear-rate viscosity:} && \eta_0 = \eta_s + nkT\,\lambda_H \\ &\text{Relaxation time:} && \lambda_1 = \lambda_H \equiv (\zeta/4H) \\ &\text{Retardation time:} && \lambda_2 = [\eta_s/(\eta_s + nkT\,\lambda_H)]\lambda_H \end{aligned} \tag{4.14}$$

This shows how the constants in an empirical constitutive equation can be given a molecular-model interpretation. However, the convected Jeffreys model gives a constant η, a constant Ψ_1, and $\Psi_2 = 0$ -- all unacceptable results.

For finitely extensible nonlinear elastic (FENE) dumbbells ($F^{(c)} = HQ/[1-(Q/Q_0)^2]$, where Q_0 is the maximum extension) a similar procedure can be used if $[1-(Q/Q_0)^2]$ is replaced by $[1-\langle(Q/Q_0)^2\rangle]$, the so-called "Peterlin approximation." This leads to the Tanner constitutive equation [19] which is nonlinear in $\tau_p = \tau - \tau_s = \tau + \eta_s\gamma_{(1)}$:

$$\begin{aligned} Z\tau_p + \lambda_H\tau_{p(1)} - \lambda_H(\tau_p - nkT\,\delta)\,D\ln Z/Dt \\ = -nkT\,\lambda_H\gamma_{(1)} \end{aligned} \tag{4.15}$$

in which $\lambda_H = \zeta/4H$, $Z = 1 + (3/b)[1 - (\mathrm{tr}\,\tau_p/3nkT)]$, and $b = HQ_0^2/kT$. The use of the Peterlin approximation for this model has been carefully studied and is known to give reasonable results. Limited testing of this "FENE-P" constitutive equation showed moderately good ability to describe rheometric data on η and Ψ_1 and nontrivial flow phenomena [20]; this is an example of how a molecular theory can suggest the form of the constitutive equation, with the constants being determined from experimental data.

Recently the elastic dumbbell model has been modified to allow for diffusion of dumbbells across streamlines [21]; this leads to a term containing $\nabla^2\tau_p$ in the constitutive equation.

4.2 Concentrated solutions of elastic dumbbells [2, p. 97]

In concentrated polymer solutions or in melts, the polymers are restricted in their motion because of the crowding of the molecules. Under such conditions, we need to retain the (anisotropic) friction tensor ζ_0 in eq. (4.2) and we also need an anisotropic Brownian motion $\mathbf{F}_\nu^{(b)} = -(kT/\Psi)[(\partial/\partial \mathbf{r}_\nu)\cdot\xi_0^{-1}\Psi]$, in which the anisotropic tensor ξ_0 accounts for skewing the Maxwellian velocity distribution. Several proposals have been made for empiricisms for ζ_0 and ξ_0, among them one by Giesekus [22]: $\zeta_0^{-1} = (1/\zeta)[\delta - (a/nkT)\tau_p]$ and $\xi_0^{-1} = \delta - (a/nkT)\tau_p$, where a is an empirical constant. These lead to the constitutive equation:

$$\tau_p + \lambda_H' \tau_{p(1)} - (a/nkT)\{\tau_p\cdot\tau_p\} = -nkT\,\lambda_H''\gamma_{(1)} \tag{4.16}$$

where $\lambda_H' = (1+2a)\lambda_H'' = [(1+2a)/(1+a)](\zeta/4H)$. This model seems capable of qualitative description of polymer melt rheometric data. By superposing equations of the form of eq. (4.16), with a spectrum of time constants, quantitative descriptions may be possible [1, p. 412].

4.3 Dilute solutions of rigid dumbbells [2, Chapter 14]

For the rigid dumbbell of rod-length L, there are only two spatial coordinates θ and ϕ (or we can use the unit vector **u** along the rod). In lieu of derivatives with respect to θ and ϕ we can use the operator $\partial/\partial\mathbf{u} = \mathbf{s}\,\partial/\partial\theta + (1/\sin\theta)\,\mathbf{t}\,\partial/\partial\phi$, where **s** and **t** are unit vectors in the θ and ϕ directions. The diffusion equation for the orientational distribution function f(**u**,t) is:

$$\frac{\partial f}{\partial t} = \frac{1}{6\lambda}\left(\frac{\partial}{\partial \mathbf{u}}\cdot\frac{\partial f}{\partial \mathbf{u}}\right) - \left(\frac{\partial}{\partial \mathbf{u}}\cdot[\kappa\cdot\mathbf{u} - \kappa:\mathbf{uuu}]f\right) \tag{4.17}$$

where $\lambda = \zeta L^2/12kT$. The stress-tensor is given by:

Kramers form: $$\tau = -\eta_s\gamma_{(1)} - 3nkT\langle\mathbf{uu}\rangle - 6nkT\,\lambda\kappa:\langle\mathbf{uuuu}\rangle \tag{4.18}$$

Giesekus form: $$\tau = -\eta_s\gamma_{(1)} + 3nkT\,\lambda\,\langle\mathbf{uu}\rangle_{(1)} \tag{4.19}$$

Because of the constraint of constant interbead distance, the forms of these equations are different from those for elastic dumbbells. Thus far, it has not been possible to get an exact constitutive equation for this model; the best we can do is to give the first few terms of the retarded-motion expansion and the memory-integral expansion (see

[1, p. 131]). Numerical solutions of eq. 4.17 have been obtained [1, p. 123] from which we know that:

$$\eta - \eta_s \doteq 0.678\ nkT\lambda(\lambda\dot{\gamma})^{-1/3} \tag{4.20}$$

$$\Psi_1 \doteq 1.20\ nkT\lambda^2(\lambda\dot{\gamma})^{-4/3};\ \Psi_2 = 0 \tag{4.21}$$

Recently the coefficient in eq. (4.21) was found exactly [23] to be $\sqrt{\pi}/[2^{2/3}\Gamma(7/6)]$. These results have also been extended to multibead rods with hydrodynamic interaction [2, p. 136].

5. Kinetic theory ideas and chain models

For flexible polymers the elastic dumbbell models are unsatisfactory because they do not contain a spectrum of relaxation times, the latter being needed in order to describe the coupled internal motions. Here we give a brief introduction to chain models.

5.1 Dilute solutions of bead-spring chains [2, Chapter 15]

The simplest model for flexible chains is the freely jointed Rouse model, a chain of N identical beads joined linearly by N-1 Hookean springs with spring constant H. The diffusion equation (the chain analog of eq. 4.10) and the stress tensor expressions (the chain analogs of eqs. 4.12 and 4.13) are easily derived and well-known. The configurational distribution function can be worked out for all flows [24; 2, p. 161], and the constitutive equation is now well known [2, p. 159]:

$$\tau = -\eta_s\gamma_{(1)} + \sum_{j=1}^{N-1} \tau_j \tag{5.1}$$

$$\tau_j + \lambda_j\tau_{j(1)} = -nkT\ \lambda_j\gamma_{(1)} \tag{5.2}$$

where the time constants λ_j are $\lambda_j = (\zeta/2H)/[4\sin^2(j\pi/2N)]$. This result is unsatisfactory since it does not describe the shear-rate dependence of η and Ψ_1. Inclusion of hydrodynamic interaction, averaged with the equilibrium configurational distribution function (the Zimm theory) is equally unsatisfactory, although useful for fitting linear viscoelastic data.

Considerable improvement is the FENE-P chain in which the spring forces are taken to be $\mathbf{F}_j^{(c)} = H\mathbf{Q}_j/(1-\langle Q_j^2/Q_0^2\rangle)$ -- that is the expression in the denominator is "consistently averaged," that is, averaged with the configurational distribution function at the instantaneous kinematic condition. Detailed calculations of material functions have been carried out for this model [25,26], and it has also been used to study "coil-stretch transitions" in unsteady elongational flows [27].

A much simpler constitutive equation, giving nearly the same results for material functions as for the FENE-P chain, is obtained for the "FENE-PM chain" [28], in which

$$\mathbf{F}_j^{(c)} = \frac{H\mathbf{Q}_j}{1 - \left[\Sigma_j \langle Q_j^2/Q_0^2 \rangle/(N-1)\right]} \tag{5.3}$$

For this choice of $\mathbf{F}_j^{(c)}$ the polymer contribution to the stress tensor is given by:

$$\tau_p = \tau + \eta_s \gamma_{(1)} = \Sigma_j \tau_j \tag{5.4}$$

$$Z\tau_j + \lambda_j \tau_{j(1)} - \lambda_j(\tau_j - nkT\,\delta)\,D\ln Z/Dt = -nkT\,\lambda_j \gamma_{(1)} \tag{5.5}$$

$$Z = 1 + 3b[1 - (\mathrm{tr}\,\tau_p)/3(N-1)nkT] \tag{5.6}$$

where the λ_j are the Rouse relaxation times given just after eq. (5.2). Note that eqs. (5.5), with $j = 1,2, \cdots N-1$, are coupled because Z depends on all of the τ_j. Although the constitutive equation in eqs. (5.4)-(5.6) for the FENE-PM chain gives about the same results for η and Ψ_1 as obtained for the FENE-P model, it does not perform as well for getting information about the details of the "chain dynamics," such as the stretching of individual links.

5.2 Dilute solutions of bead-rod chains [2, p. 217; 29]

The Kramers freely jointed bead-rod chain with N beads and N-1 rods of length a has a constant contour length of (N-1)a. Because of the internal constraints it is difficult to solve the diffusion equation. Although the second-order-fluid constants have been known for some time [2, p. 218], it was not possible to calculate the viscometric functions η and Ψ_1 until Liu's use of Brownian dynamics. Using the Langevin form of the kinetic theory, with a random-number generator to simulate the Brownian motion, he calculated the motions of Kramers chains using the classical equations of motion; the enormous amount of information generated could conveniently be represented by computer animations. Then, assuming that time averages can be equated to ensemble averages, he calculated the $\eta(\dot{\gamma})$ and $\Psi_1(\dot{\gamma})$ curves [29].

5.3 Concentrated solutions of bead-rod chains [2; Chapter 19; 30, 31, 32]

In melts and concentrated solutions the chain is not free to sample all possible configurations, because of the presence of other chains in the immediate neighborhood. To account for this one can introduce an anisotropic Stokes law to account for the fact that the frictional drag on a segment of the chain will be greater in the direction perpendicular to the chain backbone than in the direction parallel to it. One also introduces an anisotropic Brownian motion to emphasize the importance of the Brownian motion in the direction of the chain. This leads to the following constitutive relation:

$$\tau = NnkT[1/3\delta - \int_{-\infty}^{t} \mu(t-t')\mathbf{A}(t,t')dt' - \varepsilon\lambda\kappa : \int_{-\infty}^{t} \nu(t-t')\mathbf{B}(t,t')dt'] \quad (5.7)$$

in which μ and ν are something like "memory functions"

$$\mu(s) = -(\lambda/2)d\nu/ds \quad (5.8)$$

$$\nu(s) = \frac{16}{\pi^2\lambda} \Sigma_{\alpha,odd} \frac{1}{\alpha^2} e^{-\pi^2\alpha^2 s/\lambda}$$

and **A** and **B** are some kind of modified strain tensors:

$$\mathbf{A} = (1/4\pi) \int [1 + (\gamma^{[0]}:\mathbf{uu})]^{-3/2}\mathbf{uu}\, d\mathbf{u} \quad (5.9)$$

$$\mathbf{B} = (1/4\pi) \int [1 + (\gamma^{[0]}:\mathbf{uu})]^{-3/2}\mathbf{uuuu}\, d\mathbf{u}$$

There are several parameters in the constitutive equation with simple physical significance:

N = number of beads in the chain (proportional to the molecular weight)
a = the rod length (so that $(N-1)a$ is the contour length)
$\lambda = N^{3+\beta}\zeta a^2/2kT$ = the longest relaxation time
β = the "chain constraint exponent" (about 0.3 to 0.5)
ε = the "link tension coefficient" (about 0.3 to 0.5)

The parameters ε and β arise in the expression for the modified Stokes law ($\varepsilon = 1$ corresponds to an isotropic Stokes law; $\varepsilon = 0$ and $\beta = 0$ corresponds to the Doi-Edwards model) [33]. Equation 5.7 has been extensively compared with the Doi-Edwards model and with experimental data, both for monodisperse [31] and polydisperse [32] liquids.

The Doi-Edwards theory is based on the "tube" concept and also uses the stress tensor expression taken over from the theory of rubber elasticity; in addition "sliplinks" and Maxwell demons are introduced. The polymer motion is regarded as a one-dimensional stochastic problem. The Curtiss-Bird theory starts from a general phase-space kinetic theory [2, Chapter 17] and makes certain well-defined assumptions, which are necessary as successive integrations are performed (in a systematic way) to get from the system phase space down to the single-link configuration space. The principal assumptions involve making mathematical statements about the anisotropic friction tensor and the anisotropic Brownian motion. The bases for the two theories are thus quite different from one another, and it is very difficult to make a one-to-one correspondence between the elements in the two developments.

We must not forget that for concentrated liquid systems there is another totally different approach, namely the use of network theories. These theories start from the theory of rubber elasticity and then introduce assumptions regarding the destruction and reformation of network junctions during the flow [2, Chapter 20]. Generally the network theories are easier to derive and use than are the "single-chain-in-a-mean-field" theories of Doi-Edwards and Curtiss-Bird.

Acknowledgements: The author is indebted to the National Science Foundation for many years of research support, and also to the Vilas Research Fund and the John D. MacArthur Professorship.

References

[1] R.B. Bird, R.C. Armstrong and O. Hassager, *Dynamics of Polymeric Liquids: I. Fluid Mechanics* (Wiley-Interscience, New York, 1987).
[2] R.B. Bird, C.F. Curtiss, R.C. Armstrong and O. Hassager, *Dynamics of Polymeric Liquids*, Vol. 2, *Kinetic Theory*, Wiley-Interscience, New York, 1987.
[3] R.B. Bird, W.E. Stewart and E.N. Lightfoot, *Transport Phenomena*, Wiley, New York, 1960.
[4] J.O. Hirschfelder, C.F. Curtiss and R.B. Bird, *Molecular Theory of Gases and Liquids*, Wiley, New York (2nd Corrected Printing), 1964, Chapter 1.
[5] K. Walters, *Rheometry* (Chapman and Hall, London, 1975), R.W. Whorlow, *Rheological Techniques* (Wiley, New York, 1980).
[6] J.D. Huppler, E. Ashare and L.A. Holmes, Trans. Soc. Rheol. **11** (1967) 159.
[7] C.J.S. Petrie, *Elongational Flows* (Pitman, London, 1979).
[8] R.B. Bird, Chem. Engr. Communications **16** (1982) 175.
[9] H.M. Laun and H. Münstedt, Rheol. Acta **17** (1978) 415.
[10] J. Meissner, Pure and Appl. Chem. **42** (1975) 551.
[11] J.G. Oldroyd, Proc. Roy. Soc. **A200** (1950) 45, **A245** (1958) 278; J. Non-Newtonian Fluid Mech. **14** (1984) 9-46.

[12] A.S. Lodge, *Body Tensor Fields in Continuum Mechanics*, Academic Press, New York, 1975, *Elastic Liquids*, Academic Press, New York 1964.
[13] R.S. Rivlin and J.L. Ericksen, J. Rat. Mech. Anal. **4** (1955) 323. H. Giesekus, ZAMM **42** (1962) 32.
[14] W.O. Criminale Jr., J.L. Ericksen and G.L. Filbey Jr., Arch. Rat. Mech. Anal. **1** (1958) 410.
[15] A.E. Green and R.S. Rivlin, Arch. Rat. Mech. Anal. **1** (1957) 1.
[16] A. Kaye, College of Aeronautics, Cranfield, Note #134 (1962); B. Bernstein, E.A. Kearsley and L. Zapas, Trans. Soc. Rheol. **7** (1963) 391.
[17] H. Giesekus, Rheol. Acta **2** (1962) 50.
[18] H.A. Kramers, Physica **11** (1944) 1, in Dutch.
[19] R.I. Tanner, Trans. Soc. Rheol. **19** (1975) 37.
[20] L.E. Wedgewood and R.B. Bird, Ind. Eng. Chem. Research **27** (1988) 1313.
[21] A.W. El-Kareh and L.G. Leal, J. Non-Newtonian Fluid Mech. **33** (1989) 257.
[22] H. Giesekus, J. Non-Newtonian Fluid Mech. **11** (1982) 69; **12** (1983) 367.
[23] H.C. Öttinger, J. Rheol. **32** (1988) 135.
[24] A.S. Lodge and Y. Wu, Rheol. Acta **10** (1971) 539.
[25] H.C. Öttinger, J. Non-Newtonian Fluid Mech. **26** (1987) 207; L.E. Wedgewood and H.C. Öttinger, ibid. **27** (1988) 245.
[26] J.M. Wiest and R.I. Tanner, J. Rheol. **33** (1989) 281.
[27] J.M. Wiest, L.E. Wedgewood and R.B. Bird, J. Chem. Phys. **90** (1989) 587.
[28] L.E. Wedgewood, D.N. Ostrov and R.B. Bird, U. Wisc. Rheology Research Center Report, Aug. 1990.
[29] T.W. Liu, J. Chem. Phys. **90** (1989) 5826.
[30] C.F. Curtiss and R.B. Bird, J. Chem. Phys. **74** (1981) 2016, 2026.
[31] R.B. Bird, H.H. Saab and C.F. Curtiss, J. Phys. Chem. **86** (1982) 1102; R.B. Bird, H.H. Saab and C.F. Curtiss, J. Chem. Phys. **77** (1982) 4747; H.H. Saab, R.B. Bird and C.F. Curtiss, ibid. **77** (1982) 4758.
[32] J.D. Schieber, C.F. Curtiss and R.B. Bird, Ind. Eng. Chem. Fundam. **25** (1986) 471; J.D. Schieber, J. Chem. Phys. **87** (1987) 4917, 4928.
[33] M. Doi and S.F. Edwards, J. Chem. Soc., Faraday Trans. 2 **74** (1978) 1789, 1802, 1818; **75** (1979) 38.

Nomenclature

Roman Symbols

a	= parameter in Giesekus model
a	= rod length in flexible bead-rod-chain model
b_1, b_2, b_{11}, etc	= constants in the "retarded-motion expansion"
$b = HQ_0^2/kT$	= constant appearing in nonlinear spring models
$\mathbf{F}_\nu^{(h)}$	= hydrodynamic drag force on bead ν
$\mathbf{F}_\nu^{(b)}$	= Brownian (thermal) force on bead ν
$\mathbf{F}_\nu^{(\phi)}$	= spring force on bead ν
$\mathbf{F}^{(c)}$	= connector force
f	= orientational distribution function for rigid-dumbbell model
$\mathbf{g}$	= gravitational acceleration

$\hat{g}_i, \hat{g}^i$	= convected base vector and reciprocal base vector
$\hat{g}_{ij}, \hat{g}^{ij}$	= covariant and contravariant metric matrix components
H	= spring constant in bead-spring model
I_1, I_2	= invariants of the Finger strain tensor
k	= Boltzmann constant
M	= memory function in K-BKZ model
$M_1, M_2, \cdots$	= kernel functions in memory-integral expansion
m	= mass of bead
N	= number of beads in a bead-spring-chain and bead-rod-chain models
n	= number density of polymers
$\mathbf{Q} = \mathbf{r}_2 - \mathbf{r}_1$	= connector vector in a dumbbell model
Q_0	= maximum stretching of a spring
$\mathbf{Q}_j$	= connector vector for jth spring in bead-spring-chain model
$\mathbf{r}$	= position vector
$\mathbf{r}_\nu, \dot{\mathbf{r}}_\nu$	= position and velocity of bead ν
$\mathbf{r}_c$	= position of center of mass of a bead-spring model
T	= absolute temperature
t	= time (i.e., "current" time, or "present" time)
t'	= past time ($t' \leq t$)
$\mathbf{u}$	= unit vector along rod in rigid-dumbbell model
$\mathbf{v}$	= (mass-average) fluid velocity
$\mathbf{v}_\nu$	= fluid velocity at bead ν
$\mathbf{v}_\nu'$	= perturbation of fluid velocity at bead ν because of hydrodynamic interaction
$\mathbf{v}_0$	= fluid velocity at center of mass of bead-spring model [may be a function of t]
W	= potential function in the K-BKZ model
x,y,z	= cartesian coordinate
Z	= quantity appearing in nonlinear spring models

Greek Symbols

$\mathbf{A}, \mathbf{B}$	= tensors in Curtiss-Bird model
$\mathbf{B}, \mathbf{B}^{-1}$	= Finger and Cauchy strain tensors
β	= chain-constraint exponent in Curtiss-Bird model
$\dot{\gamma}$	= shear rate
$\dot{\gamma}^0$	= complex amplitude of shear-rate
$\gamma_{(1)} = \gamma^{(1)}$	= $\nabla\mathbf{v} + (\nabla\mathbf{v})^\dagger$ = rate-of-strain tensor
$\gamma_{(n)}, \gamma^{(n)}$	= nth rate-of-strain tensors [n = 1,2,3, ···] (functions of t)
$\gamma_{[n]}, \gamma^{[n]}$	= nth rate-of-strain tensors [n = 1,2,3, ···] (functions of t and t')
$\gamma_{[0]}, \gamma^{[0]}$	= relative finite-strain tensors (functions of t and t')
Δ_{ji}	= jth component of base vector $\hat{g}_i$
$\boldsymbol{\delta}$	= unit tensor (with components δ_{ij})
$\boldsymbol{\delta}_i$	= unit vector in ith cartesian coordinate direction (with components δ_{ij})
δ_{ij}	= Kronecker delta

E_{ij}	= jth component of reciprocal base vector $\hat{g}^i$
ε	= link-tension coefficient in Curtiss-Bird model
$\dot{\varepsilon}$	= elongation rate
ζ	= friction coefficient for a bead
ζ_0	= friction tensor for a bead
η	= non-Newtonian (i.e., shear-rate dependent) viscosity [defined only for steady-state shear flow!]
η_s	= solvent viscosity [here taken to be Newtonian]
η_0	= zero-shear-rate non-Newtonian viscosity
η^*	= η'-$i\eta''$ complex viscosity [a complex, frequency-dependent quantity]
$\bar{\eta}$	= elongational viscosity
η^+	= shear-stress growth function
$\kappa = (\nabla v)^\dagger$	= transpose of velocity-gradient tensor [may be a function of t]
λ_i	= time constants in bead-spring-chain models (i = 1,2, ··· N-1)
λ_i	= time constants in Oldroyd model (i = 1,2, ··· 7)
$\lambda_H = \zeta/4H$	= time constant
$\lambda_H' = \lambda_H''$	= time constants in Giesekus model
μ	= viscosity of a Newtonian fluid
μ, ν	= memory functions in Curtiss-Bird model
ξ_0	= anisotropic tensor in Brownian motion force
π	= (total) stress tensor
π_s	= solvent contribution to π
ρ	= fluid density (mass per unit volume)
τ	= extra stress tensor
$\hat{\tau}_{ij}, \hat{\tau}^{ij}$	= covariant and contravariant convected components of τ
τ_s, τ_p	= solvent and polymer contributions to τ
τ_i	= contribution of ith mode to τ in bead-spring-chain model
ϕ	= potential energy of a bead-spring system
Ψ	= configurational distribution function in terms of coordinates r_ν
Ψ_1, Ψ_2	= first and second normal-stress coefficients [defined only for steady-state shear flow!]
Ψ_{10}, Ψ_{20}	= zero-shear-rate first and second normal-stress coefficients
Ψ_1^+, Ψ_2^+	= normal-stress growth functions
ψ	= configurational distribution function in terms of connector vectors Q_k

Miscellaneous

D/Dt	= $\partial/\partial t + v \cdot \nabla$ = substantial (material) derivative
$\Phi^\dagger$	= transpose of the tensor Φ
[ ]	= average in the momentum space
< >	= average in the phase space
$(\ )_{(1)}$	= first convected derivative of ( )

$$\frac{\partial}{\partial r_\nu} = \delta_x \frac{\partial}{\partial x_\nu} + \delta_y \frac{\partial}{\partial y_\nu} + \delta_z \frac{\partial}{\partial z_\nu} = \text{a } \nabla\text{-operator in the } r_\nu\text{-space}$$

PROBLEMS ASSOCIATED WITH THE ELASTICITY OF LIQUIDS

D. D. Joseph
Department of Aerospace Engineering and Mechanics
University of Minnesota, Minneapolis, MN 55455

These lectures are in three parts:

1. Physical phenomena associated with hyperbolicity and change of type;

2. Conceptual ideas associated with effective viscosities and rigidities and the origins of viscosity in elasticity;

3. Mathematical problems associated with hyperbolicity and change of type.

The ideas which I will express in these lecture are very condensed forms of ideas which have been put forward in various papers and most completely in my recent book *Fluid Dynamics of Viscoelastic Liquids,* published in 1990 by Springer-Verlag. The mathematical theory of hyperbolicity and change of type is associated with models with an instantaneous elastic response. Basically, this means that there is no Newtonian like part of the constitutive equation. The theory for these models as it is presently known is in my book. I am persuaded that further development of this subject lies in the realm of physics rather than mathematics. The main issues are centered around the idea of the effective viscosity and rigidity and the measurements of slow speeds, topics which are discussed in this paper in a rather more discursive than mathematical manner.

1. PHYSICAL PHENOMENA ASSOCIATED WITH HYPERBOLICITY AND CHANGE OF TYPE

It is well known that small amounts of polymer in a Newtonian liquid can have big effects on the dynamics of flow. Drag reductions of the order of 80% can be achieved by adding polymers in concentrations of fifty parts per million to water. This minute addition does not change the viscosity of the liquid but evidently has a strong effect on other properties of the liquid which have as yet been inadequately identified.

We are going to consider some effects of adding minute quantities of polyethylene oxide to water on the flow over wires. The first experiments were on uniform flow with velocity U across small wires, flow over a cylinder. James and Acosta [1970] measured the heat transferred from three wires of diameter D=0.001, 0.002 and 0.006 inches. They used three different molecular weights of polymers in water (WSR 301, 205 and coagulant) in concentrations ϕ ranging from 7 parts to 400 parts per million by weight, the range of extreme dilution, in the drag reduction range. They found a critical velocity U_c in all cases except the case of most extreme dilution ϕ=6.62 ppm, as is shown in Figure 1. A brief summary of the results apparent in this figure follows.

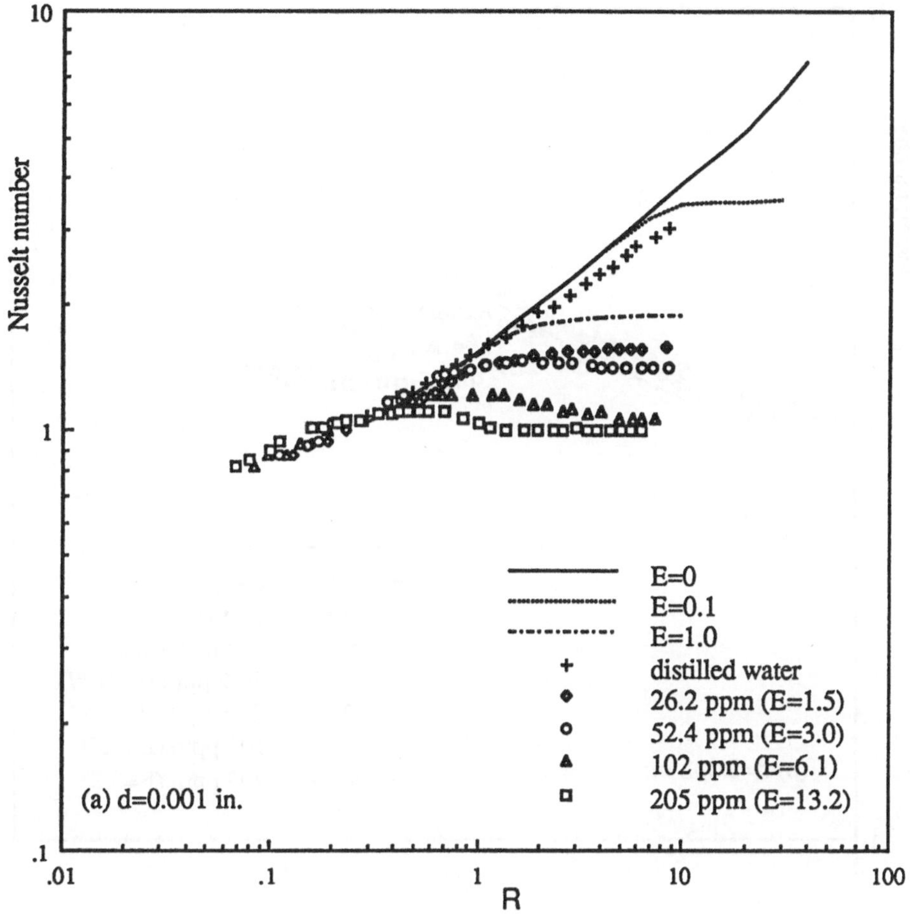
10
1
.1
Nusselt number
.01
.1
1
10
100
R
E=0
E=0.1
E=1.0
distilled water
26.2 ppm (E=1.5)
52.4 ppm (E=3.0)
102 ppm (E=6.1)
205 ppm (E=13.2)
(a) d=0.001 in.

(a)

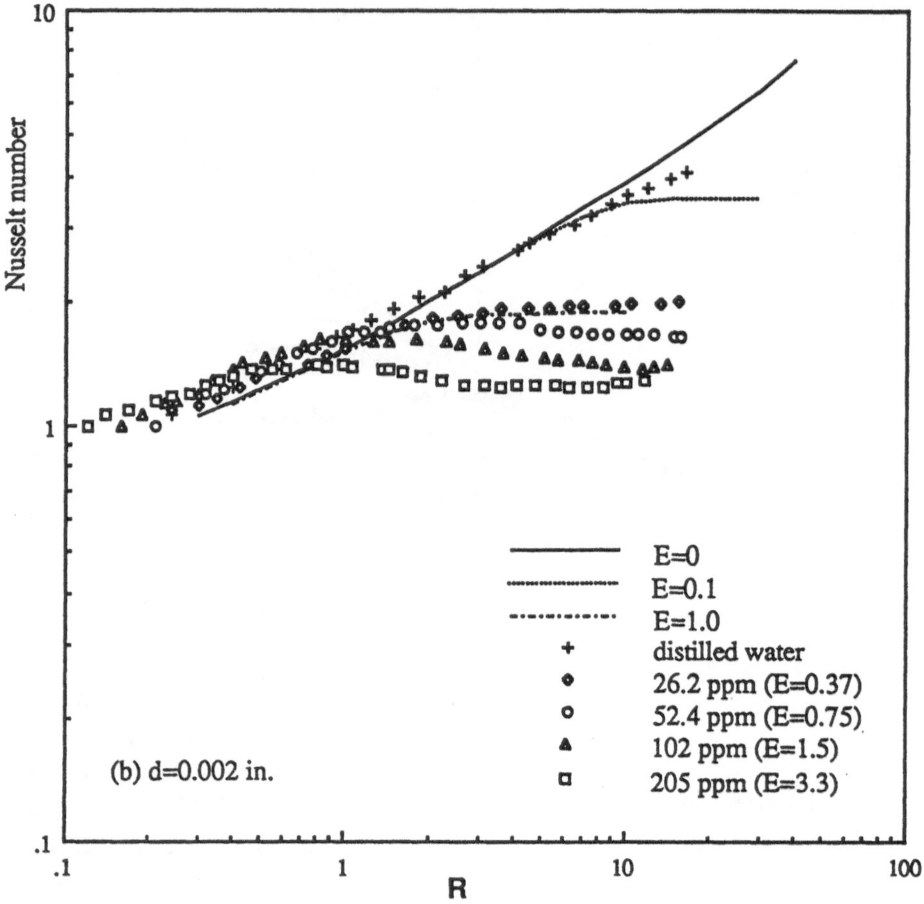
10
1
.1
Nusselt number
.1
1
10
100
R
E=0
E=0.1
E=1.0
distilled water
26.2 ppm (E=0.37)
52.4 ppm (E=0.75)
102 ppm (E=1.5)
205 ppm (E=3.3)
(b) d=0.002 in.

(b)

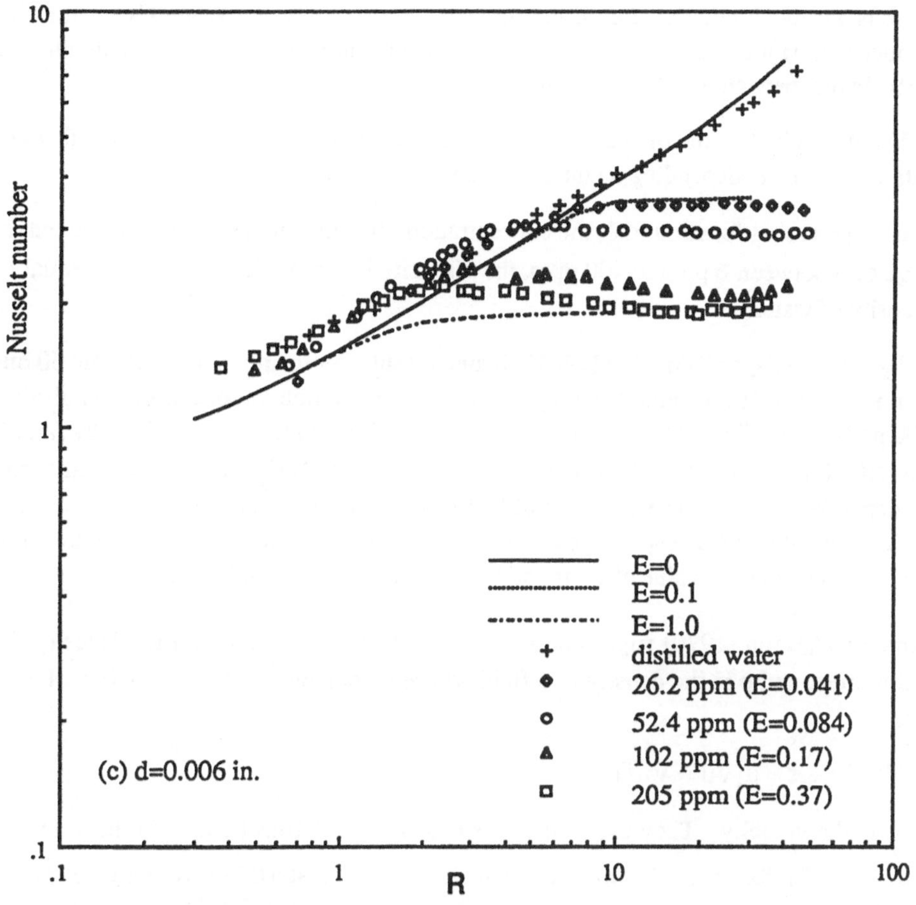

(c)

Figure 1 Heat transfer from heated wires of WSR 301 (after James and Acosta [1987]). The experimental points are dots and the lines are from computations of Hu and Joseph [1990]. (a) d=0.001 in. (b) d=0.002 in. (c) d=0.006 in.

1. There is a critical value U_c for all but the most dilute solutions: When $U<U_c$, the Nusselt number Nu(U) increases with U as in a Newtonian fluid. For $U>U_c$, the Nusselt number becomes independent of U as in Figure 1.

2. U_c is independent of the diameter of the wire. This is remarkable. It suggests that U_c is a material parameter depending on the fluid alone.

3. U_c is a decreasing function of ϕ, the concentration. It is useful to note once again that in the range of ϕ between 6 ppm to 400 ppm, the viscosity is essentially constant and equal to the viscosity of water.

Ambari, Deslouis, and Tribollet [1984] obtained results for the mass transfer from 50 micron wires in a uniform flow of aqueous polyox (coagulant) solution in concentrations of 50, 100, and 200 parts per million. Their results are essentially identical to those obtained by James and Acosta [1970]; there is a critical U_c, a decreasing function of ϕ, signalling a qualitative change for the dependence of the mass transport of U, from a Newtonian dependence when $U<U_c$, to a U independent value for $U>U_c$. Their values of U_c for the break in the mass transport curve are just about the same as the value of U_c found by James and Acosta for heat transfer.

Ultman and Denn [1970] suggested that $U_c=c=\sqrt{\eta/\lambda\rho}$ where η is the viscosity, λ the relaxation time, and ρ is the density of a fluid whose extra stress $\boldsymbol{\tau}=\mathbf{T}+p\mathbf{1}$ satisfies Maxwell's equation

$$\lambda U \partial\tau/\partial x + \tau = \mu[\nabla u = \nabla u^T] \tag{1}$$

where **u** is the velocity. They used the molecular theory of Bueche to find the value of the relaxation time λ_B for the 52.4 ppm solution and they found that a $0.7\lambda_B$ would give $\sqrt{\eta/0.7\lambda_B}$ $=U_c \sim 2.9$ cm/sec., that is, their estimate of λ_B from Bueche's theory is almost good enough to give $c=U_c$. Their calculation of the time of relaxation cannot be relevant, however, because in the Bueche theory

$$\lambda_B = \frac{12M\eta_s(10^3+12\phi)}{10^6\pi^2 RgT} \tag{2}$$

does not go to zero with the concentration ϕ. The quantities in (2) are (M, η_s, Rg, T)=(molecular weight, the viscosity of water, the gas constant, absolute temperature). The zero ϕ value of λ_B can be interpreted as a relaxation time for a single polymer in a sea of solvent. The relaxation time of one polymer cannot be the relaxation time of the solution in the limit in which the polymer concentration tends to zero, because in this limit the solution is all solvent.

Joseph, Riccius and Arney [1986] measured $c=2.48$ cm/sec in a 50 ppm, WSR 301 aqueous solution. This measurement supports the idea that $U_c=c$. We are trying now to measure wave speeds in extremely dilute solutions in the drag reduction range. We find considerable scatter in our data in these low viscosity solutions and are at present uncertain about the true value of the effective wave speed, including the values which we reported earlier.

The hypothesis that $U_c=c$ is consistent with the following argument about the dependence of the wave speed on concentration. In the regime of extreme dilution, the viscosity does not change with concentration. However, there appears to be a marked effect on the average time of relaxation which increases with concentration. It follows then that the wave speed $c=\sqrt{\eta/\rho\lambda}$ must decrease with concentration ϕ.

The shear viscosity for dilute polymers can be calculated using

$$\eta = \eta_s(1 + [\eta]\phi) \tag{3}$$

where $[\eta]=(\eta-\eta_s)/\phi$ is the intrinsic viscosity. It has a definite value as $\phi\to 0$. James and Acosta [1970] and James and Gupta [1971] developed expressions of the form $\lambda=A\phi$ from molecular theory. A is a function of the polymer properties and it can even be a slowly varying function of ϕ. They find that

$$A = \frac{2}{5}\frac{\eta_s[\eta]^2 M}{RgT} \; . \tag{4}$$

This expression shows that λ is proportional both to the largest relaxation time of polymer molecules $\lambda_m=\eta_s[\eta]n/\pi^2 RgT$ and to the concentration ϕ. James and Gupta [1971] generalized the derivation and showed significant influence of molecular weight distribution on the magnitude of the relaxation. They found that $\lambda=A\phi$, with A given by (4) could possibly underestimate the value of the relaxation time computed as a mean value from a two relaxation time model by a factor of order 10, depending on the molecular weight distribution of the polymers.

If at small concentrations ϕ, $\lambda=A\phi$ with A independent of ϕ, then the wave speeds $c=\sqrt{\eta/\lambda\rho}$ of dilute polymer solutions of two concentrations ϕ_1 and ϕ_2 are given by

$$c_2 = c_1 \sqrt{\frac{\phi_1}{\phi_2}}\sqrt{\frac{1+[\eta]\phi_2}{1+[\eta]\phi_1}} \; .$$

for extremely dilute solutions $[\eta]\phi \ll 1$ and we find that

$$c = \sqrt{\eta_s/\rho A\phi} = C\phi^{1/2} \tag{5}$$

where $\lambda=A\phi$ and η_s, A, ρ and C are independent of ϕ.

In Figure 2 we have plotted the critical velocity versus concentration for three polyox solutions and the three diameters of wires used in the experiments. We see that the line $U_c=C\phi^{1/2}$ fit the data of James and Acosta quite well. This lends support to the notion that the critical speed is equal to the shear wave speed $U_c=c$ in some approximate sense. We note that the attempt of Ultmann and Denn [1971] to fit the concentration data (their Figure 3) failed because they used the Bueche relaxation time (2) rather than the linear relation $\lambda = A\phi$, with A determined from a measurement using the wave speed meter.

We shall return to compare these observations with direct numerical simulations in part 3 of this paper.

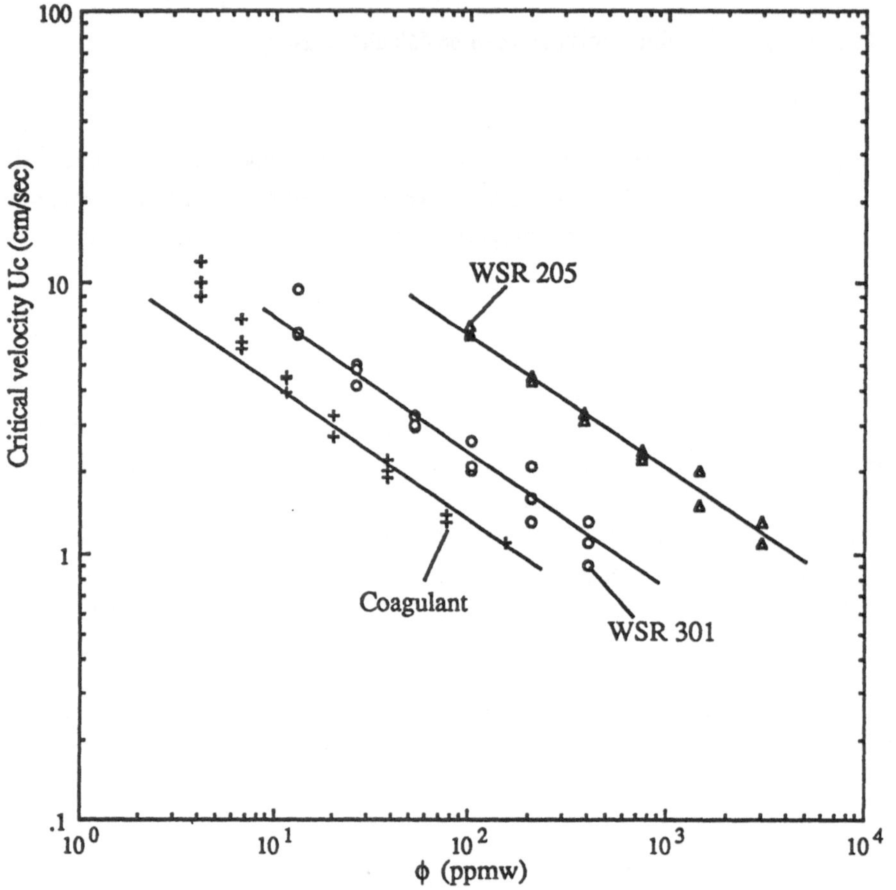

Figure 2 Critical velocity at which the Nusselt number starts to deviate from the Newtonian data versus concentration for three Polyox solutions (WSR-205, WSR-301, coagulant) and three cylindrical wires (d=0.001, 0.002, 0.006in.). Data of James and Acosta [6]. Solid lines are correlations $U_c = C\,\phi^{-1/2}$.

Konuita, Adler and Piau [1980] studied the flow around a 0.206 mm wire in an aqueous polyox solution (500 ppm, WSR-301) using laser-Doppler techniques. They found a kind of shock wave in front of the cylinder, like a bow shock. They say that the velocity of the fluid is zero in a region fluid in front of the stagnation point. Basically they say that there is no flow, or very slow flow near the cylinder. The formation of the shock occurs at a certain finite speed, perhaps U_c. This type of shock is consistent with the other observations in the sense that with a stagnant region around the cylinder, the transport of heat and mass could take place only by diffusion, without convection. This explains why there is no dependence of the heat and mass transfer on the velocity when it exceeds a critical value.

I estimated the critical speed, using the data of Konuita, Adler and Piau, and I estimated the wave speed c by extrapolating from our measurements in the polyox solutions at different concentrations. These estimates are reported in my book "Fluid Dynamics of Viscoelastic Liquids." They are consistent with the notion of a supercritical shock transition at $U_c=c$.

Another striking phenomenon which appears to be associated with a supercritical transition is delayed die swell. It is well known that polymeric liquid will swell when extruded from small diameter pipes. The swelling can be very large, four, even five times the diameter of the jet. This swelling is still not well understood even when there is no delay. Joseph, Matta and Chen [1987] have carried out experiments on 19 different polymer solutions. They found that there is a critical value of the extrusion velocity U_c such that when $U<U_c$, the swell occurs at the exit, but when $U>U_c$ the swell is delayed, as in Figure 3. If U is taken as the centerline velocity in the pipe, then the transition is always supercritical with $U_c>c$. The length of the delay increases with U. The velocity in the jet after the swell of jet has fully swelled is subcritical $U_f<U$ where U_f is the final U. This is something like a hydraulic jump with supercritical flow ahead of the delay and subcritical flow behind it.

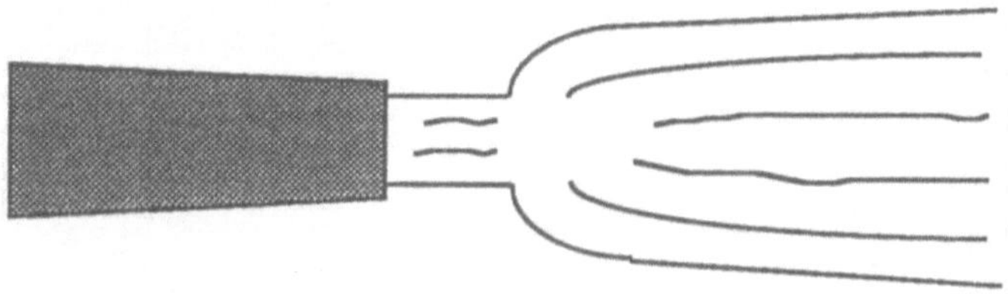

Figure 3 Delayed die swell.

Yoo and Joseph [1985] studied Poiseuille flow of an upper convected Maxwell model through a plane channel. Ahrens, Yoo and Joseph [1987] studied the same problem in a round pipe. In both cases, we get a hyperbolic region of flow in the center of the pipe when the centerline velocity U_m, equal to 2U in the Maxwell model, is greater than the wave speed c. This gives theoretical support to the idea that delayed die swell is a supercritical phenomenon.

There is a marked difference between the shape of the swell when it is delayed between different polymer solutions. The shape seems to correlate with a relaxation time

$$\lambda = \tilde{\mu}/G_c \tag{2}$$

where $\tilde{\mu}$ is the zero shear rate viscosity and G_c is the rigidity. We get G_c from measuring c

$$c^2 = G_c/\rho\ . \tag{3}$$

When λ is large, say $\lambda \geq 0(10^{-3}$ sec), the delay is sharp, as in Figure 3. When the relaxation times are small, $\lambda \leq 0(10^{-4}$ sec), the delay is smoothed; in the extreme cases it is difficult to see that the swell is actually delayed.

We can say the Newtonian fluids are fluids with very large values of λ. In the case of delayed die swell, the smoothing of the swell is probably associated by the effect of smoothing due to an effective viscosity which arises from rapidly relaxing modes which have already relaxed when the delayed swell commences. Very viscous liquids always exhibit relaxation or non-Newtonian effects because even though the relaxation is fast, there is so much to relax.

In Figure 4, we plotted the critical Mach number

$M_c = 2U_c/c$

against the diameter of the pipe. In all cases $M_c \geq 1$, nearly. The value $M_c=1$ seems to be some form of asymptote for large values of the pipe diameter d. We do not understand why different fluids have such different M_c vs. d curves. We have thought about the consequences of shear thinning, which are important for some of the test liquids, in trying to collapse the experimental curves for different liquids into one curve, but we have not been successful.

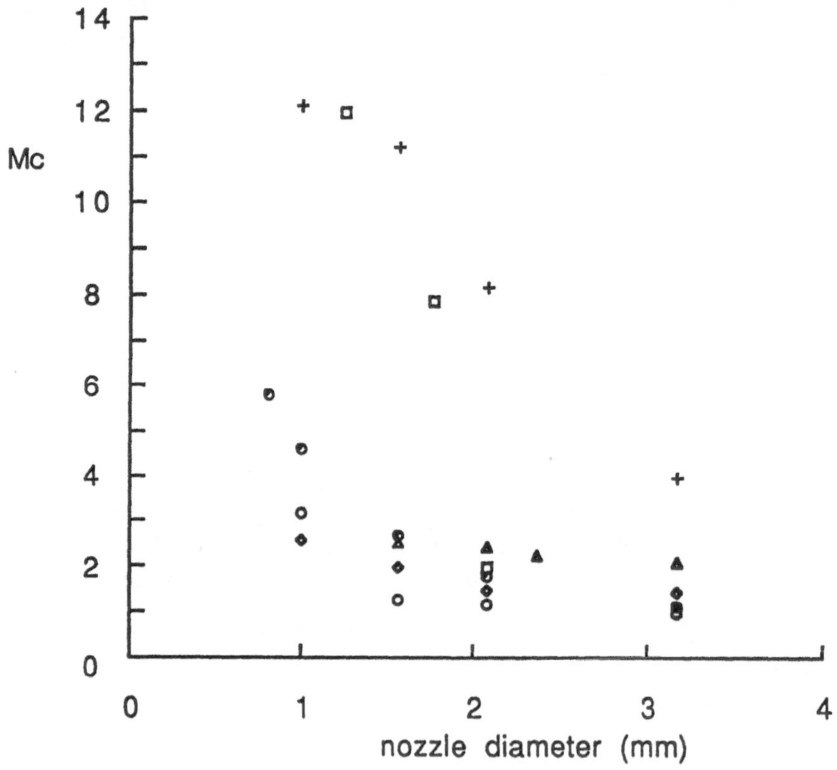

Figure 4 Mach number vs. pipe diameter.

○ 1.3% CMC □ 9.8% ELVACITE ◇ 6% PIB/D

● 2.5% POLYOX + 12.1% K-125 △ M-1

Delvaux and Crochet [1989] have done a numerical study of delayed die swell in a plane jet using an Oldroyd B model, the upper convected Maxwell model plus a very small perturbing Newtonian viscosity μ such that $\mu/(\mu+\eta)=0.05$. The results of their calculation are very interesting. They confirm the conjectures of Joseph, Matta and Chen [1987] which have been expounded, and introduce some new understandings. The main new result can be described as "the breakout of the region of the hyperbolic vorticity." At small supercritical values of the velocity (Mach numbers not too greatly in excess of one) the hyperbolic region extends slightly downstream into the jet but does not touch the jet boundary, as can be seen in panel (a) and (b) of Figure 5.

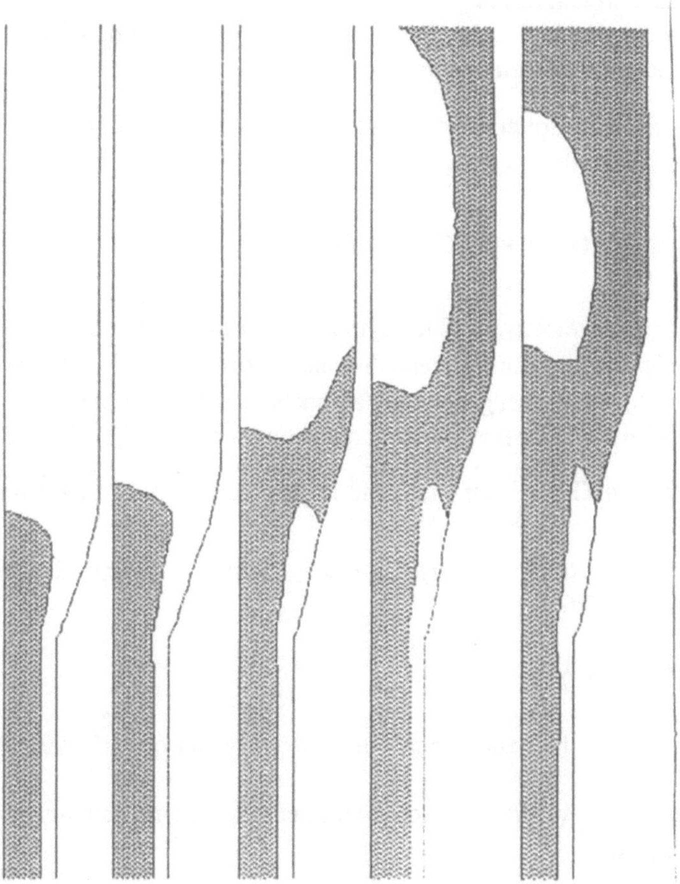

Figure 5 (after Delvaux and Crochet, 1989). Jet profile and hyperbolic regions of vorticity under different conditions in a plane jet of an Oldroyd B fluid with a very small Newtonian viscosity. (a) (M, R,W)=(2.3, 13.5, 0.39); (b) (2.9, 17, 0.49); (c) (4.1, 29.9, 0.87), (d) (5.1, 34.6, 1)

As the velocity increases, more and more of the jet is consumed by the hyperbolic region. At a certain velocity, between panel (b) and (c) of Figure 5, the hyperbolic region first touches the jet boundary, then consumes more and more of the jet boundary. Evidently the change in the curvature of the jet is associated with the breakout of the hyperbolic region. This explains why the delay is not observed at small supercritical values of the velocity but only at larger postcritical breakout values. It would be good if we could find a way to explain the way the delay depends on the jet diameter.

2. CONCEPTUAL IDEAS

Nonlinear constitutive modeling is a jungle. The possible responses of the material to stresses are too complicated to describe by one explicit expression. General expressions are too abstract to be of direct use and are always insufficiently general to describe everything. Linearizing around rest is good because many different models collapse to one. The nonlinear parameters go away. Moreover, the elasticity of liquids is preeminently associated with propagation of small amplitude waves into rest.

We start with Boltzmann's expression for the extra stress $\boldsymbol{\tau}$ which has been generalized to contain a Newtonian term

$$\boldsymbol{\tau} = 2\mu\mathbf{D}\,[\mathbf{u}(\mathbf{x}, t)] + 2\int_0^\infty G(s)\mathbf{D}[\mathbf{u}(\mathbf{x}, t-s)]ds \tag{6}$$

where $\mathbf{u}$ is the velocity, $\mathbf{D}$ is the symmetric part of grad $\mathbf{u}$ and $G(s)$ is positive, bounded and monotonically decreasing to zero. The actual stress $\mathbf{T}=-p\mathbf{1}+\boldsymbol{\tau}$ differs from $\boldsymbol{\tau}$ by a "pressure" p. Equation (6) is the most general linear functional of grad $\mathbf{u}$ in a fluid. To name a fluid, we need a Newtonian viscosity μ and a shear relaxation modulus $G(s)$. We get Jeffreys' model from (6) when we write $G(s)=\frac{\eta}{\lambda}\exp(-s/\lambda)$ and Jeffreys' model reduces to Maxwell's if also $\mu=0$.

Now we consider viscosity. In steady flow, $\mathbf{u}$ is independent of t and comes out of the integral in (6). We get

$$\boldsymbol{\tau} = 2\tilde{\mu}\mathbf{D}[\mathbf{u}(\mathbf{x})]$$

where $\tilde{\mu}=\mu+\eta$ is the static or zero shear viscosity and $\eta = \int_0^\infty G(s)ds$, the area under $G(s)$, is the elastic viscosity. We have a viscosity inequality $\tilde{\mu}\geq\eta$ with equality when there is no Newtonian viscosity $\mu=0$.

Now we consider elasticity $\mu=0$, writing

$$\mathbf{D}[\mathbf{u}(\mathbf{x}, t-s)] = -\frac{\partial}{\partial s}\mathbf{E}\,[\boldsymbol{\xi}(\mathbf{x}, t-s)]$$

where ξ is a displacement and $\mathbf{E}$ is the infinitesimal strain. If it were possible to make a step in strain without flow, and it isn't possible, we would have $\mathbf{D}[\mathbf{u}(\mathbf{x}, t)] = \mathbf{E}_0(\mathbf{x})\delta(t)$ for Dirac δ. Then, from (6), with $\mu=0$,

$$\tau = 2G(t)\,E_0(x)$$

and you can see why G(t) is called the stress relaxation function and G(0) the rigidity or shear modulus. Another way to see elasticity with $\mu=0$ is to write

$$\tau = 2\int_0^\infty -\frac{\partial}{\partial s}\{G(s)E\,[\xi\,(x, t-s)]\}ds + 2\int_0^\infty G'(s)E\,[\xi\,(x, t-s)]\,ds\,. \tag{7}$$

Now we can suppose that G(s) decays ever so slowly so that the second integral will tend to zero while the first gives rise to linear elasticity for an incompressible solid

$$\tau = 2G(0)\,E\,[(\xi\,(x, t)]. \tag{8}$$

Now we restore the Newtonian viscosity and we note that this viscosity smooths discontinuities. For example, in the problem of the suddenly accelerated plate, the boundary at $y=0$ below a semi-infinite plate is suddenly put into motion, sliding parallel to itself with a uniform speed. If $\mu=0$, this problem is governed by a telegraph equation. The news of the change in the boundary value from zero to constant velocity propagates into the interior by a damped wave with a velocity $c=\sqrt{G(0)/\rho}$. The amplitude of the velocity shock decays exponentially. A short while after the wave passes, the solution at the given y looks diffusive. If $\mu\neq0$, and is small, a sharp front cannot propagate. Instead we get a shock layer whose thickness is proportional to $\sqrt{\mu y/\bar{\mu}}$ and the solution, as in the Newtonian fluid, is felt instantly everywhere. We get a diffusive signal plus a wave. The wave could be dominant in the dynamics if μ is small.

Actually diffusion is impossible because it requires that a pulse initiated at any point be felt instantly everywhere. This same defect hold for all models with $\mu\neq0$, like Jeffreys'. Propagation should proceed as waves.

Poisson, Maxwell, Poynting and others thought that $\mu=0$ ultimately. It's all a matter of time scales. Short range forces between molecules of a liquid give rise to weak clusters of molecules which resist fast deformations elastically, then relax. Liquids are closer to solids than to gases. Liquid molecules do not bounce around with a mean free path, they move cooperatively.

So what is the difference between two liquids with the same η, one appearing viscous (Newtonian) and the other elastic? Maxwell thought that viscous liquids were actually elastic, with high rigidity and a single fast time of relaxation. To fix his idea in your mind, we compare two liquids with the same viscosity η, satisfying Maxwell's model with $G(s)=G(0)\exp(-s/\lambda)$, $G(0)=\eta/\lambda$. To have the same η the Newtonian liquid would have a relatively large G(0) and a small time λ of relaxation. The trouble with Maxwell's model, if not his idea, is that a single time of relaxation is against experiments which can never be made to fit a single time of relaxation.

There are many different times of relaxation. Experiments indicate that many liquids respond to high frequency ultrasound like a solid organic glass with

$$G(0) \sim 10^9\text{Pa},\ c=\sqrt{G(0)/\rho} \sim 10^5\ \text{cm/sec}. \tag{9}$$

This type of estimation is valid for a huge range of liquids, from olive oil to high molecular weight silicon oils. With this time of relaxation and such a high rigidity, all the liquids would look Newtonian, with t much greater than $\tilde{\mu}/G(0)$, which is of the order of 10^{-10} sec. in olive oil, and is perhaps 10^{-6} in some high viscosity silicon oils. In fact, we see much longer lasting responses which come about because there are different times of relaxation. Small molecules relax rapidly, giving rise to large rigidity G(0) and fast speed. Large molecules and polymers relax slowly, giving rise to a smaller effective rigidity $G_\mu(0)$, effective viscosity μ and slow speed

$$c=c_\mu=\sqrt{G_\mu(0)/\rho}\ . \tag{10}$$

To get this firmly in mind, we can think of a kernel with values like those given by (9), sketched in Figure 6.

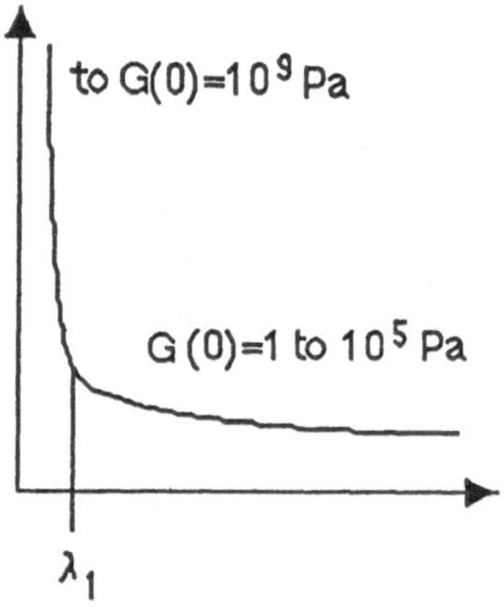

Figure 6 G(s), fast relaxation (say 10^{-10} sec) followed by a slow relaxation (say 10^{-4} sec).

We may inquire if at $t>>10^{-10}$ sec the relaxed fast modes have a dynamical effect. Yes, they give rise to an effective viscosity. We may as well collapse the glassy mode into a one-sided delta function $\mu\delta(s)$ where $\mu=G(0)\lambda_1$, or some fraction of this. This is our effective viscosity and our construction shows that is not unique. This is a very interesting concept, but it is not amenable to experiments that we know.

It is useful to define a time unit in terms of the slowest relaxation, say $\tilde{\mu}/G_c$. This gives rise to an internal clock, with a material time defined by the slowest relaxation. This time may be slow or fast on the external clock. To get this idea, think of the analog for the transport of heat. Heat is transported in solids by fast waves. The fastest wave is associated with electrons with relaxation times of 10^{-13} sec, then by lattice waves (phonons) with relaxation times of 10^{-11} sec. Both times are surpassingly short on our clock. However, at 10^{-13} sec, the electrons have all relaxed (and they give rise to diffusion) whilst the phonons have not begun to relax. Of course, it's more interesting when the slow relaxation is not too fast on our clock, as is true for viscoelastic fluids.

The notion of an external and internal clock is an appealing idea for expressing the difference between different theories of fading memory. Some theories, like Maxwell's and the more mathematical one by Coleman and Noll [1960] use an external clock; in rapid deformations the fluid responds elastically; in slow deformations the response is viscous. Fast and slow are measured in our time, on the external clock. Such theories rule out transient Newtonian responses. Models with $\mu \neq 0$, like Jeffreys', or the more mathematical one by Saut and Joseph [1983], are disallowed. To get $\mu \neq 0$ back in, even though ultimately $\mu = 0$, we need an effective μ, associated with an internal clock.

3. MATHEMATICAL THEORY

When the fluid is elastic the governing equations are partly hyperbolic. The hyperbolic theory makes sense when the Newtonian viscosity is zero or small relative to the static viscosity $\tilde{\mu}$. For very fast deformations in which the fluid responds momentarily like a glass, the equations always exhibit properties of hyperbolic response, waves and change of type. However, the glassy response takes place in times too short to notice. Hence, the hyperbolic theory is not useful where it is exact. The hyperbolic theory is useful when we get an elastic response at times we read on our clock, in the domain of the effective theory. Hence, the hyperbolic theory is useful where it is not exact.

Most of the mathematical work has been done with fluids like Maxwell's and for plane flows. These problems are governed by six quasilinear equations in six unknowns. The unknowns are two velocity components, three components of the stress, and a pressure. The continuity equation, two momentum equations and three equations for the stress govern the evolution of the six variables. The stress equations are like Maxwell's

$$\lambda\left[\frac{\partial \tau}{\partial t}+\mathbf{u}\cdot\nabla\tau+\tau\Omega-\Omega\tau-a(D\tau+\tau D)\right]=2\eta D+\ell$$

where $\mathbf{D}$ is the symmetric part and $\mathbf{\Omega}$ the antisymmetric part of $\nabla\mathbf{u}$, $-1\leq a\leq 1$ and ℓ are lower order terms, algebraic in the system variables. This system may be analyzed for type in the usual way. We get a 6th order system and it factors into three quadratic roots. Two of the roots are imaginary so that the system is not hyperbolic. The streamlines are characteristic, with double roots so that the system is not strictly hyperbolic. The third quadratic factor depends on the unknown solution, algebraically, and it can be real or complex, depending on the solution. We say that such a solution with mixed roots is of composite type. Some variables are elliptic, some are hyperbolic.

It turns out that the pair of roots which depend on the unknown solution and can change type are associated with the vorticity equation, a second order nonlinear PDE. This equation is either elliptic or it is hyperbolic, depending on the solution. It is not of composite type, but is classical, like the equation for the potential in gas dynamics.

We can think of the unsteady vorticity equation and the steady vorticity equation. The analysis of the two has greatly different consequences. The unsteady equation is ill-posed when it is elliptic and well-posed when it is hyperbolic. Ill-posed problems are catastrophically unstable to short waves, with growth rates which go to infinity with the wave number. The conditions on the stress which lead to ill-posed problems can be determined by the method of frozen coefficients, as was first done by Rutkevich [1969]. It turns out that the Maxwell models with $a=\pm 1$ cannot be ill-posed on smooth solutions, but the other models do become ill-posed for certain flows.

The problem of change of type in steady flow is different. The vorticity in steady flow can be of mixed type with elliptic and hyperbolic regions, as in transonic flow. The physical implications of these mixed "transonic" fields are not yet perfectly understood, though many examples have been calculated.

There are many models, other than those like Maxwell's, in which vorticity is the key variable. It is the only variable which is either strictly elliptic or strictly hyperbolic. The stream function satisfies Laplace's equation, the velocity and the stresses are of composite type. The stresses do not satisfy a hyperbolic equation and it is wrong to speak of the propagation of stress waves.

There are other models in which the vorticity is not the key variable. However, when these models are linearized around rest, one finds again that the steady vorticity equation is either elliptic or hyperbolic, and the unsteady vorticity equation is always hyperbolic. Hence it is precisely waves of vorticity which propagate into rest.

The mathematical consequences of composite roots are clearly evident in the recent solutions of L. E. Fraenkel [1987], H. Hu [1990], which are reviewed in Joseph's [1990] book, of the problem of linearized supercritical flow over a flat plate. The linearization here is around the uniform flow which exists at infinity, as in Oseen's problem for the Navier-Stokes equation. Fraenkel's solution shows that there is a Mach wedge of vorticity ζ centered on the leading edge of the plate. The vorticity in front of this wedge is zero and it is not zero behind the wedge [see Figure 7]. Surprisingly, the vorticity jumps from zero to infinity at the wedge, but the singularity is integrable. We have rotational flow behind the shock and irrotational flow in front of the shock. The stream function satisfies $\nabla^2\Psi=-\zeta$ where $\zeta=0$ in front of the shock. Therefore, we may write $\Psi=\Psi_1+\Psi_2$, $\nabla^2\Psi_2=-\zeta$, $\nabla^2\Psi_1=0$. To satisfy the boundary conditions on the plate, we must have a nonzero potential field Ψ_1. In fact Ψ_1 satisfies a Dirichlet problem for the region outside a strip on the positive x axis.

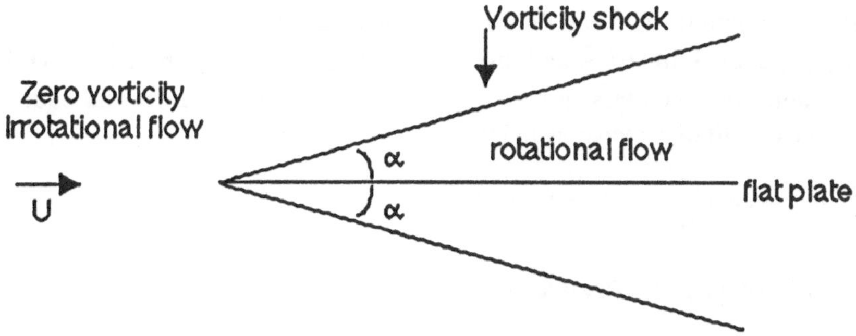

Figure 7 Mach wedge for the vorticity, $\tan \alpha = (1-M^2)^{-1/2}$.

The potential flow decays to uniform flow as one moves upstream, but the delay is slow. There is no upstream influence in the fully hyperbolic flow of a gas over a flat plate. The upstream influence of the flat plate in the flow of a Newtonian fluid is almost negligible. The persistence of Ψ_1 is a consequence of its ellipticity, ultimately to the fact that the first order system is of composite type. This type of solution may be new in mathematical physics.

The velocity and the stresses decompose into harmonic and vortical parts. Hence these fields are all of composite type. Only the vorticity is pure, strictly hyperbolic in the linearized problem of flow past bodies. The velocity and stresses are continuous across the shock. The normal derivative of the velocity, the normal and shear stress are also continuous, but the tangential derivative of the tangential components of velocity and stress are discontinuous. The elliptic component of our composite system is associated with a huge upstream influence.

Similar considerations enter into the dynamics of flow over small cylinders which we discussed in §1 of this paper. Delvaux and Crochet [1990] gave a numerical solution of the problem of flow over a cylinder using the constitutive equation of an upper convected Maxwell model. This solution is reviewed in the book of Joseph [1990]. Their solution is fully nonlinear and it supports the notion that the anomalous heat transport and drag observed in the experiments are associated with a change of type. A different numerical solution based on the algorithm SIMPLER has been given by Hu and Joseph [1990] and it agrees with the numerical solution of Delvaux and Crochet. Aspects of the solution of Hu and Joseph are discussed below.

In the present problem we wish to evaluate the effect of viscoelasticity upon the heat transfer and drag. An upper convected Maxwell model is used. We assume that the viscous heating is negligible and that temperature differences in the flow are small and such that the fluid properties (ρ, λ and η) do not change. Then the temperature field is decoupled from the velocity field, the energy equation is simply

$$c_p \rho \, (\mathbf{u} \cdot \nabla) T = \kappa \Delta T \qquad (11)$$

where T is the temperature, c_p the heat capacity and κ the thermal conductivity.

We shall scale length with the diameter of the cylinder d, velocity with the free stream velocity U, pressure with ρU^2 and stress with $\eta U/d$, and use the same symbol for the dimensional and dimensionless quantities. In the dimensionless form the equations of the momentum, the constitutive equation and the temperature are

$$(\mathbf{u}\cdot\nabla)\,\mathbf{u} = -\nabla p + \frac{1}{\Re}\,\Delta\mathbf{u} + \frac{1}{\Re}\,\nabla\cdot\tau_E\,, \tag{12}$$

$$W\left[(\mathbf{u}\cdot\nabla)\,(\tau_E + \tau_N) - \nabla\mathbf{u}\,(\tau_E + \tau_N) - (\tau_E + \tau_N)\nabla\mathbf{u}^T\right] + \tau_E = 0, \tag{13}$$

$$\Re\,\mathrm{Pr}\,(\mathbf{u}\cdot\nabla)\,T = \Delta\,T \tag{14}$$

where for the convenience of numerical treatment the extra stress is split into two parts $\tau = \tau_N + \tau_E$ with $\tau_N = 2\eta\,\mathbf{D}$ being the pseudo-Newtonian part and τ_E being the part due to the elasticity. The dimensionless temperature is taken as $(T-T_\infty)/T_\infty$ (T_∞ is the temperature of the coming fluid). In these equations the non-dimensional parameters $\Re$ (Reynolds number), W (Weissenberg number), Pr (Prandtl number) are defined as

$$\begin{aligned} \Re &= rUd/h\,, \\ W &= lU/d\,, \\ \mathrm{Pr} &= c_p r/k\,. \end{aligned} \tag{15}$$

It is helpful to introduce another two non-dimensional parameters, the viscoelastic Mach number M and the elasticity number E which are defined by

$$\begin{aligned} M &= \sqrt{W\Re} = \Re\sqrt{E} = \frac{U}{c}\,, \\ E &= \frac{W}{\Re} = \frac{hl}{rd^2}\,, \end{aligned} \tag{16}$$

where

$$c = \sqrt{\frac{\eta}{\lambda\rho}} \tag{17}$$

is the speed of shear waves in a Maxwell fluid. In the study of change of type the Mach number M is an essential parameter. The elasticity number E depends only on the fluid properties and the flow geometry. In our computation we choose the pair $(\Re, E)$ as the independent parameters, and simulate the flow in experiments by keeping E fixed and adjusting $\Re$.

In the computation we solve for the velocities, the pressure and the stresses in each iteration. Since the temperature field does not effect the velocity, the heat equation is solved after the iteration converges. Some additional quantities are also calculated. We evaluate the stream function ψ and the vorticity ω which are defined by

$$u_r = -\frac{\partial\psi}{r\partial\theta}, \qquad u_\theta = \frac{\partial\psi}{\partial r}, \tag{18}$$

$$\omega = \frac{\partial u_\theta}{\partial r} + \frac{u_\theta}{r} - \frac{\partial u_r}{r\partial\theta}. \tag{19}$$

From the computed values of the pressure and the pseudo-Newtonian stresses on the surface of the cylinder it is easy to obtain the drag force acting on the cylinder. The dimensional drag force per unit length on the cylinder is found to be

$$F_x = \rho U^2 d \int_0^\pi [\, p \cos\theta + \frac{1}{\Re}\tau_{Nr\theta} \sin\theta]_{r=d/2}\, d\theta, \tag{20}$$

which has two contributions, one from the pressure and the other from the pseudo-Newtonian shear stress. In writing (20) we noted that the contribution to the drag of the elastic part of the extra stress vanishes because $\tau_{Er\theta}=0$ on the surface of the cylinder. The drag coefficient is given by

$$C_D = \frac{F_x}{\frac{\rho}{2}U^2 d} \tag{21}$$

In our computation of heat transfer, we prescribe the upstream temperature as T_∞ (in dimensionless form T=0), and the temperature on the cylinder surface as $T_0=2T_\infty$ (in dimensionless form T=1). The dimensional average heat flux from the cylinder to the surrounding fluid is

$$Q = \frac{1}{\pi}\int_0^\pi \left[\kappa\frac{\partial T}{\partial r}\right]_{r=d/2} d\theta. \tag{22}$$

Thus the Nusselt number which characterizes the heat transfer from the cylinder to the surrounding fluid is defined as

$$Nu = \frac{Qd}{\kappa(T_0 - T_\infty)} = \frac{Qd}{\kappa T_\infty}. \tag{23}$$

We next keep the flow fixed at a certain Reynolds number and vary the elasticity number, thus we can look at the effect of the elasticity of the fluid on the flow. Figure 8 presents the streamlines in the neighborhood of the cylinder for flows at $\Re=10$ and E varying from 0 to 1.0. (a) and (b) are almost identical. Starting from (c) with M greater than one, we see an increasingly larger downstream shift of the streamlines, at the same time there is a relatively small upstream shift. The streamline pattern with viscoelastic fluids of large elasticity number differs significantly from that with Newtonian fluids. The large distortion of the streamlines creates a wide region near the cylinder where the velocity is very low, thus affects the total drag on the cylinder and the heat transfer from the cylinder to the surrounding fluid as we will see later.

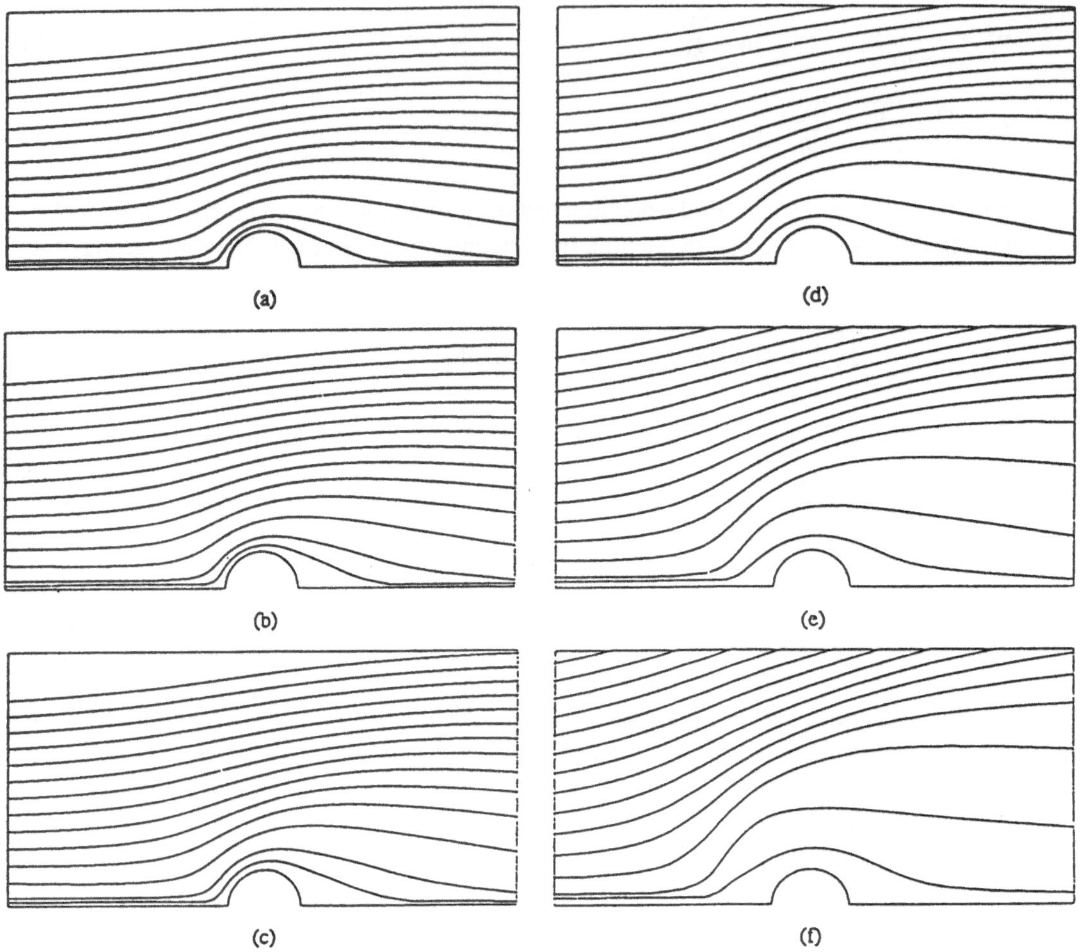

Figure 8 Streamlines in the neighborhood of the cylinder for the flow of the same Reynolds number $\mathfrak{R}$=10 and different elasticity number E. (a) E=0 (M=0). (b) E=0.01 (M=1.0). (c) E=0.1 (M=3.16). (d) E=0.25 (M=5.0). (e) E=0.5 (M=7.07). (f) E=1.0 (M=10). In the figures the values of the incoming streamlines, starting from the bottom, are 0.01, 0.05, 0.2, 0.4, 0.6, 0.8, 1.0, 1.2, 1.4, 1.6, 1.8, 2.0, 2.2 and 2.4 respectively.

Figure 8 shows the isovorticity lines at $\mathfrak{R}$=10 and E varying from 0 to 1.0. (a) is the familiar Newtonian case, where the isovorticity lines are swept downstream by the flow and the high vorticity region is at the front shoulder of the cylinder surface where the vorticity is being created. (b) is basically the same as (a) except at the front of the cylinder where the isovorticity lines are closer together signaling a sharper change of vorticity in this region. In (c), at a Mach number M=3.16, we see that the isovorticity lines jam together at the front of the cylinder thus creating a vorticity shock, like a blunt body shock in gas dynamics. As the elasticity number increases, this shock still exists and moves slightly upstream. In Figure 8(d) to 8(f), the picture of the isovorticity lines for viscoelastic fluids with large relaxation time is drastically different from that of Newtonian fluids. Besides the high vorticity zone on the front shoulder of the cylinder surface which occurs already in the Newtonian case, there exists a second high vorticity region which starts to build up and shifts away from the cylinder surface as the elasticity number increases. We find that the maximum values of the vorticity in this second region are even higher than the maximum values of the vorticity on the cylinder surface, which suggests generation of the vorticity away from the cylinder surface or behind the shock. We still do not understand the physical consequences of this build up. The existence of this second high vorticity region away from the cylinder surface was also observed in the work of Delvaux and Crochet [1990]; they found a local minimum and maximum in the vorticity plot along a path just above the cylinder ($\theta=\pi/2$). The dashed lines in Figure 9 indicate the angles, $\beta=\tan^{-1}\frac{1}{\sqrt{1-M^2}}$, ofvorticity shocks predicted in the linear theory in which the governing equations are linearized around the uniform income flow. Close to the cylinder, the vorticity shock is strong. The nonlinearity makes the shock curve around the cylinder. As E increases, the nonlinear region also increases due to the large stagnant region around the cylinder. Since the linear theory is valid far away from the cylinder, the vorticity shock, if it exists, should eventually stretch with the angle predicted in the linear theory. But because the the shock is weak and the numerical space discretization is usually coarse far away from the cylinder, it is very hard to capture this part of the shock numerically.

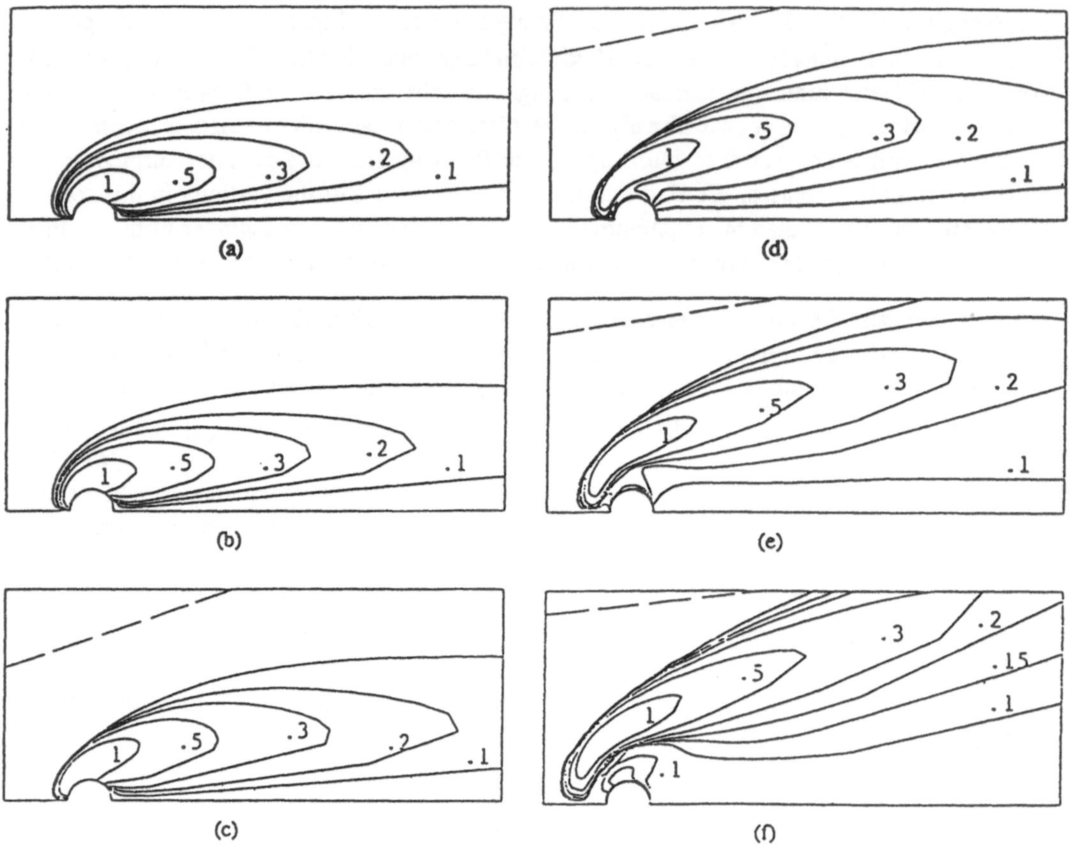

Figure 9 Isovorticity lines for the flow of the same Reynolds number $\mathfrak{R}=10$ and different elasticity number E. (a) E=0 (M=0). (b) E=0.01 (M=1.0). (c) E=0.1 (M=3.16). (d) E=0.25 (M=5.0). (e) E=0.5 (M=7.07). (f) E=1.0 (M=10). The dashed lines in the figures indicate the angle of the vorticity shocks predicted in the linearized theory.

The velocity component u in the direction of the free stream is presented in Figure 10 for $\mathfrak{R}=10$ and E=0, 0.01, 0.1, 0.25, 0.5, 1.0. Figure 10(a) gives the profile of u ahead of the cylinder along the ray $\theta=0$. 10(b) gives the profile just above the cylinder along the ray $\theta=\pi/2$. It is clear that for the flows of larger E, there is a region with small velocity close to the cylinder. This stagnant region grows with E. The diameter of this region has increased to about 3 times the cylinder diameter when E=1 as seen in Figure 10(b). Figure 10(a) also shows that there is a strong upstream influence for the viscoelastic flow with large E. In 10(b) we notice a velocity over-shoot in the region above the cylinder. This over-shoot exists for all cases with M>1 and shifts away from the cylinder as E increases. The slope of the velocity profile in (b) is consistent with the vorticity (derivatives of the velocity) distribution above the cylinder, and indicates a second high vorticity region away from the cylinder surface.

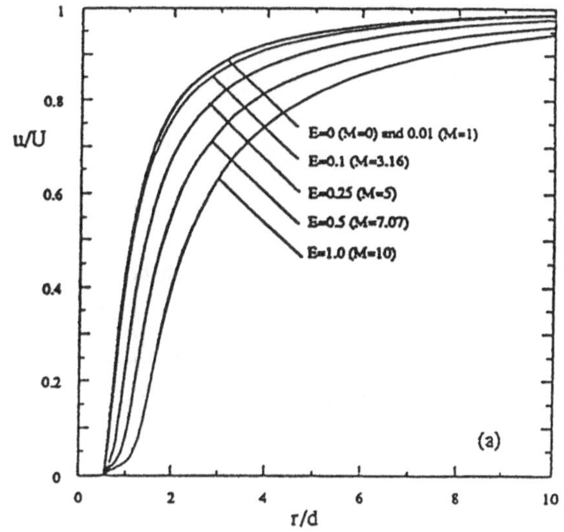

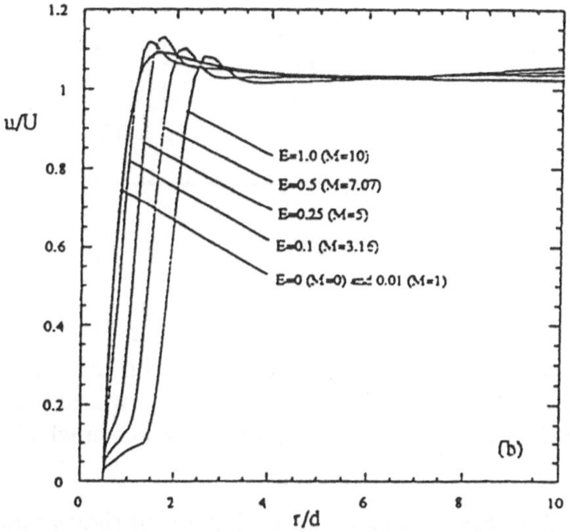

Figure 10 Effects of viscoelasticity on the velocity profile. The results are obtained with $\Re$=10 and E = 0, 0.01, 0.1, 0.25, 0.5, 1. u is the velocity component in the direction of the free stream. (a) u versus r along the path θ=0, ahead of the cylinder. (b) u versus r along the path $\theta=\pi/2$, just above the cylinder.

Numerical integration is carried out for (20) on the cylinder surface to get the drag force acting on the cylinder. The drag coefficient C_D is plotted in Figure 11 as a function of $\Re$ for four values of E= 0, 0.01, 0.1 and 1. The results for E=0, 0.01 and 0.1 are obtained using mesh No.2. For E=1 the results using the other two meshes are also presented. We see that the mesh refinement has little influence on the drag coefficient for the range of parameters in our

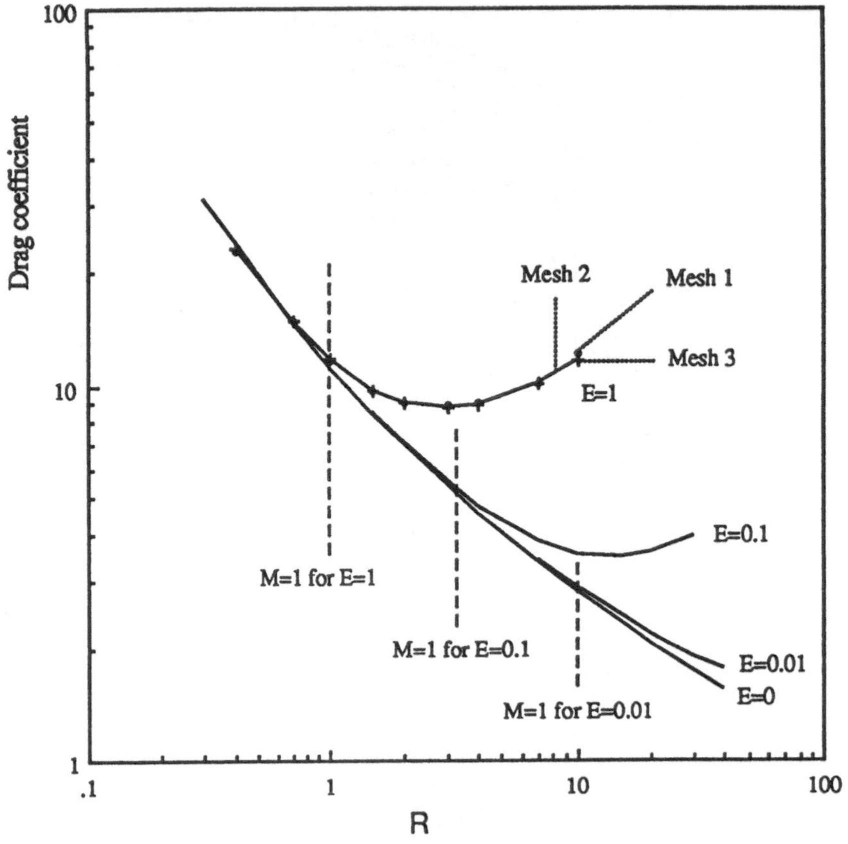

Figure 11 Drag coefficient C_D versus Reynolds number $\Re$ for elasticity number E=0 (Newtonian), 0.01, 0.1 and 1.0. Results for solid lines are obtained using mesh No.2. For E=1, the results obtained by the other two meshes are also plotted. The dashed lines indicate the values of Reynolds number at which the viscoelastic Mach number M=1.

The formula for the drag force acting on the cylinder (20) shows that the total drag can be separated into two parts, one part due to the pressure distribution around the cylinder and the other part due to the shear stress on the cylinder surface. These two contributions of the drag are plotted in Figure 12. In the figure, the drag coefficients of a viscoelastic case E=0.1 is compared with those of the Newtonian case E=0. In the Newtonian case, when $\Re$ is small, the drag coefficients due to pressure and due to shear stress are equal, as is well known. The pressure drag coefficient increases with $\Re$ because of the wake generated behind the cylinder. This is especially true in the viscoelastic case, where the drag due to pressure can be much larger than the drag due to shear stress, as we see in the figure, since we have larger wakes in viscoelastic cases. The nearly stagnant region around the cylinder is also responsible for the reduction of the drag due to the shear stress in viscoelastic flow.

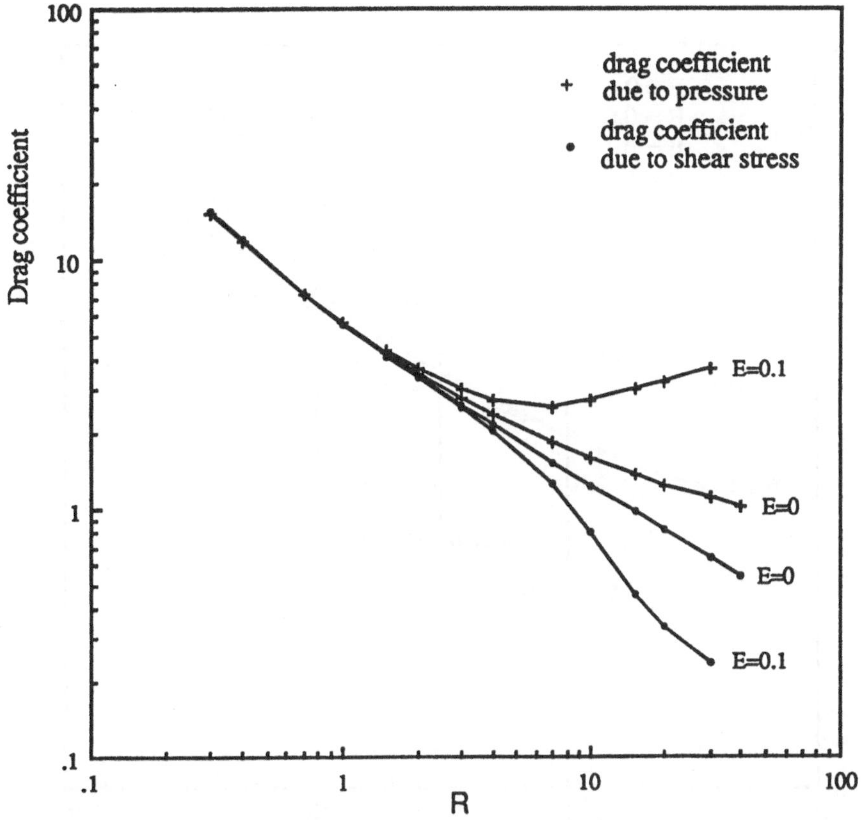

Figure 12 Effect of the elasticity of the fluids on the drag due to pressure and the drag due to shear stress on the cylinder surface.

Figure 13 presents graphs of the Nusselt number Nu versus $\mathfrak{R}$ for E=0, 0.01, 0.1 and 1 at Pr=1 and Pr=10. We checked the results for E=1 with three meshes. The results are almost identical. Again the values of $\mathfrak{R}$ at which M=1 are indicated in the figure with dashed lines. For $\mathfrak{R}$ less than these critical values, the Nusselt number for viscoelastic flow is the same as that for Newtonian flow. For $\mathfrak{R}$ greater than the critical values, the Nusselt number deviates from the Newtonian path and tends to an asymptotic value which does not depend on $\mathfrak{R}$. This deviation is more prominent for large E and Pr. We see that the effect of the viscoelasticity is to decrease the Nusselt number, or to reduce the heat transfer from the cylinder to the surrounding fluid. This can also be explained by the stagnant region which develops around the cylinder when the flow becomes supercritical (M>1).

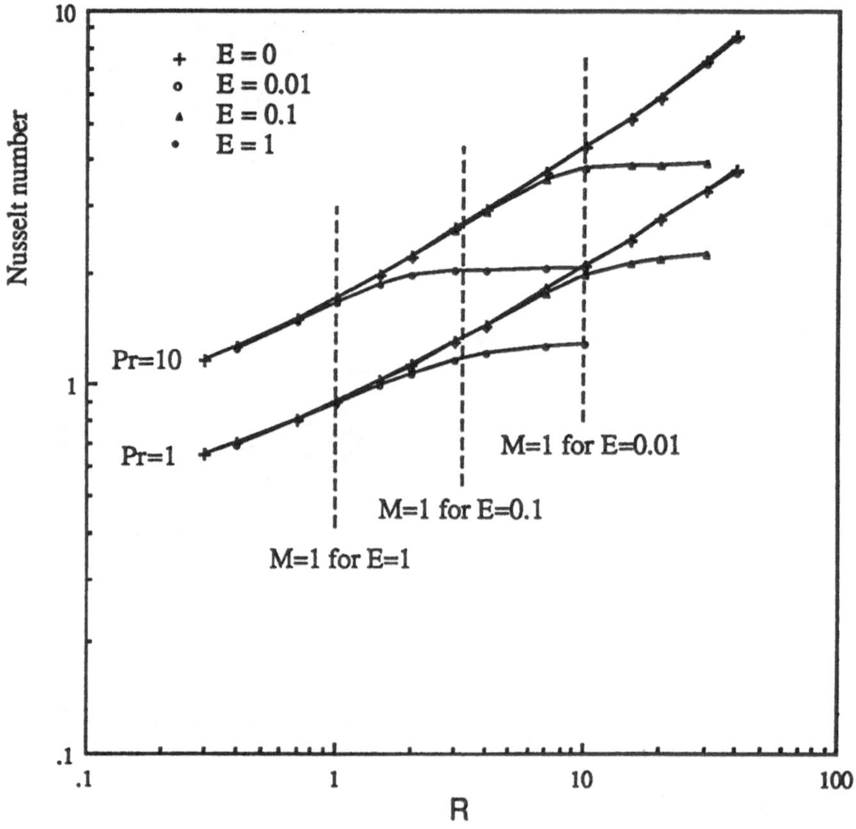

Figure 13 Nusselt number Nu versus Reynolds number $\Re$ for different elasticity numbers E=0 (Newtonian), 0.01, 0.1, 1.0, and at Prandtl number Pr=1 and 10. The dashed lines indicate the values of Reynolds number at which the viscoelastic Mach number M=1.

For the drag coefficient, the experiments of James and Acosta [6] were carried out on a wire of diameter 0.005in. in solutions of Polyox WSR-301. The intrinsic viscosity of WSR-301 was $[\eta]$=9.6 g/100ml measured in the experiments. We reproduced the data for concentration

ϕ=15.7, 30, 60, 119 and 226 ppmw in Figure 14. The shear wave speed for WSR-301 of concentration 50 ppm is about 2.48 cm/s, which is measured using a wave speed meter and listed in the tables of Joseph [1990]. Using this wave speed, we can get the shear wave speeds for the other concentrations from relation (5). Thus we estimate that elasticity numbers E=0.03 for the set of data of 15.7 ppm, E=0.07 for 30 ppmw, E=0.13 for 60 ppm, E=0.3 for 119 ppm and E=0.6 for 226 ppm as indicated in Figure 14. These values are much larger, about 50 times larger, than the values estimated in James and Gupta [1971]. As shown in Figure 14, the agreement is fair.

In the Nusselt number Figure 1 we have reproduced the experimental data for distilled water and for Polyox WSR-301 of concentrations 26.2, 52.4, 119 and 226 ppmw with three wire diameters, d=0.006in., 0.002in. and 0.001in. The elasticity numbers are similarly estimated and indicated in the figures. The experimental value of Pr is not known exactly, for distilled water at 20°C the Prandtl number is about 7. Thus the numerical results plotted in lines are obtained with Pr=7. Qualitatively, the numerical results show the same tendency of the experimental results. The differences, we think, are due to many factors. Our estimation of the elasticity number is rough, as we see from (33), a 10% error in the shear wave speed causes 20% difference in E. The heat transfer experiments were carried out with a temperature difference varying from 9-33°C. This temperature difference changes the viscosity and the shear wave speed of the solution, thus causes differences in the E. Also our choice of Maxwell model with a single relaxation time to characterize the fluid is certainly not optimal.

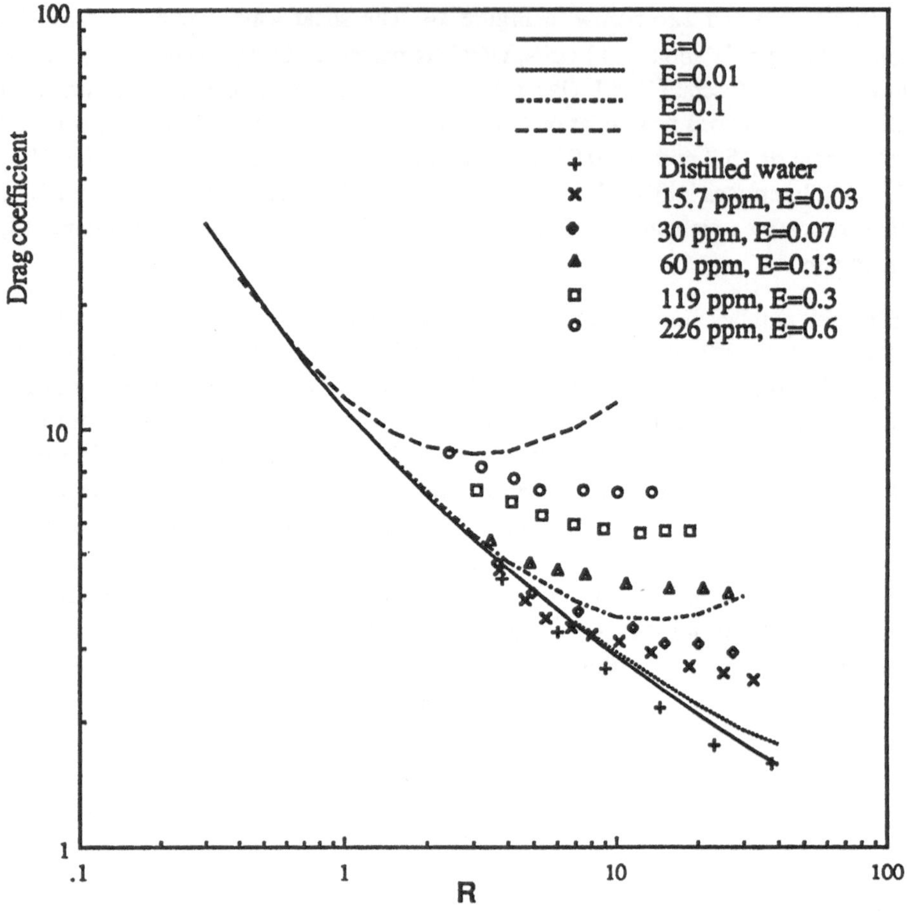

Figure 14 Comparison of the drag coefficient obtained by present computation (lines) with those measured in experiments of James and Acosta [6] (dots). The elasticity numbers for the experimental data are estimated using the shear wave speed as described.

ACKNOWLEDGEMENT

This work was supported by the Army Research Office, National Science Foundation and Department of Energy.

REFERENCES

Ahrens, M., Yoo, J. Y. and Joseph, D. D. Hyperbolicity and change of type in the flow of viscoelastic fluids through pipes, *J. Non-Newtonian Fluid Mech.*, Vol. **24**, pp. 67–83, 1987.

Ambari, A., Deslouis, C. and Tribollet, B. Coil-stretch transition of macromolecules in laminar flow around a small cylinder, *Chem. Eng. Commun.*, Vol. 29, pp. 63–78, 1984.

Coleman, B. and Noll, W. An approximation theorem for functionals with applications in continuum mechanics, *Arch. Rat. Mech. Anal.*, Vol. **6**, pp. 355–370, 1960.

Delvaux, V. and Crochet, M.J. 1990 Numerical simulation of delayed die swell. *Rheologica Acta* **29**, to appear.

Delvaux, V. and Crochet, M.J. 1990 Inertial viscoelastic flow past circular cylinder. *J. non-Newtonian Fluid Mech.*, to appear.

Fraenkel, L. E. Some results for a linear, partly hyperbolic model of viscoelastic flow past a plate, in *Material Instabilities in Continuum Mechanics and Related Mathematical Problems* (ed. J. M. Ball), Clarendon Press, Oxford, 1987.

James, D. and Acosta, A. J. The laminar flow of dilute polymer solutions around circular cylinders, *J. Fluid Mech.*, Vol. **42**, pp. 269–288, 1970.

James, D.F. and Gupta, O.P. Drag on circular cylinders in dilute polymer solutions, *Chem. Eng. Progr.* **67** Nº 111, 62–73, 1971.

Joseph, D.D. *Fluid Dynamics of Viscoelastic Liquids*. Springer, 1990.

Joseph, D. D., Matta, J. and Chen, K. Delayed die swell, *J. Non-Newtonian Fluid Mech.*, Vol. **24**, pp. 31–65, 1987.

Joseph, D.D. and Saut, J.C. Change of type and loss of evolution in the flow of viscoelastic fluids, *J. non-Newtonian Fluid Mech.* **20**, 117–141, 1986.

Konuita, A., Adler, P. M. and Piau, J. M. Flow of dilute solutions around circular cylinders, *J. Non-Newtonian Fluid Mech.*, Vol. 7, pp. 101–106, 1980.

Metzner, A.B., Uebler, E.A., and Fong, C.F.C.M. Converging flows of viscoelastic materials, *AIChE J.* **15**, 750–758, 1969.

Renardy, M., Hrusa, W. J. and Nohel, J. A. *Mathematical Problems in Viscoelasticity*, Longman, Harlow, UK, 1987.

Rutkevich, M. I. Some general properties of the equations of viscoelastic fluid dynamics, *PMM*, Vol. **33**, pp. 42–51, 1969.

Saut, J. C. and Joseph, D. D. Fading memory, *Arch. Rat. Mech. Anal.*, Vol. **81**, pp. 53–95, 1983.

Slemrod, M. Breakdown of smooth shearing flow in viscoelastic fluids for two constitutive equations; the vortex sheet vs. the vortex shock. Appendix to the paper by Joseph, D. D. Hyperbolic phenomena in the flow of viscoelastic fluids, in *Viscoelasticity and Rheology* (eds. A. S. Lodge, J. Nohel and M. Renardy), Academic Press, 1985.

Ultman, J. S. and Denn, M. M. Anomalous heat transfer and a wave phenomenon in dilute polymer solutions, *Trans. Soc. Rheol.*, Vol. **14**, pp. 307–317, 1970.

Yoo, J.Y., Ahrens, M., and Joseph, D.D. Hyperbolicity and change of type in sink flow, *J. Fluid Mech.* **153**, 203–214, 1985.

Yoo, J. and Joseph, D. D. Hyperbolicity and change of type in the flow of viscoelastic fluids through channels, *J. Non-Newtonian Fluid Mech.,* Vol. **19**, pp. 15–41, 1985.

Rheology and Shear Induced Structure of Fluids

Siegfried Hess
Institut für Theoretische Physik, Technische Universität Berlin
Hardenbergstrasse 36, D 1000 Berlin 12

Abstract

The flow behavior and its relation to shear-induced structural changes in fluids of spherical particles are treated theoretically. Results obtained by kinetic theory and by nonequilibrium molecular dynamics computer simulations are presented. Two qualitatively different flow regimes are identified: at small shear rates, changes affect the local structure; at high shear rates also a long range partial positional ordering can occur. The relevance of the results found in simple model fluids for the structural properties of flowing dispersions of spherical particles is discussed.

1 Introduction

The microscopic state of a fluid as described by one- and two- particle distribution-functions is affected by a viscous flow. The flow birefringence looked for and detected by Maxwell [1] is an early experimental evidence for a shear-flow-induced orientation of nonspherical particles. Also over a century ago, changes in the relative positional order of particles were conjectured by Reynolds [2] who demonstrated his point by a model experiment. Deviations of the velocity distribution function from a (local) Maxwellian due to transport processes were predicted by the kinetic theory of gases founded by Maxwell and Boltzmann [3], [4].

Of course, additional insight into the interrelation between the rheological behavior of fluids and their microscopic properties has been gained during the last century although the progress has not been steady. An increased interest in this topic during the last decade is reflected by several special issues of journals [5] – [7]. In this field, there is a vivid interplay between theory, nonequilibrium molecular dynamics computer simulations, experiments and applications [8].

In fluids composed of spherical particles, a viscous flow induces an anisotropy of the velocity distribution function [9] and of the local structure as characterized by the pair-correlation function or of the static structure factor which is measured by scattering techniques [10] – [12].

In fluids of nonspherical particles, a viscous flow causes a partial orientation which leads to the afore mentioned flow birefringence. In dense fluids, the axes of the particles are aligned [13], [14]; in molecular gases an aligment of the rotational angular momenta [15], [16] occurs. The orientation of segments plays an important role in the viscous porperties of polymeric liquids [14], [17].

This article is mainly concerned with the kinetic theory and the nonequilibrium molecular dynamics (NEMD) computer simulation studies on fluids of spherical particles. Some general remarks on rheological properties and notations are made first.

2 Pressure Tensor, Viscosity Coefficients

2.1 General Remarks

Point of departure for the theoretical treatment of the flow behavior of a fluid with mass density ρ and the flow velocity $\mathbf{v}$ is the local conservation equation for the linear momentum. In the absence of external forces, this equation can be written as

$$\frac{\partial}{\partial t}(\rho \mathbf{v}) + \nabla \cdot (\rho \mathbf{v}\mathbf{v}) + \nabla \cdot \mathbf{p} = 0 \tag{1}$$

or equivalently,

$$\rho \frac{d}{dt}\mathbf{v} + \nabla \cdot \mathbf{p} = 0 \tag{2}$$

where

$$\frac{d}{dt} = \frac{\partial}{\partial t} + \mathbf{v} \cdot \nabla \tag{3}$$

is the substantial derivative and $\mathbf{p}$ is the pressure tensor of the fluid. The tensor $\mathbf{p}$ can be decomposed into its "irreducible" parts associated with tensors of ranks $l = 0, 1, 2$ according to

$$\mathbf{p} = p\,\delta + \mathbf{p}^a + \overleftrightarrow{\mathbf{p}}, \tag{4}$$

δ is the unit tensor, p is one third of the trace of the tensor $\mathbf{p}$. In thermal equilibrium, p equals the hydrostatic pressure p_{eq} and the antisymmetric part $\mathbf{p}^a$, as well as the symmetric traceless part $\overleftrightarrow{\mathbf{p}}$ vanish. For a fluid composed of spherical, structureless particles to be considered in the following, $\mathbf{p}^a$ is always zero. The symbol $\overleftrightarrow{\cdots}$ indicates the symmetric traceless part of a tensor, e.g.

$$\overleftrightarrow{\mathbf{a}\,\mathbf{b}} = \frac{1}{2}(\mathbf{a}\,\mathbf{b} + \mathbf{b}\,\mathbf{a}) - \frac{1}{3}\mathbf{a} \cdot \mathbf{b}\,\delta \tag{5}$$

for the dyadic constructed from components of two vectors $\mathbf{a}$ and $\mathbf{b}$. In the following, the quantity $\overleftrightarrow{\mathbf{p}}$ is referred to as the "friction pressure tensor".

The gradient of the velocity field $\mathbf{v}$, viz. $\nabla\mathbf{v}$ can be decomposed in analogy to (4). The scalar part ($l = 0$) is the divergence $\nabla \cdot \mathbf{v}$, the vectorial ($l = 1$) and tensorial ($l = 2$) parts are characterized by the vorticity

$$\omega = \frac{1}{2}\nabla \times \mathbf{v} \tag{6}$$

and by the deformation rate (shear rate) tensor

$$\overleftrightarrow{\gamma} = \overleftrightarrow{\nabla\mathbf{v}}\,. \tag{7}$$

In the linear flow regime, the constitutive laws linking the components of the pressure tensor with the velocity gradient are Newton's law

$$\overleftrightarrow{\mathbf{p}} = -2\eta\,\overleftrightarrow{\gamma} \tag{8}$$

and

$$p - p_{eq} = -\eta_V \, \nabla \cdot \mathbf{v}, \tag{9}$$

with the shear viscosity η and the bulk (volume) viscosity η_V. These coefficients depend on the particular state of the system (density, temperature, ...) but not on the shear rate. Notice that the constitutive laws (8, 9) do not involve the vorticity ω. In the nonlinear flow regime, eqs. (8, 9) are to be generalized in several respects. Firstly, the shear viscosity η has to be replaced by a 4-th rank viscosity tensor and one may introduce two 2nd rank viscosity tensors describing the nonlinear coupling between the scalar $p - p_{eq}$ and the 2nd rank tensor p. The directional properties of the nonlinear viscosity tensors depend on directions specified by ω and $\overleftrightarrow{\gamma}$; their components, as well as η_V, depend on the magnitudes of $\nabla \cdot \mathbf{v}$, ω and $\overleftrightarrow{\gamma}$. In contradistinction to (8, 9), nonlinear constitutive laws depend on the specific flow geometry. In the following, a plane Couette arrangement is considered where the tensorial analysis can be somewhat simplified.

2.2 Special Geometry: Plane Couette Flow

For a flow in x-direction between flat plates parallel to the x-z-plane, one has

$$\mathrm{v}_x = \gamma y, \quad \mathrm{v}_y = 0, \quad \mathrm{v}_z = 0 \tag{10}$$

with the (constant) shear rate

$$\gamma = \frac{\partial \mathrm{v}_x}{\partial y}. \tag{11}$$

In this case, the divergence $\nabla \cdot \mathbf{v}$ vanishes, the vorticity and the deformation rate tensor are given by

$$\omega = -\frac{1}{2}\gamma \mathbf{e}^z, \qquad \overleftrightarrow{\gamma} = \gamma \, \overleftrightarrow{\mathbf{e}^x \mathbf{e}^y}, \tag{12}$$

where $\mathbf{e}^{x,y,z}$ are the unit vectors parallel to the coordinate axes. In matrix notation, the tensor $\overleftrightarrow{\gamma}$ reads

$$\overleftrightarrow{\gamma} = \frac{1}{2}\gamma \begin{pmatrix} 0 & 1 & 0 \\ 1 & 0 & 0 \\ 0 & 0 & 0 \end{pmatrix}. \tag{13}$$

Notice that the magnitudes of both the vorticity ω and of the deformation rate tensor $\overleftrightarrow{\gamma}$ are determined by the shear rate γ. For a plane Couette symmetry, only 3 of the 5 independent components of the friction pressure tensor are nonzero; they are denoted by p_+, p_-, p_0 and introduced via the ansatz

$$\overleftrightarrow{\mathsf{p}} = 2p_+ \, \overleftrightarrow{\mathbf{e}^x \mathbf{e}^y} + p_- \, (\mathbf{e}^x \mathbf{e}^x - \mathbf{e}^y \mathbf{e}^y) + 2p_0 \, \overleftrightarrow{\mathbf{e}^z \mathbf{e}^z} \, . \tag{14}$$

In matrix notation, (14) reads

$$\overleftrightarrow{\mathsf{p}} = p_+ \begin{pmatrix} 0 & 1 & 0 \\ 1 & 0 & 0 \\ 0 & 0 & 0 \end{pmatrix} + p_- \begin{pmatrix} 1 & 0 & 0 \\ 0 & -1 & 0 \\ 0 & 0 & 0 \end{pmatrix} + p_0 \frac{2}{3} \begin{pmatrix} -1 & 0 & 0 \\ 0 & -1 & 0 \\ 0 & 0 & 2 \end{pmatrix}. \tag{15}$$

Clearly, p_+ is the x-y-component of the pressure tensor, p_- and p_0 are associated with normal pressure differences, viz.

$$p_+ = p_{xy}, \tag{16}$$
$$p_- = \frac{1}{2}(p_{xx} - p_{yy}) \tag{17}$$
$$p_0 = \frac{1}{2}\left(p_{zz} - \frac{1}{2}(p_{xx} + p_{yy})\right). \tag{18}$$

The quantities p_-, p_+ are essentially the real and the imaginary parts of the spherical $m = \pm 2$ components of the tensor p, p_0 ist proportional to the $m = 0$ component; the $m = \pm 1$ components vanish for this special geometry. Notice that the p_+-term of (15, 16) has the same directional property as the $\overleftrightarrow{\gamma}$-tensor of (12, 13). The p_- and p_0 terms involve tensors which are proportional to

$$\omega \times \overleftrightarrow{\gamma} \quad \text{and} \quad \overleftrightarrow{\gamma} \cdot \overleftrightarrow{\gamma}, \tag{19}$$

respectively. Thus for a vorticity free flow field (e.g. 4 roller geometry), p_- vanishes due to symmetry arguments.

For a plane Couette geometry, the generalization of the constitutive laws (8, 9) to the nonlinear flow regime are written as

$$p_k = -\eta_k \gamma, \quad k = +,\ -,\ 0, \tag{20}$$

and

$$p - p_{eq} = -\eta_{02}\gamma \tag{21}$$

with the viscosity coefficients η_+, η_-, η_0 and η_{02}. In the linear flow regime, i.e. for $\gamma \to 0$, η_+ reduces to the Newtonian viscosity and the η_-, η_0, η_{02} vanish. In the nonlinear regime, the non-Newtonian viscosity η_+ as well as the coefficients η_-, η_0, η_{02} are functions of the shear rate γ. The entropy production is proportional to $\eta_+\gamma^2$, thus $\eta_+ > 0$ but η_-, η_0 and η_{02} may have either sign. The viscometric functions $\Psi_{1,2}$ defined by

$$p_{xx} - p_{yy} = -\Psi_1\gamma^2, \quad p_{yy} - p_{zz} = -\Psi_2\gamma^2 \tag{22}$$

are related to the coefficients η_-, η_0 by

$$\gamma\Psi_1 = 2\eta_-, \quad \gamma\Psi_2 = -(2\eta_0 + \eta_-). \tag{23}$$

The coefficient η_{02} describes the change of the hydrostatic pressure in an nonlinear shear flow, $\eta_{02} < 0$ implies an increase of p at constant density or an equivalent increase of the volume at constant pressure (shear dilatancy). So far, the generalized transport coefficients needed to describe the viscous behavior of a fluid in the nonlinear flow regime have been introduced. General symmetry arguments were used, however, no microscopic explanation for the nonlinear behavior has been given. As a first step in this direction, the pressure tensor is related to microscopic quantities.

2.3 Microscopic Expressions for the Pressure Tensor

The pressure tensor p of a fluid of N particles in a volume V can be decomposed into "kinetic" and "potential" parts accoring to

$$\mathsf{p} = \mathsf{p}^{kin} + \mathsf{p}^{pot}, \tag{24}$$

with

$$\mathsf{p}^{kin} = V^{-1} m \sum_i \mathbf{c}^i \mathbf{c}^i. \tag{25}$$

Here, m is the mass of a particle and

$$\mathbf{c}^i = \dot{\mathbf{r}}^i - \mathbf{v}(\mathbf{r}^i) \tag{26}$$

is the peculiar velocity of particle "i", $\mathbf{r}^i$ is its velocity with respect to the laboratory frame. The potential contribution to p is

$$\mathsf{p}^{pot} = V^{-1} \frac{1}{2} \sum \sum\nolimits_{i \neq j} \mathbf{r}^{ij} \mathbf{F}^{ij}. \tag{27}$$

Here $\mathbf{r}^{ij} = \mathbf{r}^i - \mathbf{r}^j$ is the difference between the position vectors of particles "i" and "j", $\mathbf{F}^{ij} = \mathbf{F}(\mathbf{r}^{ij})$ with $\mathbf{F}(\mathbf{r}) = -\partial \phi / \partial \mathbf{r}$ is the force between them; $\phi(\mathbf{r})$ is the binary interaction potential. For spherical particles, one has $\mathbf{F} = -\mathbf{r} r^{-1} \phi'$ where the prime denotes differentiation with respect to $r = \|\mathbf{r}\|$. The N-particle averages (25) and (27) provide the prescription for the evaluation of the pressure in a molecular dynamics simulation. For many theoretical considerations, it is advantageous to write p^{kin} and p^{pot} as integrals over 1- and 2-particle distribution functions, viz.

$$\mathsf{p}^{kin} = m \int \mathbf{c}\mathbf{c} f(\mathbf{c}) \, d^3c, \tag{28}$$

$$\mathsf{p}^{pot} = \frac{1}{2} n^2 \int \mathbf{r}\mathbf{F} g(\mathbf{r}) \, d^3r. \tag{29}$$

Here, $f(\mathbf{c})$ with the normalization $\int f \, d^3c = n$ is the velocity distribution function, n is the number density of the fluid. The quantity $g(\mathbf{r})$ with $g \to 1$ for $r \to \infty$ is the pair-correlation function, $\mathbf{r}$ is the position of any (other) particle of the fluid with respect to an arbitrary reference particle.

In terms of N-particle averages, the distribution functions are given by

$$f(\mathbf{c}) = V^{-1} \sum_i \delta\left(\mathbf{c} - \mathbf{c}^i\right), \tag{30}$$

$$n g(\mathbf{r}) = N^{-1} \sum \sum\nolimits_{i \neq j} \delta\left(\mathbf{r} - \mathbf{r}^{ij}\right) \tag{31}$$

with $n = N/V$.

Both p^{kin} and p^{pot} as given either by (25, 27) or by (28, 29) can be decomposed analogously to (4). Similary, the viscosity coefficients η_k and η_{02} introduced in (20, 21) are the sums of kinetic and potential contributions which are expected to be the dominating ones in dilute gases and in dense fluids, respectively. In thermal equilibrium, the velocity distribution function $f(\mathbf{r})$ depends on the velocity $\mathbf{c}$ via $c = \|\mathbf{c}\|$ only and the pair correlation function is a function of $r = \|\mathbf{r}\|$. Due to this isotropy, the symmetric traceless part $\overleftrightarrow{\mathrm{p}}$ of p vanishes. Furthermore, p reduces to the equilibrium pressure p_{eq}. Thus in a nonequilibrium state, the molecular distribution functions $f(\mathbf{c})$ and $g(\mathbf{r})$ deviate from their equilibrium values. Before this point is discussed in detail in part 3, some results on the (nonlinear) viscosity coefficients as obtained from molecular dynamics simulations are presented next.

2.4 Examples from NEMD

In a molecular dynamics (MD) computer simulation the equations of motion of N particles interacting with a given force law are integrated numerically. Macroscopic quantities such as the pressure tensor, as well as one- and two-particle distribution functions are evaluated from the known positions and momenta of the particles as N-particle averages, see eqs. (25, 27, 30, 31). These quantities then are averaged over many time steps. The dynamical system studied with the help of the computer mimics a physical system; the extraction of data can be looked upon as a "measurement" on a model system.

Periodic boundary conditions are used (together with the standad minimum image convention) if one is interested in bulk properties rather than boundary layer effects. Typical values for N are a few hundered to a few thousand.

The control of constraints which guarantee thermal equiblibrium or a stationary nonequilibrium state (nonequilibrium molecular dynamics: NEMD) as well as the construction of the "measuring devices" which allow the extraction of the desired data are of crucial importance (just as in a real experiment). Of course, relaxation phenonema can also be studied by NEMD.

Methods to simulate a plane Couette (simple shear) flow have been developed and tested, for details see refs. [18], [19], [20]. For gases excellent agreement (without any adjustable parameter) between results obtained by the kinetic theory based on the Boltzmann equation and NEMD results are found for the viscosity coefficients and the shear-flow induced distortion of the velocity distribution function even in the nonlinear flow regime [9].

As typical examples, data for the viscosity coefficients of a liquid composed of Lennard-Jones (LJ) particles with the interaction potential

$$\phi^{LJ}(\mathbf{r}) = 4\varepsilon_{LJ}\left(\left(\frac{s}{r}\right)^{12} - \left(\frac{s}{r}\right)^{6}\right) \tag{32}$$

are presented. Reduced variables are introduced with the help of the characteristic energy $\varepsilon \equiv \varepsilon_{LJ}$, the characteristic length s and with the mass m of a particle, e.g. densities are expressed in units of s^{-3}, temperatures T in units of εk_B^{-1} (k_B is the Boltzmann constant), times and shear rates in units of $t_0 = s(m/\varepsilon)^{1/2}$ and of t_0^{-1}, respectively. The simulation was perforned for a system of $N = 8^3 = 512$ particles in a volume V determined by the (constant) number density $n = 0.84$, the temperature was kept constant at $T = 1$ by rescaling the magnitude of the (peculiar) velocities. For comparison, the triple point is at $n \approx 0.84$ and $T \approx 0.72$. The potential is cut off at $r = r_C = 2.5$. Periodic boundary conditions are used with the appropriate modifications for a plane Couette flow. The equations of motion are integrated with a 5th-order predictor-corrector method (Gear), the Couette flow is simulated by the "homogeneous shear" algorithm [21].

In Figs. 1–3, the pressure components p_+, p_-, p_0 as well as p and $\delta p = p - p_{eq}$ and the pertaining viscosity coefficients are displayed as functions of the shear rate γ (in reduced units). The kinetic contribution to the friction pressure and to η_+ is only about one tenth of the potential contribution. In an experiment the sum of both contributions is measured. In Figs. 2 and 3, only the potential contributions are shown. The dominating features are a shear-thinning and an increase of the pressure corresponding to a volume dilatancy mentioned in the introduction. At high values of γ, there is an indication of shear thickening.

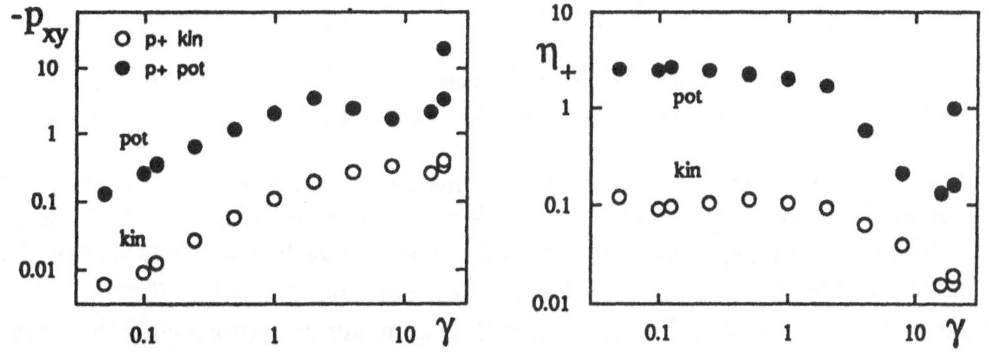

Figure 1: The kinetic and potential contributions to the x-y-component of the pressure tensor (p_+) and the shear viscosity η_+ as functions of the shear rate γ from a NEMD simulation for a Lennard-Jones-liquid at $n = 0.84$ and $T = 1.0$ (reduced LJ-units; number of particles: $N = 512$).

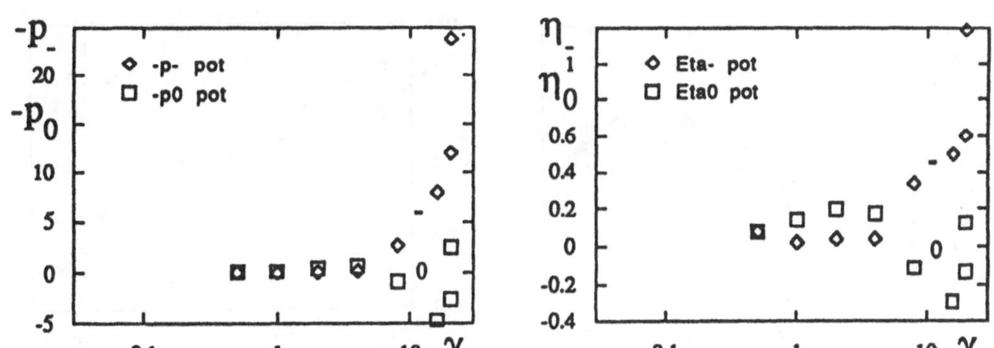

Figure 2: The normal pressure differences p_- and p_0 and the pertaining viscosity coefficients (cf. eqs 17, 18, 20) from the same NEMD simulation as in Fig.1.

From practically all curves one inferes that a crossover from one functional dependence on the shear rate γ to another one occurs between $\gamma = 2$ and $\gamma = 4$. Two values (both averaged over 16 000 time steps) are shown for the highest γ-value ($\gamma = 20$), one of them appears like an extrapolation of the small shear rate behavior.

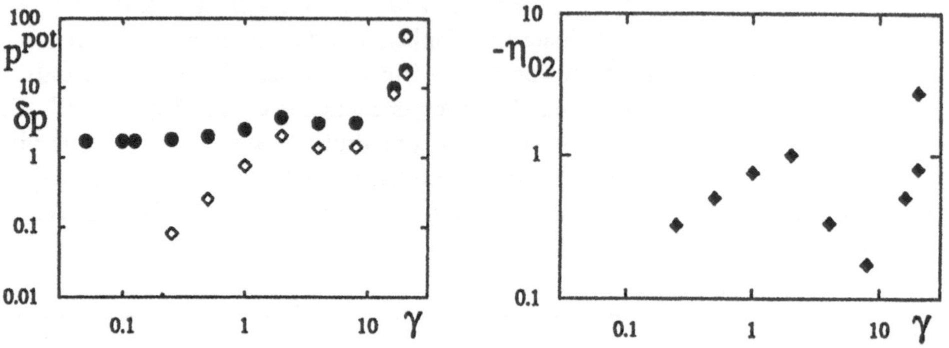

Figure 3: The potential contribution to the pressure p, the difference $\delta p = p - p_{eq}$ and the viscosity coefficient η_{02} (cf. eq. 21) from the same NEMD simulation as in Fig.1.

At high densities, the viscous properties are mainly determined by the repulsive part of the potential. Results obtained for dense fluids of "soft spheres" (SS) with the interaction potential $\phi^{SS} = \varepsilon_{SS}(s/r)^{12}$ or $\phi^{SS} = r^{-12}$ (in SS-units) are rather similar to those of dense LJ-liquids. In Fig. 4, p_{xy} and η_+ are shown for the potential $\phi = r^{-12} + Br^{-1}\left(1 - \frac{1}{2}\kappa r\right)^2$, cut off at $r_c = 2\kappa^{-1} = 2.5$. The term with the parameter B approximates the screened Coulomb interaction $r^{-1}e^{-\kappa r}$. The values $B = 1$ (circles) and $B = 10$ (squares) were chosen; in both cases $n = 0.84$ and $T = 0.25$ (SS-units). A plastic or pseudo-plastic behavior is found for small shear rates. Again a crossover in the shear rate dependence occurs, here for γ between 1 and 2. The structural difference in these two regimes are discussed in section 3.

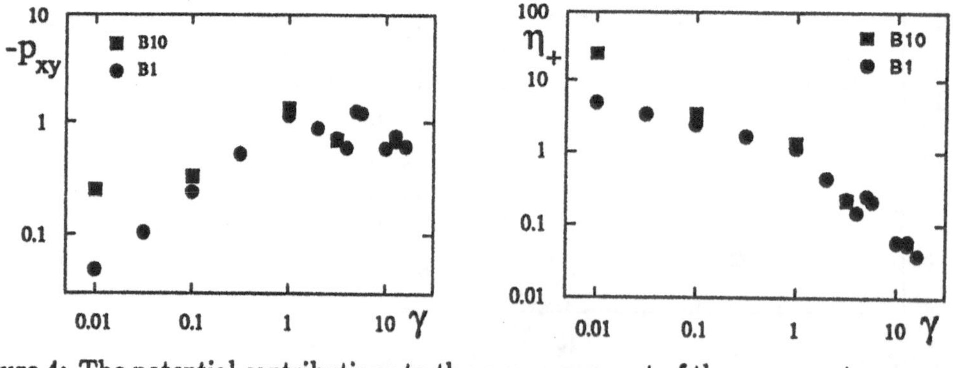

Figure 4: The potential contributions to the x-y-component of the pressure tensor p_+ and the shear viscosity η_+ as functions of the shear rate γ for a soft sphere like model fluid (r^{-12}-potential plus screened Coulomb interaction) at $n = 0.84$ and $T = 0.25$ (reduced SS-units, $N = 128$).

2.5 Kinetic Theory

As the point of departure for the theory of the (nonlinear) viscous behavior of a dense fluid, a kinetic equation for the pair-correlation function $g(\mathbf{r})$ is used [22]. The pressure tensor p (in the following the subscript "pot" is omitted for simplicity) is obtained according to (29). The kinetic equation is written as [23]

$$\frac{\partial g}{\partial t} + \mathbf{r}\cdot(\nabla\mathbf{v})\cdot\frac{\partial}{\partial \mathbf{r}}g + D(g) = 0. \tag{33}$$

It is recalled that $\mathbf{r} = \mathbf{r}_1 - \mathbf{r}_2$ is the difference of the position vectors between two particles located at $\mathbf{r}_1$ and $\mathbf{r}_2$. For the Couette geometry, the flow term can be written as $\gamma y \frac{\partial}{\partial x}$ where x and y are the components of $\mathbf{r}$. In general, a decomposition of $\nabla\mathbf{v}$ into its irreducible parts cf. (12) yields

$$\mathbf{r}\cdot(\nabla\mathbf{v})\cdot\frac{\partial}{\partial \mathbf{r}} = \omega\cdot\mathcal{L} + \overleftrightarrow{\gamma} : \mathsf{L} \tag{34}$$

with the differential operators

$$\mathcal{L} = \mathbf{r}\times\frac{\partial}{\partial \mathbf{r}}, \qquad \mathsf{L} = \overleftrightarrow{\mathbf{r}\frac{\partial}{\partial \mathbf{r}}} \tag{35}$$

which are the generators of SU(3). The "damping" term $D(g)$ of (33) guarantees that in the absence of a perturbing flow g relaxes to its equilibrium value g_{eq}. General properties are

$$\int D(g)\, d^3r = 0, \qquad D\left(g_{eq}\right) = 0. \tag{36}$$

A special form for $D(g)$ has been proposed by Kirkwood [21] (diffusion in the potential of the mean force), modifications have been derived more recently [23], [24]. In order to highlight the effects which stem from the flow term one models $D(g)$ by a simple relaxation time approximation [25], [26]

$$D(g) \simeq \tau^{-1}\left(g - g_{eq}\right) \tag{37}$$

involving a relaxation time τ.

Now, multiplication of (33) by $\mathbf{r}\,\mathbf{F}$, cf. (29), with (37) and use of (34) leads to the relaxation equation

$$\frac{\partial}{\partial t}\overleftrightarrow{\mathsf{p}} - 2\,\overleftrightarrow{\omega\times\mathsf{p}} + 2\sigma_{22}\overleftrightarrow{\overleftrightarrow{\gamma}\cdot\mathsf{p}} + \ldots + \tau^{-1}\overleftrightarrow{\mathsf{p}} = -2G\overleftrightarrow{\gamma}. \tag{38}$$

In (38)

$$G = \frac{2\pi n^2}{15}\int_0^\infty \left(r^4\phi'\right)' g_S\, dr \tag{39}$$

is the shear modulus which in thermal equilibrium, where $g_S = g_{eq}$ agrees with the expression of Born and Green [22] obtained by a rather different method. The quantity σ_{22} is given by $\sigma_{22} = -\frac{6}{7}\left(1 + \frac{2}{3}\nu\right)$ for a simple power law potential $\phi = r^{\nu}$. For this case, incidentally, G is proportional to p^{pot}, i.e. $G = -\frac{1}{5}(\nu+3)\,p^{pot}$. The dots in (38) stand for terms which couple the 2nd rank tensor p via $\overleftrightarrow{\gamma}$ with tensors of rank $l = 0$ and $l = 4$; these terms are disregarded in the following.

For a stationary situation ($\partial p/\partial t = 0$), and in the linear flow regime (38) reduces to Newtons law (8) with the Newtonian shear viscosity

$$\eta = G\tau. \tag{40}$$

In the nonlinear regime, the terms involving ω and $\overleftrightarrow{\gamma}$ on the l.h.s. of (38) have to be taken into account. For a plane Couette flow, the ansatz (14) or (15) may be inserted into (38) to obtain equations for the quantities p_+, p_- p_0. The inhomogeneity with G only shows up in the p_+ equation. The ω-term of (38) leads to a coupling between p_+ and p_-, the $\overleftrightarrow{\gamma}$ involving σ_{22} couples p_+ and p_0. If these latter contributions are taken into account in lowest order only, the resulting expressions for the viscosity coefficients are

$$\eta_+ = \eta\left(1+\tau^2\gamma^2\right)^{-1}, \tag{41}$$

$$\eta_- = \gamma\tau\eta_+, \tag{42}$$

$$\eta_0 = \frac{1}{4}\sigma_{22}\gamma\tau\eta_+, \tag{43}$$

for η see (40).

Analogously to (38), an equation for the trace p of the pressure tensor can be obtained by a multiplication of the kinetic equation by $\mathbf{r}\cdot\mathbf{F}$. With the same approximation as discussed above one finds

$$\frac{\partial}{\partial t}p + \tau^{-1}\left(p - p_{eq}\right) = -\sigma_{02}\,\overleftrightarrow{\gamma} : \mathsf{p} \tag{44}$$

where σ_{02} is given by $-\nu/3$ for an interaction potential $\phi = r^\nu$, i.e. $\sigma_{02} = 4$ for $\nu = 12$.

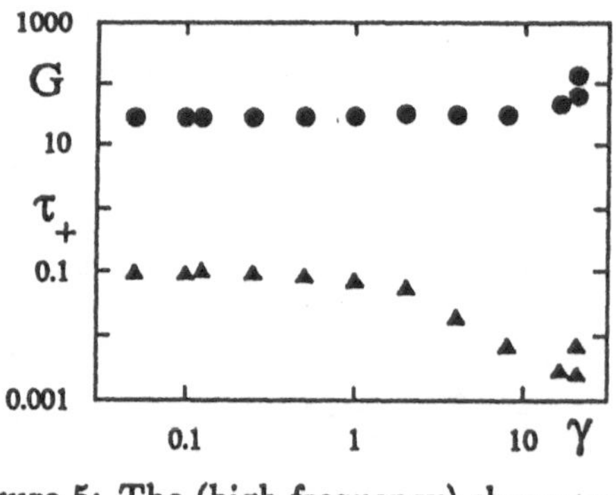

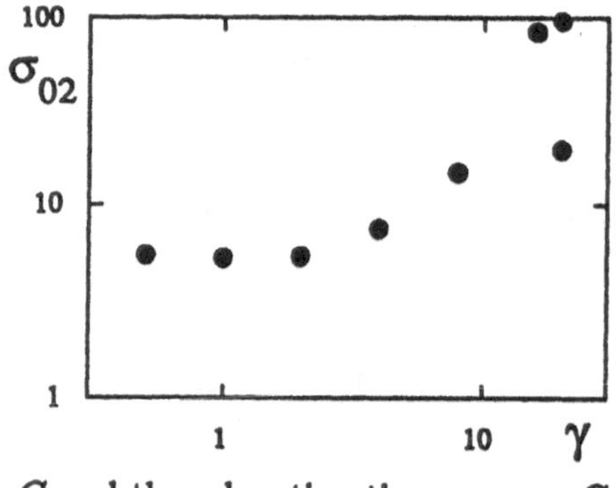

Figure 5: The (high frequency) shear modulus G and the relaxation time $\tau_+ = \eta_+ G^{-1}$ as functions of the shear rate for the LJ-liquid at $n = 0.84$, $T = 1.0$).

Figure 6: The quantity $\sigma_{02} = \delta p\left(\gamma\tau_+ P_+\right)^{-1}$ as function of the shear rate γ for the LJ-liquid at $n = 0.84$ and $T = 1.0$.

Thus in a steady state one finds a relation of the form (21) with the viscosity coefficient

$$\eta_{02} = -\sigma_{02}\gamma\tau\eta_+. \tag{45}$$

Clearly, $\eta_+ \to \eta$ and η_-, η_0, $\eta_{02} = 0$ for $\gamma \to 0$, as expected.

How about a comparison of the theoretical results (41 – 43, 45) with the computer simulation data? An estimate for the relaxation time τ can be obtained by dividing η_+ by the shear modulus G which can be calculated in the NEMD simuation according to $G = V^{-1}\frac{1}{30}\sum\sum_{i\neq j} r_{ij}^{-2}\left(r_{ij}^4 \phi_{ij}'\right)'$ with $r_{ij} = \|\mathbf{r}_{ij}\|$ and $\phi_{ij} = \phi(\mathbf{r}_{ij})$. In Fig.5, G and the ratio $\tau_+ = \eta_+/G$ are plotted for the LJ liquid at $n = 0.84$, $T = 1.0$. For $\gamma \leq 2$, both quantities show a rather weak dependence on γ; $\tau \approx 0.1$ is a reasonable value. Qualitatively, and for $\gamma < 4$, one finds the expected behavior: the non-Newtonian viscosity η_+ decreases with increasing shear rate γ, the normal pressure coefficients η_- and η_0 are nonzero and have the correct sign; the same holds true for the shear dilatancy coefficient η_{02}. Quantitatively, however, there are significant differences: the computer values for η_+ show a stronger decrease with increasing γ than expected due to (41), the other viscosity coefficients are smaller than predicted by (42, 43). Moreover, there seems to be a nonanalytic functional dependences of the form

$$\eta_+ = \eta\left(1 + A_1\gamma^{1/2} + \ldots\right)^{-1}. \tag{46}$$

The quantitative discrepancy between the computer results and the theory may be caused by the approximations made to derive (41 – 43) and (45). Firstly, the tensorial coupling generated by the $\overleftrightarrow{\gamma}$: L-term of (33), (34) has been taken into account in lowest order only. Secondly, and probably more important, the damping term and consequently the relaxation processes to be incorporated into eqs. (38), (44) are more complicated than implied by (37): $D(g)$, in general, possesses a whole spectrum of relaxation times and $\overleftrightarrow{\mathbf{r}\,\mathbf{F}}$ as well as $\mathbf{r} \cdot \mathbf{F}$ are certainly not eigenfunctions of $D(g)$. On the other hand, relation (45) is reasonably well obeyed in the LJ-liquid for $\gamma < 4$ (and $\gamma > 0.1$ where the statistical accuracy of δp is good enough); see Fig. 6 where $\sigma_{02} = -\eta_{02}/(\gamma\tau_+\eta_+)$, displayed as a function of γ, assumes a constant value for small γ. The simple theory presented here is a first step towards a microscopic understanding of the nonlinear flow behavior of dense fluids; this applies also to the nonequilibrium structure to be discussed next.

2.6 Remarks on Colloidal Suspensions

From the theoretical considerations presented so far follows that a non-Newtonian behavior can be observed when the product $\gamma\tau$ of the shear rate γ and the structural relaxation time τ are of the order of 1 (or at least about 10^{-2}). Experimentally accessible shear rates are smaller than about $10^5\ s^{-1}$. In a real simple liquid like Argon, τ is typically shorter than $10^{-10}\,s$ and consequently $\gamma\tau < 10^{-5}$, thus there is no chance to see a non-Newtonian behavior.

The situation is different in (dense) colloidal dispersions of spherical particles where one has much longer relaxation times (e.g. in the order of $10^{-3}\ s$ to $1\ s$). Of course, a proper theoretical treatment of dispersions has to take the solvent-particle interaction and long range hydrodynamic effects into account. However, in many cases where the particle-particle interactions are the dominating ones, the rheological behavior and the structural properties are quite analogous to those of simple fluids [27]–[29]. Some colloidal dispersions may even be looked upon as model "macro-fluids" where the phenomena calculated for simple fluids can be observed experimentally.

3 The Structure of Streaming Fluids

3.1 Expansion of the Pair-Correlation Function for a Plane Couette Flow

Nonequilibrium processes affect the structure of a fluid; the pair-correlation function $g(\mathbf{r})$ deviates from its equilibrium value. The flow-induced distortion of $g(\mathbf{r})$ is not only a feature occurring in the kinetic theory for the viscosity but its spatial Fourier transform is also directly measurable in colloidal systems. Firstly, some general remarks on the functional dependence of $g(\mathbf{r})$ on $\mathbf{r}$, more precisely on $r = \|\mathbf{r}\|$ and $\hat{\mathbf{r}} = r^{-1}\mathbf{r}$, are made.

In general, $g(\mathbf{r})$ can be expanded with respect to the spherical harmonics $Y_{lm}(\hat{\mathbf{r}})$, the expansion coefficients depend on r. Since $g(\mathbf{r})$ equals $g(-\mathbf{r})$ (replacing $\mathbf{r}$ by $-\mathbf{r}$ corresponds to the relabelling of the names of the two particles of a pair), only even l occur in the expansion. Cartesian tensors, equivalent to the spherical harmonics, are easily adapted to the Couette geometry considered here. Symmetry arguments similar to those used for the friction pressure tensor, cf. (14, 15) allows one to write $g(\mathbf{r})$ as

$$g(\mathbf{r}) = g_S(r) + g_+(r)\hat{x}\hat{y} + g_-(r)\frac{1}{2}\left(\hat{x}^2 - \hat{y}^2\right) + g_0(r)\left(\hat{z}^2 - \frac{1}{3}\right) + \ldots, \tag{47}$$

$\hat{x}$, $\hat{y}$, $\hat{z}$ are the Cartesian components of $\hat{\mathbf{r}}$. The scalar or "isotropic" part of $g(\mathbf{r})$ is

$$g_S = (4\pi)^{-1} \int g(\mathbf{r})\, d^2\hat{r}. \tag{48}$$

The three 2nd rank tensorial coefficients $g_k(r)$ are given by

$$g_k = (4\pi)^{-1} \int Y_k(\hat{\mathbf{r}})\, g(\mathbf{r})\, d^2\hat{r} \tag{49}$$

with

$$Y_+ = 2\hat{x}\hat{y}, \qquad Y_- = \hat{x}^2 - \hat{y}^2, \qquad Y_0 = \frac{3}{2}\left(\hat{z}^2 - \frac{1}{3}\right). \tag{50}$$

Note that the $Y_{+,-}$ are linear combinations of the spherical harmonics $Y_{2\pm 2}$. The dots in (47) stand for terms involving tensors of ranks $l = 4, 6, \ldots$.

One of the $l = 4$ terms in (47) is $g_4(r)K_4(\mathbf{r})$ with

$$K_4 = \frac{5}{4}\sqrt{21}\left(\hat{x}^4 + \hat{y}^4 + \hat{z}^4 - \frac{3}{5}\right); \tag{51}$$

g_4 is defined in analogy to (49). For the full set of the $l = 4$ terms see [30].

In thermal equilibrium, g_S reduces to the equilibrium radial distribution function g_{eq} and the quantities $g_{+,-,0,4}$ vanish.

Insertion of the expansion (47) into (29) and use of (14) yields

$$p_k = -\frac{2\pi}{15} n^2 \int_0^\infty r\phi'\, g_k(r)\, r^2\, dr, \qquad k = +,\, -,\, 0 \tag{52}$$

$$p - p_{eq} = -\frac{2\pi}{3} n^2 \int_0^\infty r\, \phi'\, (g_S(r) - g_{eq})\, r^2\, dr. \tag{53}$$

These relations show that the coefficients η_k, η_{02} as introduced by (20, 21) can only be nonzero if $g_k \neq 0$ and $g_S \neq g_{eq}$. In this sense, the viscosity coefficients provide an indirect evidence for the flow-induced distortion of the structure of a fluid.

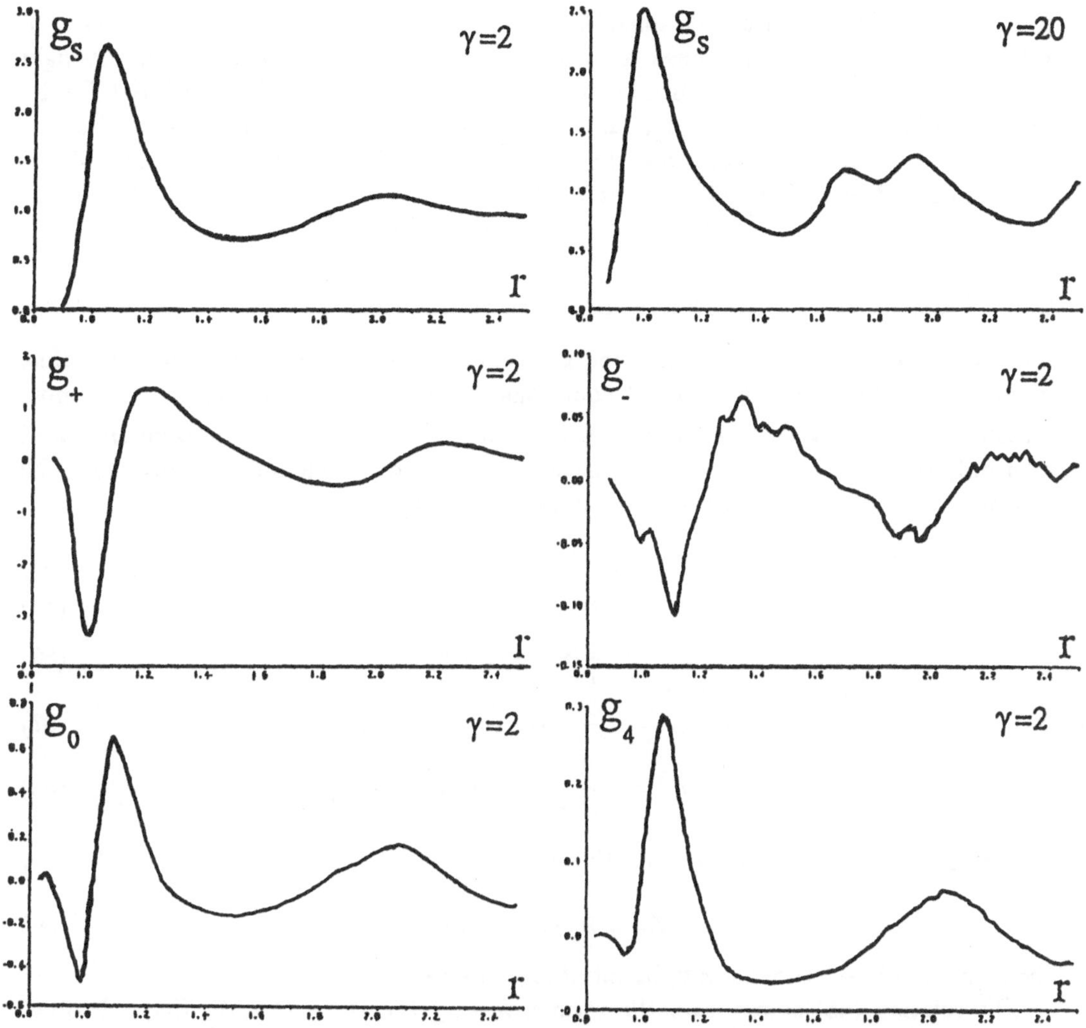

Figure 7: The partial distribution functions $g_{S,+,-,0,4}$ as functions of the interparticle distance r for the LJ-liquid at $n = 0.84$, $T = 1.0$ and for the shear rates $\gamma = 2$. For $\gamma = 20$ in the highly ordered state g_S is also shown. The vertical axes extend from 0 to 3.0 and 0 to 2.5 for g_S, from -4 to 2 for g_S, from -0.15 to 0.1 for g_-, from -0.5 to 0.8 for g_0 and -0.1 to 0.3 for g_4; the range of the r-values is from 0.8 to 2.5.

In molecular dynamics simulations, the functions g_S, g_+ and more recently also g_-, g_0 and higher order terms have been extracted, an example is shown in Fig.7. By light scattering techniques, the structure function $S(\mathbf{k})$ which is essentially the spatial Fourier transform of $g(\mathbf{r})$ has been investigated for systems of interacting colloidal particles [10], [11]. It has been demonstrated that the $g_{S,+,-,0}$ inferred from simulation of soft spheres (at the (reduced) density of 0.8) leads to a $S(\mathbf{k})$ which is qualitatively very similar to that one observed experimentally; see the pictures presented in [31].

3.2 Generalized Stokes-Maxwell Relations

Insertion of the expansion (47) into the kinetic equation (33) with the relaxation time approximation (37) leads to coupled equations for the functions g_S, g_+, ... If those contributions which stem from the $\overleftrightarrow{\gamma}$: L term of (34) are taken into account in lowest order only (tensors of rank $l \geq 4$ are neglected) the following generalized Stokes-Maxwell relations between the functions $g_{...}$ are found: [32], [33]

$$
\begin{aligned}
g_+ &= -B_1 r g_S', && (54)\\
g_- &= B_2 g_+, && (55)\\
g_0 &= \frac{1}{7} B_3 \left(\frac{3}{2} g_+ + r g_+'\right), && (56)\\
g_S - g_{eq} &= -\frac{1}{15} B_{02} \left(3 g_+ + r g_+'\right), && (57)\\
g_4 &= \frac{2}{15}\frac{1}{\sqrt{21}} B_4 \left(r\, g_+' - 2 g_+\right) && (58)
\end{aligned}
$$

with coefficients $B_{...}$ which depend on the shear rate γ but not on r. In lowest order in γ one finds:

$$B_1 = B_2 = B_3 = B_{02} = B_4 = \gamma\tau. \qquad (59)$$

The prime in (54 – 58) denotes differentiation with respect to r.

The relation (55) stems from the term involving the vorticity ω in (34), all other relations are "generated" by the term involving the deformation rate tensor $\overleftrightarrow{\gamma}$. In the linear flow regime, i.e. for $\gamma \to 0$, the quantities g_-, g_0, g_4 vanish, g_S approaches g_{eq}, and (54) reduces to the relation

$$g_+ = -\gamma\tau r g_{eq}'. \qquad (60)$$

Stimulated by a remark made by Poisson, both Stokes and Maxwell proposed a physical picture which, via a rather different route, leads to (60): distort the equilibrium structure g_{eq} by a small deformation which equals the deformation rate tensor $\overleftrightarrow{\gamma}$ times the "lifetime" τ. It is for this reason that (60) and its generalized versions (54 – 57) are referred to as Stokes-Maxwell relation. In molecular dynamics simulations, relation (60) has been tested some time ago, the nonlinear relations (54–57) and additional ones have been derived and tested more recently [30], [32], [33]. In Fig. 8, results for a $N = 512$ particles soft sphere system are presented, the temperature T and the density n (in reduced units) are 0.25 and 0.7, respectively. The data shown are for the shear rate $\gamma = 1$. The full curves on of Fig.8 are the functions $g_{+,-,0,4}$ extracted from the simulation, the dashed curves are obtained from the "measured" g_S and g_+ functions according to (54, 56, 59). The vertical scales are different in the various graphs. Clearly, the relations (54–59) seem to hold to a much better degree than expected in view of the approximations involved in their derivation and on account of the discussion presented on the shear rate dependence of the viscosity coefficients in the previous section.

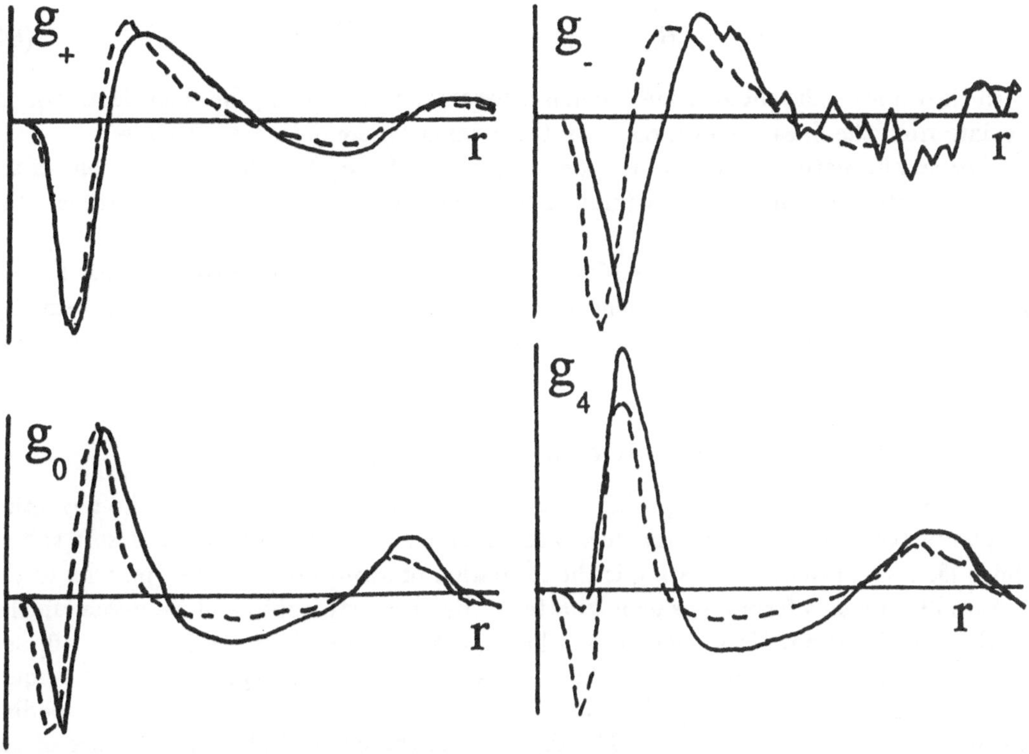

Figure 8: Test of the Stokes-Maxwell relations (54–56, 58) for a r^{-12}-soft sphere fluid at $n = 0.7$, $T = 0.25$ (SS-units). The full curves are the functions g_+, g_-, g_0, g_4 as extracted from the NEMD simulation, the dashed curves are evaluated from the computed g_S and g_+ according to the r.h.s. of eqs. (54–56, 58) with adjustable factors B_1, B_2,.... The vertical scale is in arbitrary units, the (horizontal) r-values range from 0.8 to 2.5 just as in Fig.7.

Notice however, that Fig.8 shows data for one fixed value of γ. How about the γ-dependence of the coefficients $B_{1,...}$? Experimental values for them are obtained, e.g. by comparison of the height of the first extrema of the functions given on the left hand and on the right hand sides of eqs. (54–56). Alternatively, one may equate integrals over these curves, in particular insertion of the relations (54–57) into (29) and use of (14), (20, 21) yields relations between the viscosity coefficients involving the $B_{...}$ coefficients:

$$\gamma\eta_+ = B_1 G \tag{61}$$

$$\eta_- = B_2 \eta_+ \tag{62}$$

$$\eta_0 = \frac{3}{2} B_3 \eta_+, \tag{63}$$

where (63) is only valid for the r^{-12}-potential. The "experimental" B values obtained by the two methods are consistent with each other. The theoretical expressions (59), however, are too simple. More precisely, B_1 can be fitted by

$$B_1 = \gamma\tau \left(1 + A_1\gamma^{1/2} + \ldots\right)^{-1} \tag{64}$$

with the same coefficient A_1 as in the non-Newtonian viscosity η_+, the coefficients B_2, B_3 deviate from the linear expression $\gamma\tau$ in the γ-range where data are available.

For an alternative solution of the kinetic equation for $g(\mathbf{r})$ see [34]; calculations of the static structure factor $S(\mathbf{k})$ in a fluid undergoing a viscous flow were also presented in [35]–[38].

At high shear rates, the Stokes-Maxwell relations break down for two reasons. Firstly, higher ($l \geq 4$) rank tensor components of $g(\mathbf{r})$ can no longer be disregarded and secondly, a long range (partial) positional ordering takes place. This phenomenon is described in the next section.

3.3 Shear-induced Positional Ordering

At high shear rates, a partial positional ordering of the particles occurs. The formation of planes parallel to the stream lines was observed for dense LJ [39] and soft sphere fluids [33]. Even more fascinating is the formation of strings of particles (parallel to the stream lines) which form a hexagonal pattern when projected onto a plane perpendicular to the flow direction. This phenomenon has first been noticed for hard spheres [40] and independently for soft spheres [41]. Fig.9 shows such a snapshot picture for a LJ-liquid under pressure (T=1.0, n=0.84) at the shear rate $\gamma = 20$ in the state corresponding to the smaller viscosity in Fig.1. The diameter of the circles indicating the size of the particles is equal to the distance where the LJ-potential is zero. Notice that the strings are actually "tubes" which can accommodate up to 9 particles (within one periodicity box) but which contains as few as 2 particles. Thus the fluid is by far not as densely packed as the figure suggested at a first glance. There is room in the flow direction and this room is needed since the particles occassionally have to jump from one tube to one below or above in order to transport momentum. In Fig.9 also the projections of the velocities of the particles on the plane normal to the flow direction are depicted by arrows. In some tubes, the particles have very little transverse momentum whereas in others one notices fast particles obviously moving towards adjacent tubes. It must be stressed that the configuration seen in Fig. 9 changes with time. Pair-correlation functions $g(\mathbf{r})$, on the other hand, when averaged over many thousand time steps will reveal those features which persist. There are marked qualitative differences noticed for $g(\mathbf{r})$ in a sheared "amorphous fluid" and in a fluid with "string" or "tube" ordering.

The partially ordered states at high shear rates can be looked upon as a new nonequilibrium (shear-induced) phase; the transition from one state into the other occurs in a rather narrow range of values for the shear rate γ. The crossover in the shear rate dependence of rheological properties seen in Figs. 1–4 is associated with this transition.

The formation of the long range order takes place over time considerably longer than the relaxation time τ mentioned above; stress-overshoot is observed.

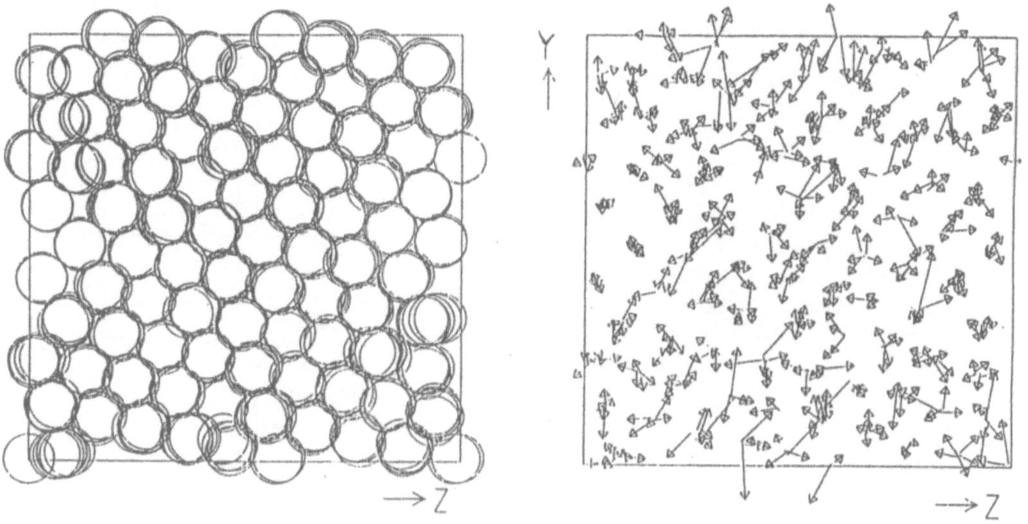

Figure 9: Snapshots of the positions and the velocities of 512 particles of a LJ-liquid at $n = 0.84$, $T = 1.0$, $\gamma = 20$ in a well ordered state. The projection of the 3-dimensional system is on the y-z-plane which is perpendicular to the flow direction. The particles are marked by circles with the radius $r = 1$; the velocities are indicated by arrows.

At still higher shear rates there is a tendency for the particles to organize themselves in blocks of several layers parallel to the flow velocity. Thus the actual velocity profile deviates from the assumed linear Couette profile. This was already noticed some time ago (with soft sphere fluids) indirectly via a break down of the "thermostat" which was not due to the thermostating procedure as such but was caused by the failure to enforce a linear flow profile above a certain shear rate. This prompted us to change the simulation algorithm such that the system has the freedom to choose its velocity profile [42]. The shear rate γ is prescribed only as a velocity gradient averaged over the height (direction of the velocity gradient) of the periodicity box. It turned out that for small and intermediate shear rates the system "voluntarily" adjusts to the linear Couette profile and no differences are noticed between the results obtained with the newer and with the older NEMD simulations until the partial long range positional ordering sets in at a critical value $\gamma = \gamma_c$. Beyond that point, a plug-like flow profile as shown in Fig.10 is found (LJ-liquid, $T = 1.0$, $n = 0.84$, $\gamma = 4.0$). The arrows indicate the mean flow velocities in 32 layers (perpendicular to the velocity gradient) averaged over several thousand time steps; the slope of the dashed line is equal to the mean velocity gradient, i.e. to the prescribed shear rate γ. A snapshot projection of the positions into a plane perpendicular to the flow direction is also shown in Fig. 10 with the particles of two neighboring noncubic periodicity boxes. Clearly, there are amorphous regions and layers with a relatively high degree of spatial ordering. The rheological properties extracted from the simulations with the old and the new algorithms are identical for $\gamma < \gamma_c$; for $\gamma > \gamma_c$ (i.e. in the positionally

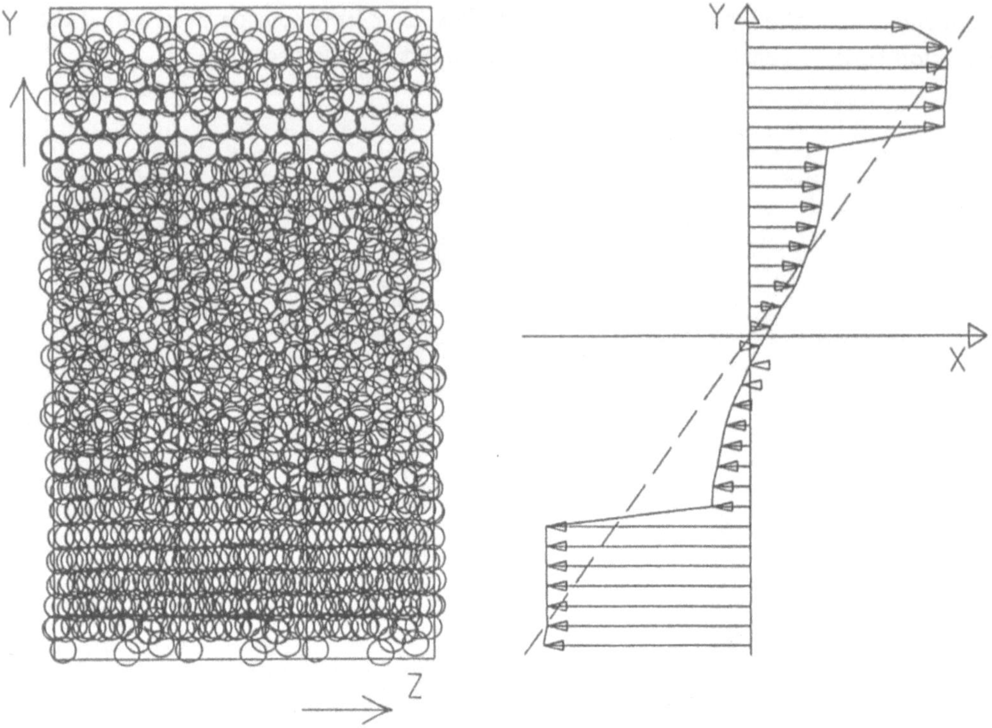

Figure 10: Snapshot projection of the positions of the particles on the y-z-plane and the velocity profile in the x-y-plane for a LJ-liquid at $n = 0.84$, $T = 1.0$ and $\gamma = 4$. The periodicity box has been stretched in the y-direction; only the mean velocity gradient γ is controlled, resulting in a plug-like flow (W. Loose, TU Berlin).

ordered state) the viscosities found with the new method (profile-unbiased thermostat: PUT [43]) are somewhat smaller than the old ones displayed in Fig.1.

The problem of thermostatting and the formation of spatial structures in NEMD simulation has been discussed in detail in [44]. The occurence of this nonequilibrium phase transition can also be inferred from a stability analysis [42]. The positional ordering also occurs in boundary driven flows devoted to the study of small slip phenomena [45]. In an experiment where the pressure rather than the volume is kept constant, it is expected that the transition from amorphous to the long range ordered state occurs gradually.

3.4 Light Scattering and Small Angle Neutron Scattering

The intensity $I(\mathbf{k})$ measured in a light scattering or small angle neutron scattering (SANS) experiment is proportional to the product $P(\mathbf{k})S(\mathbf{k})$ of the relevant form factor $P(\mathbf{k})$ and the static structure factor $S(\mathbf{k})$. Here $\mathbf{k}$ is the scattering wave vector, i.e. the difference between the wave vectors of the incident and the scattered radiation. As mentioned before,

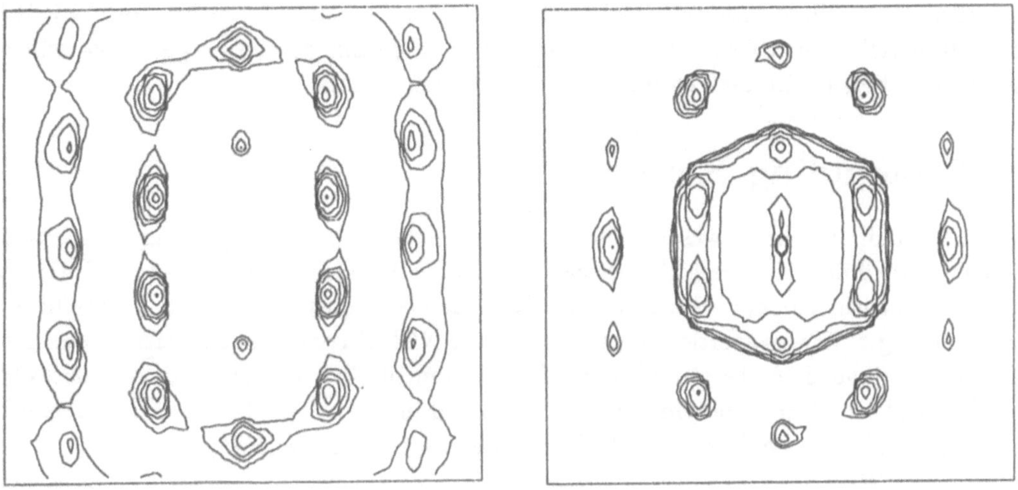

Figure 11: Contour plots of the structure factor $S(\mathbf{k})$ calculated in a NEMD simulation of a soft sphere like model system [42] ($n = 0.84$, $T = 0.25$) in a state similar to that one shown in Fig.9 and the intensity $I(\mathbf{k}) = P(\mathbf{k})S(\mathbf{k})$ (with the theoretical form factor $P(\mathbf{k})$ for a sphere with radius 0.55) in the x-z-plane corresponding to a small angle scattering experiment with the incident beam along the direction of the velocity gradient (W. Loose, TU Berlin).

$S(\mathbf{k})$ is essentially the spatial Fourier transform of $g(\mathbf{r})$, more specifically, for $\mathbf{k} \neq 0$,

$$S(\mathbf{k}) = 1 + n \int e^{i\mathbf{k}\cdot\mathbf{r}} \left(g(\mathbf{r}) - 1\right) \, d^3r. \tag{65}$$

In the computer simulation, $S(\mathbf{k})$ can be extracted as a N-particle average according to

$$S(\mathbf{k}) = N^{-1} \left(\left(\sum_j \cos \mathbf{k} \cdot \mathbf{r}^j \right)^2 + \left(\sum_j \sin \mathbf{k} \cdot \mathbf{r}^j \right)^2 \right). \tag{66}$$

The $\mathbf{k}$ vectors have to be chosen such that they "fit" into the periodicity box. The small angle scattering with a detector field can be mimiced by the choice of a large set of $\mathbf{k}$ values in a plane perpendicular to the direction of the incident beam. In [46], intensity plots of a sheared model fluid are shown for three scattering geometries.

Light scattering experiments [10], [11], [47] on the shear flow induced distortion of the structure have stimulated a good part of the theoretical work described here. SANS experiments on sheared dense colloidal dispersions have also been devoted to shear-induced melting [12], [48]. More recently, the shear-induced positional ordering was found [49] in the low viscosity state of shear-dilatant dispersions [50].

In Fig.11, contour plots of $S(\mathbf{k})$ and $P(\mathbf{k})\, S(\mathbf{k})$ are shown for a structure factor extracted from a NEMD simulation of a sheared model fluid [51] (soft spheres with a screened Coulomb interaction, cf. section 2.4; $n = 0.84$, $T = 0.25$, $B = 1$, SS -units) in a highly ordered state similar to that one shown in Fig.9. For the form factor P the theoretical expression for a sphere with radius 0.55 was used. The theoretical intensity $I \sim P(\mathbf{k})S(\mathbf{k})$ perfectly matches the SANS-intensity observed for the same scattering geometry, where the incident beam is parallel to the gradient direction (y-axis).

4 Concluding Remarks

Here, the main emphasis was on the interrelation between rheological and structural properties in fluids of spherical particles as studied by kinetic theory and NEMD. It should be mentioned that general thermodynamic considerations [52] can be useful for the modelling of flow properties. Furthermore, the NEMD computer simulations can and have been extended to the study of the flow alignment, the anisotropy of the viscosity and the local structure in polymeric liquids [53] and in anisotropic fluids such as nematic ane nematic discotic liquid crystals [54] and oriented ferro-fluids [55], [56].

Acknowledgements

Many thanks are due to W. Loose for providing Figs. 8–11 and to O. Hess for substantial help in preparing this manuscript. Fruitful discussions on the theoretical and computational aspects, on experiments and applications of rheology and the structure of fluids with B. J. Ackerson (Oklahoma State Univ.), H. J. M. Hanley (NIST, Boulder, Colorado), H. M. Laun (BASF, Ludwigshafen), and P. Lindner (ILL, Grenoble) are gratefully acknowledged. This work is conducted under the auspices of the Sonderforschungsbereich "Anisotrope Fluide" (SFB 335) of the Deutsche Forschungsgemeinschaft.

References

[1] J. C. Maxwell, Proc. Roy. Soc. London (A), **22**, 46 (1873)

[2] O. Reynolds, Phil. Mag. **20**, 469 (1885)
Reynolds filled a leather bag with marbels, topped it with water and then twisted it, thereby inducing a flow. The water level drops because the close packing of the marbels is disrupted as layers of marbels slide over each other during the twisting motion. Consequently, the marbels are further apart, on average, and the water fills the additional space between them.

[3] C .G .D. Cohen and W. Thirring (eds.), *The Boltzmann Equation, Theory and Applications*, Springer, Wien 1983

[4] S. Chapman and T. G. Cowling, *The mathematical theory of non-uniform gases*, Cambridge Univ. Press, Cambridge 1939;
J. O. Hirschfelder, C. F. Curtiss and R .B. Bird, *Molecular theory of gases and liquids*, Wiley, New York (1954)

[5] H. J. M. Hanley (ed.), *Nonlinear Fluid Behavior*, North Holland, Amsterdam 1983, Physica **118 A** (1983)

[6] Physics Today **27**, Special issue, Jan. 1984

[7] D. Quemada (ed.), *Nonlinear behavior of dispersive media*, J. Mécanique Théor. Appl., Numéro spécial 1985

[8] S. Hess and W. Loose, Ber. Bunsenges., Phys. Chem **94**, 216 (1990)

[9] W. Loose and S. Hess, Phys. Rev. Lett.**58**, 2443 (1987); Phys. Rev. A **37**, 2099 (1988)

[10] N. A. Clark and B .J. Ackerson, Phys. Rev. Lett.**44**, 2844 (1980);
B. J. Ackerson and N. A. Clark, Physica **118 A**, 221 (1983);
B. J. Ackerson and P. N. Pusey, Phys. Rev. Lett. **61**, 1033 (1988)

[11] B .J. Ackerson, J. van der Werff, and C. G. de Kruif, Phys. Rev. A **37**, 4819 (1988)

[12] B. J. Ackerson, J. B. Hayter, N. A. Clark, and L. Cotter, J. Chem. Phys. **84**, 2344 (1986);
J. C. van der Werff, B. J. Ackerson, R. P. May, and C. G. de Kruif, Physica **165 A**, 375 (1990)

[13] A. Peterlin and H. A. Stuart, in Hand- und Jahrbuch d. Chem Phys. (ed. Eucken-Wolf) **8**, 113 (1943)

[14] H. Janeschitz-Kriegel, *Polymer melt rheology and flow birefringence*, Springer, Berlin 1983

[15] S. Hess, Phys. Lett. **30 A**, 239 (1969); Springer Tracts in Mod. Phys. **54**, 136 (1970)

[16] F. Baas, Phys. Lett. **36A**, 107 (1971);
F. Baas, J. N. Breunese, H. F. P. Knaap, and J. J. M. Beenakker, Physica **88 A**, 1 (1977)

[17] R. B. Bird, R. C. Armstrong, and O. Hassager, *Dynamics of polymeric liquids*, J. Wiley, New York 1977

[18] W. G. Hoover, Annu. Rev. Phys. Chem. **34**, 103 (1983);
D. J. Evans and G. P. Morriss, Comp. Phys. Rep. **1**, 287 (1984);
D. J. Evans and W. G. Hoover, Ann. Rev. Fluid Mech. **18**, 243 (1986);
M. P. Allen and D. J. Tildesley, *Computer Simulation of Liquids*, Clarendon, Oxford (1987)

[19] B. D. Holian and D. J. Evans, J. Chem. Phys. **78** 5157 (1983)

[20] S. Hess and W. Loose, in D. Axelrad and W. Muschik (eds.) *Constituive laws and microstructure*, Springer, Berlin 1988, p. 92

[21] D. J. Evans, Mol. Phys. **37**, 1745 (1979)

[22] J. G. Kirkwood, J. Chem. Phys. **14**, 180 (1946);
R. Eisenschitz, *Statistical theory of irreversible processes*, Oxford Univ. Press, Oxford 1960;
H. S. Green, *Handbuch der Physik*, vol. 10, S. Flügge (ed.), Springer, Berlin 1960

[23] S. Hess, in [5], p. 79

[24] T. Oktsuki, Physica **108A**, 441 (1981)

[25] J. F. Schwarzl and S. Hess, Phys. Rev. A **33**, 4277 (1986)

[26] S. Hess, in [7], p. 1

[27] J. B. Hayter and J. Penfold, J. Chem. Soc. Faraday Trans. I, **77**, 851 (1981);
Mol. Phys. **4**, 109 (1981);
J. P. Hansen and J. B. Hayter, Mol. Phys. **46**, 651 (1982)

[28] P. N. Pusey, J. Phys. **A8**, 1433 (1976)

[29] P. N. Pusey, H. N. W. Lekkerkerker, E. G. D. Cohen, and I. M. de Schepper, Physica **164 A**, 12 (1990)

[30] H. J. M. Hanley, J. C. Rainwater, and S. Hess, Phys. Rev. A **36**, 1795 (1987)

[31] H. J. M. Hanley, J. C. Rainwater, N. A. Clark, and B. J. Ackerson, J. Chem. Phys. **79**, 4448 (1983);
J. C. Rainwater, H. J. M. Hanley, and S. Hess, Phys. Lett **116A**, 450 (1988)

[32] S. Hess and H. J. M. Hanley, Phys. Lett. **98A**, 35 (1983)

[33] S. Hess, J. de Physique **46C3**, 191 (1984)

[34] H.-M. Koo and S. Hess, Physica **145A**, 361 (1987)

[35] S. Hess, Phys. Rev. A **22**, 2844 (1980)

[36] D. Ronis, Phys. Rev. A **29**, 1453 (1984)

[37] J. K. G. Dhont, J. Fluid Mech. **204**, 421 (1989)

[38] N. J. Wagner and W. B. Russel, Physica **155 A**, 475 (1989)

[39] D. M. Heyes, J. J. Kim, C. J. Montrose, and T. A. Litovitz, J. Chem. Phys. **73**, 3979 (1980)

[40] J. J. Erpenbeck, Phys. Rev. Lett. **52**, 1333 (1984)

[41] S. Hess, Int. J. Thermophys. **6**, 657 (1985)

[42] W. Loose and S. Hess, Rheol. Acta **28**, 91 (1989)

[43] D. J. Evans and G. P. Morriss, Phys. Rev. Lett. **56**, 2172 (1986)

[44] W. Loose and S. Hess, in *Microscopic simulation of complex flows*, M. Mareschal (ed.), ASI series, Plenum Press (1990)

[45] S. Hess and W. Loose, Physica A **162**, 138 (1989)

[46] O. Hess, W. Loose, T. Weider, and S. Hess, Physica B **156/157**, 505 (1989)

[47] B. J. Ackerson, Physica A ... (1991)

[48] P. Lindner, *Annual report of the Institute Laue-Langevin (ILL)*, p. 76, Grenoble 1987

[49] E. Hädicke, S. Hess, H. M. Laun, P. Lindner, B. Schmitt, F. Schmidt, R. Bung, ILL-experiment (D11, Dec. 1989), preliminary results

[50] H. M. Laun, Angew. Makromol. Chem. **123/124**, 335 (1984);
Proc. X-th Internat. Congr. Rheol., Sydney, Aug. 1988

[51] T. Weider, U. Stottut, W. Loose, and S. Hess, Physica A ... (1991)

[52] C. Péres-Garcia and D. Jou, J. Non-Equilib. Thermodyn. **7**, 191 (1982);
Phys. Lett **95A**, 23 (1983)

[53] S. Hess, J. Non-Newtonian Fluid Mech. **23**, 305 (1987)

[54] D. Baalss and S. Hess, Phys. Rev. Lett. **57**, 86 (1986);
Z. Naturforsch. **43a**, 662 (1988);
H. Sollich, D. Baalss, and S. Hess, Mol. Cryst. Liq. Cryst. **168**, 189 (1989)

[55] S. Hess, J. B. Hayter, and R. Pynn, Mol. Phys. **53**, 1527 (1984);
T. Weider, Diplomarbeit, TU Berlin 1988;
S. Hess, J. Non-Equilib. Thermodyn. **11**, 176 (1986)

[56] S. Hess, J. F. Schwarzl, and D. Baalss, J. Physics (1991)

On the dynamics of polymers in solution

J.M. Rubí, J. Bonet Avalos
Departament de Física Fonamental, Facultat de Física
Universitat de Barcelona, Diagonal 647, 08028 Barcelona (Spain)
and D. Bedeaux
Department of Physical and Macromolecular Chemistry
Gorlaeus Laboratoria, University of Leiden
P.O. Box 9502, 2300 RA Leiden (The Netherlands)

1 Introduction

It is well-known that the dynamics of objects in fluids plays a very important role in the study of transport phenomena or in the rheology of suspensions and polymer solutions [1,2,3]. In the formulation of the problem, the fluid is usually assumed to be Newtonian while objects may have a defined shape (spheres, spheroids, rods, fibers,...), they may be deformable (drops, bubbles,...) and they can adopt complicated structures (chains, coils, rings, branched structures,...).

It is clear that the perturbation caused by the motion of the objects through the fluid will depend on their geometry. If they have a well-defined and simple form (for example, spherical) one can readily compute the force exerted by the fluid on them and, from this, one gets the expression for the friction or mobility tensors [4]. In many cases, however, the geometry may be complicated or the configuration may depend on time. For this reason it may be convenient to introduce simple models finding a compromise between reality and solvability.

The case of polymers is particulary interesting. In fact, macromolecules have conformations ranging from the flexible chain to the rigid rod. It has become costumary to assume that coils behave as hard or permeable spheres with an effective radius. These models where introduced by Kirkwood [5] and Debije, Bueche and Brinkman [6]. Dumbbell models have also been used to describe the dynamics of polymer chains. In these models, one assumes that the chain reduces to two spheres (beads) joined by a rigid link [7,8] or an spring [9,10]. They have been extensively discussed in ref. [11] and in Prof. Bird's lectures. A generalization of this model can be introduced by considering the polymer constituted by several beads connected by springs (bead-spring chain model). For stiff chains, the so-called worm-like chain model has also been proposed [12,13]

To describe the dynamics of a chain, we must consider the forces acting on each bead or segment. Hydrodynamic forces are due to the friction of the beads with the fluid and to the presence of hydrodynamic interactions between them [1]. Brownian forces appear as a consequence of the colloidal nature of the beads and originate in the presence of fluctuations in the fluid [14]. There are also internal forces between segments of the chain and external forces related to the presence of external fields. The simplest dynamics was studied by Rouse [15] who considered that the internal forces were elastic. The interaction of the beads with the fluid was introduced through the friction force of the single beads given by Stokes law. The model was generalized by Zimm [16] incorporating hydrodynamic interactions between the beads.

Our purpose in this paper is to develop a general formalism including retardation effects of the hydrodynamic interactions to analyze the dynamics of polymers in solution. To this end we have organized the paper as follows. We plan to discuss the derivation of the equation of motion of the polymer in section 2 for both flexible and rigid chains. In the first case we study the influence of the nonstationary motion of the fluid on the dynamics of the chains [17]. For this purpose we develop a general formalism whose starting point is the Navier-Stokes equation with random and induced force sources [18,19]. The random forces account for the existence of fluctuations in the fluid while the induced forces are related to the perturbations caused by the motion of the segments. Crucial in our analysis is the obtention of the nonstationary mobility kernel which incorporates time-dependent hydrodynamic interactions. In the second case, we show that the induced forces method also enables us to obtain the equation of motion of rod-like polymers. In section 3 we study the resulting dynamics of the polymer. We then get information about the motion of the centre of mass and the confomational changes of flexible chains. For rigid rods our theory enables us to reproduce the translational and rotational diffusion coefficients, given by Yamakawa, in a simple manner.

2 Equation of motion of a polymer

Our purpose in this section is to derive the equation of motion of a polymer. We start from the Navier-Stokes equation in which we have included stochastic and induced forces. The properties of the former follow from fluctuating hydrodynamics [20] while the latter account for the presence of particles in the fluid. The cases of flexible chains and rigid rods will be examined.

2.1 Flexible chain

We will consider a flexible polymer composed by N beads, moving through an incompressible Newtonian fluid of density ρ and viscosity η. The motion of the fluid is governed by the Navier-Stokes equation

$$\rho \frac{\partial \vec{v}(\vec{r},t)}{\partial t} = -\nabla \cdot \vec{\vec{P}}(\vec{r},t) \tag{2.1}$$

which is valid in the linearized regime, together with the incompressibility condition

$$\nabla \cdot \vec{v} = 0 \tag{2.2}$$

In eqs. (2.1) and (2.2) $\vec{v}(\vec{r},t)$ is the velocity field and $\vec{\vec{P}}(\vec{r},t)$ is the pressure tensor. This quantity is given by

$$P_{\alpha\beta} = p\delta_{\alpha\beta} - \eta \left(\frac{\partial v_\beta}{\partial r_\alpha} + \frac{\partial v_\alpha}{\partial r_\beta} \right) + \Pi^R_{\alpha\beta} \tag{2.3}$$

where p is the hydrostatic pressure and $\Pi^R_{\alpha\beta}$ is the random pressure tensor. This quantity defines a Gaussian white noise stochastic process whose correlations are given by

$$< \vec{\vec{\Pi}}^R(\vec{r},t) > \; = \; 0 \tag{2.4}$$

$$< \Pi^R_{\alpha\beta}(\vec{r},t)\Pi^R_{\gamma\mu}(\vec{r}\,',t') > \; = \; 2k_BT\eta\, \Delta_{\alpha\beta\gamma\mu}\, \delta(\vec{r}-\vec{r}\,')\, \delta(t-t') \tag{2.5}$$

where

$$\Delta_{\alpha\beta\gamma\mu} = \delta_{\alpha\gamma}\delta_{\beta\mu} + \delta_{\alpha\mu}\delta_{\beta\gamma} - \frac{2}{3}\delta_{\alpha\beta}\delta_{\gamma\mu} \tag{2.6}$$

To compute the mobility of the particles we will use the induced force method [18,19,21]. One then reformulates the problem by assuming that the fluid field is also defined inside the particles and the perturbations caused by their motion are introduced through induced forces. The Navier-Stokes equation is now written

$$\rho \frac{\partial \vec{v}(\vec{r},t)}{\partial t} = -\nabla p(\vec{r},t) + \eta \nabla^2 \vec{v}(\vec{r},t) + \sum_i \vec{F}_i(\vec{r},t) - \nabla \cdot \vec{\vec{\Pi}}^R(\vec{r},t) \tag{2.7}$$

where $\vec{F}_i$ is the induced force density on the ith particle. Of course, this induced force density is only different from zero inside and on the surface of the particle and is chosen so that

$$p(\vec{r},t) = 0; \qquad |\vec{r}-\vec{R}_i| < a \tag{2.8}$$
$$\vec{v}(\vec{r},t) = \vec{u}_i(t); \qquad |\vec{r}-\vec{R}_i| \leq a \tag{2.9}$$

where $\vec{R}_i(t)$ are the position vectors of the centre of the spheres. Equation (2.8) follows from the fact that the particles are rigid while eq. (2.9) is simply the stick boundary condition at the surface of the spheres.

It follows from these expressions together with eq. (2.7) that the induced force density is given by

$$\begin{aligned}\vec{F}_i(\vec{r},t) &= \hat{n}(\vec{r},t)\cdot\vec{\vec{P}}^{s}(\vec{r},t)\,\delta(|\vec{r}-\vec{R}_i(t)|-a)\\ &+ \left[\rho\frac{d\vec{u}_i}{dt} - \nabla\cdot\Pi^R(\vec{r},t)\right]\theta(a-|\vec{r}-\vec{R}_i(t)|)\end{aligned} \tag{2.10}$$

where $\hat{n}(\vec{r},t) = (\vec{r}-\vec{R}_i)/|\vec{r}-\vec{R}_i|$.

Now we will formally solve eq. (2.7) for the velocity field $\vec{v}(\vec{k},t)$. One gets

$$\begin{aligned}\vec{v}(\vec{k},t) &= \vec{v}_0(\vec{k},t) + \int_{-\infty}^{t} dt' e^{-\nu k^2(t-t')}\,\frac{(\vec{\vec{1}}-\hat{k}\hat{k})}{\rho}\,\cdot\\ &\left[\sum_i \vec{F}^i(\vec{k},t') - i\vec{k}\cdot\vec{\vec{\Pi}}^R(\vec{k},t')\right]\end{aligned} \tag{2.11}$$

where $\nu = \eta/\rho$ is the kinematic viscosity, $\vec{\vec{1}}$, the unit matrix, $\hat{k} = \vec{k}/k$ and $\vec{v}_0(\vec{k},t)$ is the velocity field due to the presence of sources at infinity.

Since we will consider length scales larger than the radius of the beads and times so that $t \gg a^2/\nu$, we will assume that the induced force density is uniformly distributed at the surface of the beads. In such a case, eq. (2.10) may be replaced by

$$\vec{F}_i(\vec{k},t) \simeq e^{-i\vec{k}\cdot\vec{R}_i(t)}\left(\frac{\sin ka}{ka}\right)\vec{F}_i(\vec{k}=0,t) \tag{2.12}$$

where $\vec{F}_i(\vec{k}=0,t)$ is the total force exerted for the ith-particle on the fluid. The drag force experienced by the bead is $\vec{F}_i^H(t) \equiv -\vec{F}_i(\vec{k}=0,t)$. Substituting the value of the induced force in eq. (2.11) we get

$$\begin{aligned} \vec{v}(\vec{k},t) &= \vec{v}_0(\vec{k},t) - \int_{-\infty}^{t} dt' e^{-\nu k^2(t-t')} \frac{(\vec{\vec{1}} - \hat{k}\hat{k})}{\rho} \\ &\cdot \left[\sum_i e^{-i\vec{k}\cdot\vec{R}_i(t)} \left(\frac{\sin ka}{ka}\right) \vec{F}_i^H(t) + i\vec{k}\cdot \vec{\vec{\Pi}}^R(\vec{k},t') \right] \end{aligned} \tag{2.13}$$

Using stick boundary conditions, we can obtain the velocity of the ith-bead, $\vec{u}_i(t)$, by averaging the velocity field on the surface of the bead. One then obtains

$$\begin{aligned} \vec{u}_i(t) &= \frac{1}{4\pi a^2} \int d\vec{r}\, \vec{v}(\vec{r},t)\, \delta(|\vec{r} - \vec{R}_i(t)| - a) \\ &= \frac{1}{(2\pi)^3} \int d\vec{k}\, e^{i\vec{k}\cdot\vec{R}_i} \left(\frac{\sin ka}{ka}\right) \vec{v}(\vec{k},t) \end{aligned} \tag{2.14}$$

Substituting now eq. (2.13) into eq. (2.14) we finally get

$$\vec{u}_i(t) \equiv \vec{R}_i \cdot \vec{\vec{\beta}} - \sum_j \int_{-\infty}^{t} dt' \vec{\vec{\mu}}_{ij}(t-t'; \vec{R}_i(t) - \vec{R}_j(t')) \cdot \vec{F}_j^H(t') + \vec{u}_i^R(t) \tag{2.15}$$

where incident velocity field was assumed to be given by $\vec{v}_o(\vec{r},t) = \vec{r}\cdot\vec{\vec{\beta}}$. Here $\vec{\vec{\beta}}$ is a traceless matrix. The time-dependent mobility kernels were found to be given by

$$\vec{\vec{\mu}}_{ij}(\tau; \vec{R}_{ij}) \equiv \int \frac{d\vec{k}}{(2\pi)^3} e^{-\nu k^2 \tau}\, e^{i\vec{k}\cdot\vec{R}_{ij}} \left(\frac{\sin ka}{ka}\right)^2 \frac{(\vec{\vec{1}} - \hat{k}\hat{k})}{\rho} \tag{2.16}$$

with $\tau = t - t'$ and $\vec{R}_{ij} = \vec{R}_i(t) - \vec{R}_j(t)$. Furthermore, the random term is

$$\begin{aligned} \vec{u}_i^R(t) &= -\int_{-\infty}^{t} dt' \int \frac{d\vec{k}}{(2\pi)^3} e^{-\nu k^2(t-t')} \left(\frac{\sin ka}{ka}\right) e^{i\vec{k}\cdot\vec{R}_i(t)} \\ &\quad \frac{(\vec{\vec{1}} - \hat{k}\hat{k})}{\rho} \cdot \left[i\vec{k}\cdot \vec{\vec{\Pi}}^R(\vec{k},t') \right] \end{aligned} \tag{2.17}$$

We have considered in eq. (2.16) that $\vec{R}_i(t) - \vec{R}_j(t') \simeq \vec{R}_i(t) - \vec{R}_j(t)$ in view of the linearization of the Navier-Stokes equation.

On one hand, the explicit expression for the two-particle mobility kernel may easily be evaluated if $|\vec{R}_{ij}| \gg a$. In that case, we can take $ka = 0$ (point particle approximation) in eq. (2.16). Then, the integration over $\vec{k}$-space can be performed, arriving at

$$\vec{\vec{\mu}}_{ij}(\tau; \vec{R}_{ij}) = \frac{e^{-R_{ij}^2/4\nu\tau}}{\rho(4\pi\nu\tau)^{3/2}} \vec{\vec{1}} + \frac{1}{4\pi\rho} \nabla_{\vec{R}_{ij}} \nabla_{\vec{R}_{ij}} \frac{1}{R_{ij}} \Phi\left(\frac{R_{ij}}{(4\nu\tau)^{1/2}}\right) \tag{2.18}$$

$\Phi(x)$ being the error function defined as

$$\Phi(x) \equiv \frac{2}{\sqrt{\pi}} \int_0^x d\xi e^{-\xi^2} \tag{2.19}$$

This result is valid if $t \gg a^2/\nu$ and $R_{ij} \gg a$. The single-particle mobility kernel can also be obtained from eq. (2.16), setting $\vec{R}_{ij} = 0$. For our purposes, however, only the two-particle mobility kernel is required. After integration in time, eq. (2.18) reduces to the Oseen tensor $\vec{\vec{T}}(\vec{R}_{ij})$ defined as

$$\vec{\vec{T}}(\vec{R}_{ij}) = \frac{1}{8\pi\eta R_{ij}} \left(\vec{\vec{1}} + \hat{R}_{ij}\hat{R}_{ij}\right) \tag{2.20}$$

where $\hat{R}_{ij}$ is the unit vector $\vec{R}_{ij}/R_{ij}$.

On the other hand, the random parts of the velocities given in eq. (2.17) may be shown to satisfy

$$\begin{aligned} < \vec{u}_i^R(t) > &= 0 & (2.21) \\ < \vec{u}_i^R(t)\vec{u}_j^R(t') > &= k_B T \vec{\vec{\mu}}_{ij}(t-t'; \vec{R}_{ij}) & (2.22) \end{aligned}$$

using the properties of the stochastic part of the viscous pressure tensor given through eqs. (2.4) and (2.5).

The equation of motion of the ith-particle is given by

$$M\frac{d\vec{u}_i(t)}{dt} = \vec{F}_i^H(t) + \vec{F}_i^{int}(t) + \vec{F}_i^{ext}(t) \tag{2.23}$$

where M is the mass of a segment, $\vec{F}_i^H$ is the total force exerted by the fluid on the ith-particle, $\vec{F}_i^{int}$ is the force due to internal interactions with other segments and $\vec{F}_i^{ext}$ includes the forces due to external fields. We can then solve for $\vec{F}_i^H$ from eq. (2.23) and replace it into eq. (2.15), yielding

$$\begin{aligned}\vec{u}_n(t) &= \vec{R}_i \cdot \vec{\vec{\beta}} + \int_{-\infty}^{t} dt' \int_0^N dm \, \vec{\vec{\mu}}_{nm}(t-t'; \vec{R}_{nm}) \cdot \\ &\left[\vec{F}_m^{int}(t') + \vec{F}_m^{ext}(t') - M\frac{d\vec{u}_m}{dt'}\right] + \vec{u}_n^R(t)\end{aligned} \tag{2.24}$$

where the continuous chain limit has been taken. Here, the mobility kernel is given by eq. (2.18) which is valid for $|\vec{R}_{nm}| \gg a$. If $|\vec{R}_{nm}| \simeq a$, we will still use eq. (2.18) while in the case where $|\vec{R}_{nm}| < a$ we will assume that the mobility is zero. This cutoff is necessary in order to avoid the divergences which originate from the use of a continuous chain [22].

Notice that eq. (2.24) may be transformed into a generalized Langevin equation [23]. In fact, by multiplying this expression by the friction tensor $\vec{\vec{\xi}}_{mn}(t)$ (defined as a generalized inverse of the mobility kernel $\vec{\vec{\mu}}_{nm}(t)$) and integrating over t and m one gets the generalized Langevin equation

$$\begin{aligned}M\frac{d\vec{u}_n}{dt} = -\int_{-\infty}^{\infty} dt' \int_0^N dm \, \vec{\vec{\xi}}_{nm}(t-t'; \vec{R}_{nm}) \cdot (\vec{u}_m(t') &- \vec{R}_m(t') \cdot \vec{\vec{\beta}}) \\ &+ \vec{F}_n^{ext}(t) + \vec{F}_n^R(t)\end{aligned} \tag{2.25}$$

where the random term $\vec{F}_n^R(t)$ is

$$\vec{F}_n^R(t) = \int_{-\infty}^{\infty} dt' \int_0^N dm \, \vec{\vec{\xi}}_{nm}(t-t'; \vec{R}_{nm}) \cdot \vec{u}_m^R(t') \tag{2.26}$$

This random force has zero mean and satisfies the fluctuation-dissipation theorem

$$< \vec{F}_n^R(t)\vec{F}_m^R(t') >= k_B T \vec{\vec{\xi}}_{nm}(t-t'; \vec{R}_{nm}) \tag{2.27}$$

which follows from eqs. (2.21) and (2.22).

Now we want to introduce an explicit form for $\vec{F}_n^{int}$. Let us consider the case where no external fields are present. If $\mathcal{H}$ is the interaction potential, the forces between segments are given by

$$\vec{F}_n^{int} = -\frac{\delta \mathcal{H}}{\delta \vec{R}_n} \tag{2.28}$$

According to the elastic chain model with short-range interactions between segments, we can assume that $\mathcal{H}$ is given by

$$\mathcal{H} = \int_0^N dn \left\{ \frac{H}{2} \left(\frac{\partial \vec{R}_n}{\partial n} \right)^2 + \int_0^N dm \, U_{int}(\vec{R}_n - \vec{R}_m) \right\} \tag{2.29}$$

where

$$H = \frac{3k_BT}{b^2} \tag{2.30}$$

b being the effective bond length. In the case in which

$$U_{int}(\vec{R}_n - \vec{R}_m) = \frac{k_BT\tilde{v}}{2} \delta(\vec{R}_n - \vec{R}_m) \tag{2.31}$$

$\tilde{v}$ being the excluded volume, we have the usual model for the excluded volume interaction [24,25]. In such a case, we can finally write

$$\begin{aligned} \frac{\partial \vec{R}_n(t)}{\partial t} &= \vec{R}_n(t) \cdot \vec{\vec{\beta}} + \int_{-\infty}^t dt' \int_0^N dm \, \vec{\vec{\mu}}_{nm}(t - t'; \vec{R}_n - \vec{R}_m) \cdot \\ &\left[H \frac{\partial^2 \vec{R}_m}{\partial m^2} - \frac{k_BT\tilde{v}}{2} \frac{\partial}{\partial \vec{R}_m} \int_0^N dm' \, \delta(\vec{R}_m - \vec{R}_{m'}) - M \frac{\partial^2 \vec{R}_m}{\partial t'^2} \right] + \vec{u}_n^R(t) \end{aligned} \tag{2.32}$$

where use has been made of the definition $(\partial \vec{R}_n(t)/\partial t) \equiv \vec{u}_n(t)$. A similar equation has also been obtained by Edwards and Freed [26] using a variational principle.

2.2 Rigid rod

Another interesting situation is the case where the polymer is rigid. In this case, we have a considerable reduction of the number of degrees of freedom of the macromolecule. Here we will use the quasistatic approximation for the Navier-Stokes equation because, due to the linear configuration of the polymer, the mass of fluid moving with the macromolecule is so small that the inertial effects disappear in a very short time. We will consider that the polymer has cylindrical shape with its length L larger than its radius a, instead of a line of beads of the same radius ("shish kebab" model). Following a procedure analogous to the one of the preceding section, we will derive a Langevin equation for every longitudinal element of the rod which will be used to obtain the equation of motion of the centre of mass and its orientation.

We will consider a rigid rod suspended in a Newtonian fluid. The position of the centre of mass of the rod will be denoted by $\vec{R}$ and the unit vector along the cylindre's axis by $\hat{s}$. We will take cylindrical coordinates with the origin placed at the centre of mass of the rod and the revolution axis coinciding with the cylindre's axis. A point in space will be then located by the position vector

$$\vec{r} = s\,\hat{s} + \vec{r}_\perp \tag{2.33}$$

where $\vec{r}_\perp$ is a vector orthogonal to $\hat{s}$. The points on the surface of the rod satisfy $|\vec{r}_\perp| = a$ and $|s| \leq L/2$

The Navier-Stokes equation in the creeping flow approximation is given by

$$0 = -\nabla p(\vec{r},t) + \eta\nabla^2\vec{v}(\vec{r},t) + \vec{F}(\vec{r},t) - \nabla\cdot\vec{\vec{\Pi}}^R(\vec{r},t) \tag{2.34}$$

where $\vec{F}(\vec{r},t)$ stands for the induced force density introduced in the same way as in subsection 2.1. In this case the induced force density is only nonzero at the surface of the cylindre. For the flexible chain case this was only true approximately. We can then write

$$\vec{F}(\vec{r},t) = \vec{f}(\vec{r},t)\,\delta(|\vec{r}_\perp| - a)\theta(L/2 - |s|) \tag{2.35}$$

where $\vec{f}(\vec{r},t)$ is given by $\hat{n}(\vec{r},t)\cdot\vec{\vec{P}}^s(\vec{r},t)$ at the surface, where now $\hat{n} = \vec{r}_\perp/|\vec{r}_\perp|$. Notice that in eq. (2.35) we have neglected the induced forces due to the end surfaces, since the corrections introduced by presence of these surfaces are of the order a/L.

We can formally solve eq. (2.34) in Fourier space, yielding

$$\vec{v}(\vec{k},t) = \vec{v}_0(\vec{k},t) + \frac{(\vec{\vec{1}} - \hat{k}\hat{k})}{\eta k^2}\cdot\left[\vec{F}(\vec{k},t) - i\vec{k}\cdot\vec{\vec{\Pi}}^R(\vec{k},t)\right] \tag{2.36}$$

The Fourier transform of the induced force density is obtained from eq. (2.35) giving

$$\begin{aligned} \vec{F}(\vec{k},t) &= \int d\vec{r} e^{-i\vec{k}\cdot\vec{r}} \vec{f}(\vec{r},t)\,\delta(|\vec{r}_\perp| - a)\,\theta(L/2 - |s|) \\ &= \int_{-L/2}^{L/2} ds e^{-ik_\parallel s} \int_0^{2\pi} d\varphi e^{-ik_\perp a \cos\varphi}\, \vec{f}(s,\varphi;t) \end{aligned} \tag{2.37}$$

where $k_\parallel$ and $k_\perp$ are the parallel and radial components of the vector $\vec{k}$ with respect to the $\hat{s}$-axis and φ is the azimutal angle. For a very thin rod as the one we are dealing with, we can take $\vec{f}(s,\varphi;t)$ independent of φ as a good approximation, leaving only its dependence in the axial coordinate. We can then write

$$\vec{F}(\vec{k},t) = J_0(k_\perp a) \int_{-L/2}^{L/2} ds e^{-ik_\parallel s}\, \vec{f}(s,t) \tag{2.38}$$

where $\vec{f}(s,t) = 2\pi a\, \vec{f}(s,\varphi;t)$ and $J_0(x)$ is the Bessel function of zeroth order. The formal solution given in eq. (2.36) can also be written in real space as

$$\vec{v}(\vec{r},t) = \vec{v}_0(\vec{r},t) + \int d\vec{r}'\overleftrightarrow{T}(\vec{r}-\vec{r}')\cdot\left[\vec{F}(\vec{r}',t) - \nabla\cdot\overleftrightarrow{\Pi}^R(\vec{r}',t)\right] \tag{2.39}$$

where $\overleftrightarrow{T}$ is the usual Oseen tensor defined in eq. (2.20)

Following the same line of reasoning as in sec. 2.1 we use stick boundary conditions. Neglecting rotations around $\hat{s}$-axis and due to the fact that the $a \ll L$, the velocity of a surface element is approximately given by the velocity of the corresponding point on the $\hat{s}$-axis. If we denote a point on the axis by $\vec{\rho}(s,t)$, we can write

$$\frac{\partial\vec{\rho}(s,t)}{\partial t} \simeq \frac{1}{2\pi a}\int d\vec{r}'\,\delta(|\vec{r}_\perp{}'| - a)\theta(L/2 - |s|)\,\delta(s-s')\,\vec{v}(\vec{r},t) \tag{2.40}$$

which is simply the average of the fluid velocity over the contour of the transverse section of the cylindre. Using now the formal solution given in eq. (2.39) one arrives at

$$\begin{aligned} 2\pi a\frac{\partial\vec{\rho}(s,t)}{\partial t} &\simeq \int d\vec{r}'\,\delta(|\vec{r}_\perp{}'| - a)\theta(L/2 - |s|)\,\delta(s-s') \\ &\left\{\vec{v}_0(\vec{r},t) + \int d\vec{r}'\overleftrightarrow{T}(\vec{r}-\vec{r}')\cdot\left[\vec{F}(\vec{r}',t) - \nabla\cdot\overleftrightarrow{\Pi}^R(\vec{r}',t)\right]\right\} \end{aligned} \tag{2.41}$$

By assuming that the flow in the absence of polymer is homogeneous and given by $\vec{v}_0(\vec{r},t) = (\vec{r}+\vec{R})\cdot\vec{\vec{\beta}}$, the first integral on the right hand side of eq. (2.41) simply gives $2\pi a\,\vec{\rho}(s,t)$. The term containing the induced force density transforms into

$$2\pi a \int_{-L/2}^{L/2} ds' \left\{ \int \frac{d\vec{k}}{(2\pi)^3}\, e^{ik_{\parallel}(s-s')}\, J_0^2(k_\perp a)\, \frac{\vec{\vec{1}} - \hat{k}\hat{k}}{\eta k^2} \right\} \cdot \vec{f}(s') \tag{2.42}$$

from which we can identify the mobility kernel

$$\vec{\vec{\mu}}(s-s') = \int \frac{d\vec{k}}{(2\pi)^3}\, e^{ik_{\parallel}(s-s')}\, J_0^2(k_\perp a)\, \frac{\vec{\vec{1}} - \hat{k}\hat{k}}{\eta k^2} \tag{2.43}$$

Finally, the random part of the velocity is defined by

$$\begin{aligned} &2\pi a \vec{v}^R(\vec{r},t) = \\ &\quad -\int d\vec{r}'\, \delta(|\vec{r}_\perp{}'| - a)\theta(L/2 - |s|)\, \delta(s-s') \int d\vec{r}''\vec{\vec{T}}(\vec{r}' - \vec{r}'') \cdot \left[\nabla \cdot \vec{\vec{\Pi}}^R(\vec{r}'',t)\right] \\ &\quad 2\pi a \int \frac{d\vec{k}}{(2\pi)^3}\, e^{ik_{\parallel}(s-s')}\, J_0(k_\perp a)\, \frac{\vec{\vec{1}} - \hat{k}\hat{k}}{\eta k^2} \cdot \left\{ i\vec{k} \cdot \vec{\vec{\Pi}}^R(\vec{k},t) \right\} \end{aligned} \tag{2.4}$$

where, in deriving the last line we have omited the function $\theta(L/2-|s|)$, implicitly considering that the functions of s are defined only in the interval $-L/2 \leq s \leq L/2$. Let us calculate the averages of the random part of the velocty. Clearly, we have

$$< \vec{v}^R(s,t) >= 0 \tag{2.45}$$

After some algebra, we can obtain

$$< \vec{v}^R(s,t)\vec{v}^R(s',t') >= 2k_B T \vec{\vec{\mu}}(s-s')\, \delta(t-t') \tag{2.46}$$

As far as the longitudinal induced force density $\vec{f}(s,t)$ is concerned, it can be considered as minus the total force per unit legth $\vec{f}^H(s,t)$ exerted by the fluid. After neglecting inertial effects, we can then write the force balance

$$0 = \vec{f}^H(s,t) + \vec{f}^{int}(s,t) + \vec{f}^{ext}(s,t) \tag{2.47}$$

where $\vec{f}^{int}$ is the force per unit legth due to interactions with neighbouring segments and, consequently, its direction is parallel to $\hat{s}$-axis. $\vec{f}^{ext}$ is the force per unit legth due to external fields.

In view of these considerations, eq. (2.41) can be written in the final form

$$\frac{\partial \vec{\rho}(s,t)}{\partial t} = \vec{\rho}(s,t) \cdot \vec{\vec{\beta}} + \int_{-L/2}^{L/2} ds' \, \vec{\vec{\mu}}(s-s') \cdot \left[\vec{f}^{int}(s',t) + \vec{f}^{ext}(s',t)\right] + \vec{v}^R(s,t) \tag{2.48}$$

in which it will be used in the following section

3 Polymer dynamics

Once we have obtained the equation of motion of the polymer we then proceed to analyze the dynamics. In particular, we will focus our attention on the motion of the centre of mass for both, flexible and rigid chains, and on the conformations of the polymer for flexible chains.

3.1 Dynamics of a flexible chain in equilibrium

We will start our analysis by introducing the Gaussian chain. The equation of motion for such a chain follows from eq. (2.32), simply by neglecting excluded volume interactions. As is well-known, the distribution function for the segments follows Gaussian statistics and the radius of gyration R_G is given by [27]

$$R_G = \frac{N^{1/2} b}{\sqrt{6}} \tag{3.1}$$

To simplify our analysis we will use the so-called "preaveraging" approximation first introduced by Zimm in the static case which consists of substituting the configurational-dependent mobility kernel given in (2.18) by its equilibrium average

$$\begin{aligned} < \vec{\vec{\mu}}_{nm}(\tau; \vec{R}_{nm}) >_{eq} &= \frac{2}{3\rho(4\pi\nu)^{3/2}} \frac{\vec{\vec{1}}}{(b^2|n-m|/6\nu + \tau)^{3/2}} \\ &\equiv \mu_{nm}(\tau)\vec{\vec{1}} \end{aligned} \tag{3.2}$$

where the equilibrium average is computed by using Gaussian statistics for the distribution of the vectors $\vec{R}_{nm}$. The steady-state mobility can be obtained from eq. (3.2) simply by integrating in time

$$\mu_{nm}^{st} = \int_0^\infty d\tau \, \mu_{nm}(\tau) = \frac{1}{\eta\pi b(6\pi|n-m|)^{1/2}} \tag{3.3}$$

which agrees with the result obtained after averaging the Oseen tensor [27]. The average and the correlations of $\vec{u}_n^R$ are obtained from eqs. (2.21) and (2.22) by also averaging over the equilibrium distribution function. The final form of the equation of motion then follows from eq. (2.32) by neglecting the excluded volume term. One then obtains

$$\frac{\partial \vec{R}_n}{\partial t} = \int_{-\infty}^t dt' \int_0^N dm \, \mu_{nm}(t-t') \left[H\frac{\partial^2 \vec{R}_m}{\partial m^2} - M\frac{\partial^2 \vec{R}_m}{\partial t'^2} \right] + \vec{u}_n^R \tag{3.4}$$

The analysis of collective motions may be carried out by using the development of $\vec{R}_n$ in normal modes

$$\vec{R}_n = \vec{X}_0 + 2\sum_{p=1}^N \vec{X}_p(t)\, \varphi_p(n) \tag{3.5}$$

Here $\varphi_p(n) = \cos(p\pi n/N)$, which satisfy the orthogonality condition. The normal coordinates $\vec{X}_p(t)$ are defined as

$$\vec{X}_p(t) = \int_0^N \frac{dn}{N} \vec{R}_n(t)\, \varphi_p(n) \tag{3.6}$$

where $\vec{X}_o(t)$ describes the motion of the centre of mass and $\vec{X}_p(t)$ the internal rearrangements with characteristic length scales of the order of R_G/p.

According to eq. (3.6), it is possible to obtain from eq. (3.4) the equation of motion for the normal coordinates

$$\frac{\partial \vec{X}_p}{\partial t} = \sum_{q=0}^N \int_{-\infty}^t dt' \, \mu_{pq}(t-t') \left[K_q \vec{X}_q - NM\frac{\partial^2 \vec{X}_q}{\partial t'^2} \right] + \vec{w}_q^R \tag{3.7}$$

where use has been made of the definitions

$$K_p = \frac{Hp^2\pi^2}{N} = \frac{6k_BT\,p^2\pi^2}{Nb^2} \tag{3.8}$$

$$\mu_{pq}(\tau) = \frac{1}{N^2}\int_0^N dn\,dm\,\mu_{nm}(\tau)\,\varphi_p(n)\,\varphi_q(m) \tag{3.9}$$

$$\vec{w}_p^R(t) = \frac{1}{N}\int_0^N dn\,\vec{u}_n^R(t)\,\varphi_p(n) \tag{3.10}$$

Here, K_p is the p-mode effective elastic constant, μ_{pq} is the normal mobility matrix and $\vec{w}_p^R$ is the normal stochastic source satisfying the properties

$$\begin{aligned} < \vec{w}_p^R(t) > &= 0 \\ < \vec{w}_p^R(t)\vec{w}_q^R(t') > &= k_BT\mu_{pq}(t-t')\vec{\vec{1}} \end{aligned} \tag{3.11}$$

which follows from eqs. (2.22) and (2.21)

Explicit values for the mobility matrix elements μ_{pq} can be obtained by using the mobility given by eq. (3.2) together with eq. (3.9). One gets

$$\begin{aligned} \mu_{pq}(\tau) &= \frac{2}{3(4\pi\nu)^{3/2}N^2}\left(\frac{6\nu}{b^2}\right)^{3/2} \\ &\int_0^N dn\,[\cos(\frac{p\pi n}{N})\cos(\frac{q\pi m}{N})\int_{-n}^{N-n} dm\,\frac{1}{(|m|+6\nu\tau/b^2)^{3/2}}\cos\left(\frac{q\pi m}{N}\right) - \\ &\cos(\frac{p\pi n}{N})\sin(\frac{q\pi m}{N})\int_{-n}^{N-n} dm\,\frac{1}{(|m|+6\nu\tau/b^2)^{3/2}}\sin\left(\frac{q\pi m}{N}\right)] \end{aligned} \tag{3.12}$$

As an approximation, we will consider that, in the case where $p \neq 0$ and $q \neq 0$, the limits of the integral over m can be extended to infinity [27]. Contrarily, when $p = 0$ and $q = 0$ we will calculate the integral exactly. Consequently, on one hand we have

$$\mu_{pq}(t_p) \equiv \mu_p(t_p)\delta_{pq} = \delta_{pq}\frac{1}{\tau_p\xi_p}\phi_p(t_p); \tag{3.13}$$

Here, $\tau_p \equiv R_G^2/(\nu\pi p)$ is the relaxation time of the perturbation caused by the motion of a blob of radius $R_G/(\pi p)^{1/2}$, $t_p \equiv t/\tau_p$ is a dimensionless time and $\xi_p \equiv (12\pi^3)^{1/2}\eta p^{1/2}N^{1/2}b$ is the effective friction coefficient for the p-mode. Moreover, the functions $\phi_p(t_p)$ have been defined as

$$\phi_p(t_p) = \cos t_p\,(1 - 2S(\sqrt{t_p})) - \sin t_p\,(1 - 2C(\sqrt{t_p})) - \left(\frac{2}{\pi t_p}\right)^{1/2} \tag{3.14}$$

where the functions $C(x)$ and $S(x)$ are the Fresnel integrals.

On the other hand, the element μ_{oo} reduces to

$$\mu_{oo}(\tau) = \frac{1}{N^2}\int_0^N dn\,dm\,\mu_{nm}(\tau) = \frac{3}{4\xi_o\tau_o}\,\phi_o(t_o) \tag{3.15}$$

where we have defined the relaxation time $\tau_o \equiv R_G^2/\nu$, the dimensionless time $t_o \equiv t/\tau_o$ and the effective friction coefficient $\xi_o \equiv 3(6\pi^3)^{1/2}\eta N^{1/2}b/8$, whose meaning is analogous to their corresponding quantities introduced previously, and the function $\phi_o(t_o)$ is

$$\phi_o(t_o) = t_o^{-1/2} - 2[(1+t_o)^{1/2} - t_o^{1/2}] \tag{3.16}$$

Using the dimensionless quantities we can finally rewrite eq. (3.7) in the more suitable form

$$\frac{\partial \vec{X}_o}{\partial t_o} = -\int_{-\infty}^{t_o} dt_o'\,\frac{3NM}{4\xi_o\tau_o}\,\phi_o(t_o - t_o')\frac{\partial^2 \vec{X}_o}{\partial {t_o'}^2} + \tau_o\vec{w}_o^R \tag{3.17}$$

for the zeroth order mode and

$$\frac{\partial \vec{X}_p}{\partial t_p} = \int_{-\infty}^{t_p} dt_p'\,\frac{\tau_p}{\xi_p}\,\phi_p(t_p - t_p')\left[K_p\vec{X}_p - \frac{NM}{\tau_p^2}\frac{\partial^2 \vec{X}_q}{\partial {t_p'}^2}\right] + \tau_p\vec{w}_q^R \tag{3.18}$$

for higher order modes.

Our analysis will mainly be focused on the description of the motion of the centre of mass. The inertial effects due to the finite mass of the polymer are very small; however, for flexible chains the inertial effects due to the amount of mass of fluid partially moving along with the polymer are much more important. To justify our assertion it is convenient to define the dimensionless parameter γ_o, comparing the relaxation time of the inertial effects due to the mass of the polymer, NM/ξ_o, with the relaxation time of fluid perturbations, τ_o

$$\gamma_o = \frac{3NM}{4\xi_o \tau_o} \sim \frac{\rho^*}{\rho}\frac{a^3}{b^3}\frac{1}{N^{1/2}} \ll 1 \tag{3.19}$$

ρ^* being the polymer density which is usually comparable to the density of the solvent.

We now proceed to the calculation of the centre of mass velocity correlation function $< \dot{\vec{X}}_o(t_o)\dot{\vec{X}}_o(0) >$. This quantity may be obtained from eq. (3.17) using the fact that the random term satisfies $< \dot{\vec{X}}_o(0)\vec{w}_o^R(t_o) >= 0$ [23]. One then obtains

$$< \dot{\vec{X}}_o(t_o)\dot{\vec{X}}_o(0) >= -\gamma_o \int_0^{t_o} dt'_o\, \phi_o(t_o - t'_o) \frac{d}{dt'_o} < \dot{\vec{X}}_o(t'_o)\dot{\vec{X}}_o(0) > \tag{3.20}$$

This equation should be solved using the initial condition

$$< \dot{\vec{X}}_o(0)\dot{\vec{X}}_o(0) >= \frac{k_B T}{NM}\vec{\vec{1}} \tag{3.21}$$

dictated by the law of equipartition of energy. The velocity correlation function can be obtained simply by using Laplace transforms which, for an unspecified function $A(t)$, reads

$$A(s) = \int_0^\infty dt\, e^{-st}\, A(t) \tag{3.22}$$

Then, Laplace transforming eq. (3.20) we get

$$< \dot{\vec{X}}_o(s_o)\dot{\vec{X}}_o(0) >= \frac{\gamma_o\, \phi_o(s_o)}{1 + \gamma_o\, s_o \phi_o(s_o)} < \dot{\vec{X}}_o(0)\dot{\vec{X}}_o(0) > \tag{3.23}$$

where $s_o = s\tau_o$. In the time domain we are considering $t_o \sim 1$, we can set $s_o \sim 1$ in eq. (3.23). In that case, $\gamma_o s_o \ll 1$ consequently, eq. (3.23) transforms into

$$< \dot{\vec{X}}_o(t_o)\dot{\vec{X}}_o(0) > \simeq \gamma_o\, \phi_o(t_o) < \dot{\vec{X}}_o(0)\dot{\vec{X}}_o(0) > \tag{3.24}$$

In expression (3.24) the dynamics of the fluid is still present due to the time dependence of the memory kernel. However, the approximation we have introduced neglects the inertial effects due to the polymer mass according to (3.19). In view of the expression of $\phi_o(t_o)$ given in eq. (3.16), we have that for $t_o \ll 1$, the correlation function behaves as $t^{-1/2}$ while, in the long-time limit, one has

$$< \dot{\vec{X}}_o(t_o \to \infty)\dot{\vec{X}}_o(0) > \to \frac{2}{3}\frac{k_BT}{\rho(4\pi\nu t)^{3/2}}\vec{\vec{1}} \tag{3.25}$$

Notice that this behaviour is the same as the one for a Brownian particle [28], showing the hydrodynamic long-time tail $t^{-3/2}$. This time behaviour is in fact independent of the coupling of modes we have neglected, due to the fact that no reference to characteristic lengths of the system remains in this limit. In that sense the behaviour of the velocity correlation function for long times is universal. However, the short-time behaviour we have obtained may in principle be modified when the coupling between modes through off-diagonal terms in the mobility matrix is taken into account.

The diffusion coefficient is obtained from eq. (3.16) using the corresponding Green-Kubo formula

$$\begin{aligned} D_G &= \frac{1}{3}\int_0^\infty dt < \dot{\vec{X}}_o(t) \cdot \dot{\vec{X}}_o(0) > \\ &= \frac{k_BT}{\xi_o} = \frac{8}{3}\frac{k_BT}{(6\pi^3)^{1/2}\eta b N^{1/2}} \end{aligned} \tag{3.26}$$

which is precisely the expression given in the continuous Zimm model [27].

From our analysis we can also compute the mean-square displacement

$$\delta\vec{X}_o(t) \equiv \vec{X}_o(t) - \vec{X}_o(0) = \int_0^t dt' \dot{\vec{X}}_o \tag{3.27}$$

In view of eq. (3.24) one gets

$$< \delta X_o^2(t) >= 6\frac{k_BT}{\xi_o}\tau_o\left\{t_o + t_o^{3/2} - \frac{2}{5}\left[(1+t_o)^{5/2} - t_o^{5/2} - 1\right]\right\} \tag{3.28}$$

which, in the long-time limit, reduces to the well-known result $6D_Gt$ for the Zimm model. From expression (3.28) we conclude that, for times $t \sim 60\tau_0$, our model still introduces corrections to the Zimm model of the order of 10 %. This is due to the fact that the behaviour of the mean-square displacement slowly approaches the diffusive regime [29,30].

As far as the motion of the internal degrees of freedom is concerned, there appears a new time-scale introduced by the effective elastic constant K_p. It is then convenient to introduce the dimensionless parameter

$$\alpha_p = \frac{\tau_p K_p}{\xi_p} \sim \frac{k_B T p^{1/2}}{\rho \eta^2 b N^{1/2}} \ll 1 \tag{3.29}$$

which compares the relaxation time due to fluid perturbations τ_p, caused by the motion of a blob of size $R_G/\sqrt{\pi p}$, with the relaxation time associated to elastic forces K_p/ξ_p.

· Following the same line of reasoning as in the analysis of the centre of mass motion, we can obtain the velocity correlation function for the p-mode

$$< \dot{\vec{X}}_p(t_p) \cdot \dot{\vec{X}}_p(0) > \simeq -\frac{3k_BT}{\xi_p \tau_p} \phi_p(t_p) \tag{3.30}$$

where use has been made of the law of equipartition of energy

$$< \dot{\vec{X}}_p(0) \cdot \dot{\vec{X}}_p(0) > = \frac{3k_BT}{NM} \tag{3.31}$$

and we have neglected the inertia of the polymer mass, as before. Notice that the elastic force does not appear in our result (3.30) because, in view of the value of α_p, we have considered that the configuration of the polymer remains frozen in the time in which the velocity correlation function takes significant values.

According to eq. (3.14), in the short-time limit the velocity correlation function behaves as $t^{-1/2}$ while, in the long-time limit, one gets a decay $t^{-5/2}$ for this quantity. We then conclude that in the long-time limit, the behaviour of this correlation function is the same as the one for the rotational motion of a Brownian particle [31].

Concerning the correlation function $< \vec{X}_p(t_p) \cdot \vec{X}_p(0) >$, one should realize that the time-scales in which the configuration of the polymer changes appreciably are governed by the elastic forces. This means that the effects of the fluid enter only through the static quantity ξ_p giving the exponential behaviour [27]

$$< \vec{X}_p(t) \cdot \vec{X}_p(0) > = \frac{k_BT}{K_p} e^{-\frac{K_p}{\xi_p}t} \tag{3.32}$$

Up to now we have analyzed the dynamics of the Gaussian chain. When excluded volume interactions are present, the study of the dynamics is much more complex. However, we can rather easily obtain qualitative information about the motion of the centre of mass of the chain, which is precisely the case where inertial effects are more important. If we perform the equilibrium average of the mobility kernel and introduce the same approximation as we used in eq. (3.13), we can proceed as in the ideal case to get

$$\begin{aligned} < \dot{\vec{X}}_o(t)\dot{\vec{X}}_o(0) > &\simeq \gamma_o \phi_o^{(\bar{v})}(t) < \dot{\vec{X}}_o(0)\dot{\vec{X}}_o(0) > \\ &\sim \mu_{oo}^{(\bar{v})}(t)\vec{\vec{1}} \end{aligned} \tag{3.33}$$

where the superscript v refers to the fact that the equilibrium average has been computed by using the equilibrium distribution function corresponding to the chain with excluded volume interactions. Now, by means of eqs. (2.16) and (3.15), interchanging the order of integration and averaging with respect to the equilibrium distribution function, we can see that

$$\mu_{oo}^{(v)}(t)\vec{\vec{1}} = \int \frac{d\vec{k}}{(2\pi)^3} e^{-\nu k^2 \tau} \left[\int_0^N \frac{dn\, dm}{N^2} < e^{i\vec{k}\cdot\vec{R}_{nm}} >^{(\bar{v})} \right] \frac{(\vec{\vec{1}} - \hat{k}\hat{k})}{\rho} \tag{3.34}$$

Here, the term between brackets is $g(\vec{k})/N$ where $g(\vec{k})$ is the static structure factor which, due to the isotropy of the system, is a function only of $k \equiv |\vec{k}|$. After integration over the solid angle one gets

$$\mu_{oo}^{(v)}(t) \sim \int_0^\infty dk\, k^2\, e^{-\nu k^2 t} g(k) \tag{3.35}$$

If $R_G \sim bN^\nu$ (here ν refers to the Flory exponent), scaling arguments lead us to $g(k) \sim k^{-1/\nu}$ in the case $kR_G \gg 1$ [27,31,32,33], while $g(k) \sim 1$ when $kR_G \ll 1$. Introducing these expressions into eq. (3.35) we can finally arrive at the conclusion that the decay of the centre of mass velocity correlation function is $t^{-(3\nu-1)/2\nu}$ in the limit $t \ll R_G^2\rho/\eta$, while in the limit $t \gg R_G^2\rho/\eta$ it behaves as $t^{-3/2}$ as we expected. In the case of the ideal chain, the value of the Flory exponent is $\nu = 1/2$ which gives a short-time decay $t^{-1/2}$ recovering the value obtained above. For the chain with excluded volume, the value of the exponent is approximately given by $\nu = 3/5$ yielding a short-time decay $t^{-2/3}$. These asymptotic decays ensure that the centre of mass velocity correlation function is integrable in time which means that the diffusion coefficient exists.

3.2 Dynamics of a rigid rod

As we have shown in subsection 2.2, the dynamics of the rigid rod is described by eq. (2.48). In this expression one should notice that $\vec{f}^{int}$ cannot be given in terms of any potential energy as in the case of the flexible chain (cf. eq. 2.29). The existence of this force is related to the fact that the length of the rod should be constant.

To proceed, it is convenient to expand the variables of eq. (2.48) in terms of Legendre polynomials. For an unspecified quantity $\phi(x)$ one has

$$\phi(x,t) = \sum_{l=0}^{\infty} \phi_l(t)\, P_l(x) \tag{3.36}$$

where $x \equiv 2s/L$, $P_l(x)$ is the Legendre polynomial of order l and

$$\phi_l = \int_{-1}^{1} dx\, \phi(x) P_l(x) \tag{3.37}$$

If we expand $\vec{f}^{int}(x,t)$ and $\vec{f}^{ext}(x,t)$ in eq. (2.48), multiplying both sides of this equation by $P_l(x)$ and integrating over x, we finally get

$$\dot{\vec{\rho}}_l(t) = \vec{\rho}_l(t) \cdot \vec{\vec{\beta}} + \sum_{q=0}^{\infty} \vec{\vec{\mu}}_{lq} \cdot \frac{L}{2} \left[\vec{f}_q^{int}(t) + \vec{f}_q^{ext}(t) \right] + \vec{v}_l^R \tag{3.38}$$

where use has been made of the orthogonality property of the Legendre polynomials. In this last equation we have defined the mobility matrices $\vec{\vec{\mu}}_{lq}$

$$\vec{\vec{\mu}}_{lq} \equiv \int_{-1}^{1} dx\, dx'\, P_l(x)\, \vec{\vec{\mu}}(x-x')\, P_q(x') \tag{3.39}$$

Moreover, the random term $\vec{v}_l^R(t)$ has zero mean and satisfies

$$< \vec{v}_l^R(t) \vec{v}_q^R(t') >= 2k_B T \vec{\vec{\mu}}_{lq}\, \delta(t-t') \tag{3.40}$$

as follows from eqs. (2.45) and (2.46). Notice that for symmetry reasons the matrices $\vec{\vec{\mu}}_{lq}$ are different from zero only if l and q are both even or odd.

Since $\vec{\rho}(s,t) = \vec{R}(t) + s\hat{s}$, we can easily compute $\vec{\rho}_o = \sqrt{2}\vec{R}$ and $\vec{\rho}_1 = L/\sqrt{6}\,\hat{s}$ with $\vec{\rho}_q = 0$ for $q \geq 2$. Deriving $\vec{\rho}(s,t)$ with respect to time we simply have $\dot{\vec{\rho}}_o = \sqrt{2}\dot{\vec{R}}$ and $\dot{\vec{\rho}}_1 = L/\sqrt{6}\,\dot{\hat{s}} = L/\sqrt{6}\,\vec{\omega} \times \hat{s}$, where $\vec{\omega}$ is the angular velocity. Equation (3.38) then gives

$$\dot{\vec{\rho}}_o(t) = \vec{\rho}_o(t) \cdot \vec{\vec{\beta}} + \vec{\vec{\mu}}_{oo} \cdot \frac{L}{2} \left[\vec{f}_o^{int}(t) + \vec{f}_o^{ext}(t) \right] + \vec{v}_o^R \tag{3.41}$$

$$\dot{\vec{\rho}}_1(t) = \vec{\rho}_1(t) \cdot \vec{\vec{\beta}} + \vec{\vec{\mu}}_{11} \cdot \frac{L}{2} \left[\vec{f}_1^{int}(t) + \vec{f}_1^{ext}(t) \right] + \vec{v}_1^R \tag{3.42}$$

in which, to obtain the values of diffusion coefficients, we have only considered contributions due to the matrices $\vec{\vec{\mu}}_{oo}$ and $\vec{\vec{\mu}}_{11}$. These matrices can be computed by means of eqs. (2.43) and (3.39). One then obtains

$$\vec{\vec{\mu}}_{oo} = \frac{1}{2\pi\eta L} \left\{ 2 \left[\ln \frac{L}{2a} + 2\ln 2 - 3/2 \right] \hat{s}\hat{s} + \left[\ln \frac{L}{2a} + 2\ln 2 - 1/2 \right] (\vec{\vec{1}} - \hat{s}\hat{s}) \right\} \tag{3.43}$$

$$\vec{\vec{\mu}}_{11} = \frac{1}{2\pi\eta L} \left\{ 2 \left[\ln \frac{L}{2a} + 2\ln 2 - 17/6 \right] \hat{s}\hat{s} + \left[\ln \frac{L}{2a} + 2\ln 2 - 11/6 \right] (\vec{\vec{1}} - \hat{s}\hat{s}) \right\} \tag{3.44}$$

In eq. (3.41), $\vec{f}_o^{int}$ vanishes due to the action-reaction principle. Moreover, $\vec{f}_1^{int}$ has only longitudinal contributions and, as a consequence, will not affect rotational motions. The random terms $\vec{v}_o^R$ and $\vec{v}_1^R$ are not correlated since $\vec{\vec{\mu}}_{01} = \vec{\vec{\mu}}_{10} = 0$.

Let us analyze first of all, the translational diffusion of the rod in the absence of flow ($\vec{\vec{\beta}} = 0$) and of external fields ($\vec{f}^{ext} = 0$). In this case, eq. (3.41) simply gives

$$\dot{\vec{\rho}}_o(t) = \vec{v}_o^R(t) \tag{3.45}$$

Consequently, one obtains for the centre of mass velocity correlation function

$$< \dot{\vec{\rho}}_o(t)\dot{\vec{\rho}}_o(t') > = < \vec{v}_o^R(t)\vec{v}_o^R(t') > = 2k_B T \vec{\vec{\mu}}_{oo} \delta(t - t') \tag{3.46}$$

In the isotropic case, the diffusion coefficient is given by the Green-Kubo formula

$$D_G = \frac{1}{3} \int_0^\infty dt' < \dot{\vec{R}}(t) \cdot \dot{\vec{R}}(t') > \tag{3.47}$$

Using the relation $\dot{\vec{\rho}}_o = \sqrt{2}\dot{\vec{R}}$ and eqs. (3.43) and (3.46), one finally gets

$$D_G = \frac{k_B T}{3\pi\eta L}\left[\ln\frac{L}{2a} + 2\ln 2 - 1\right] \tag{3.48}$$

which agrees with the result obtained by Yamakawa et al. [34].

We can now proceed to the calculation of the rotational diffusion coefficient D_r for a rigid rod in equilibrium. Setting $\vec{\vec{\beta}} = 0$ and $\vec{f}^{ext} = 0$ in eq. (3.42) we get

$$\frac{L}{\sqrt{6}}\vec{\omega}(t) \times \hat{s}(t) = +\vec{\vec{\mu}}_{11} \cdot \frac{L}{2}\vec{f}_1^{int}(t) + \vec{v}_1^R(t) \tag{3.49}$$

where use has been made of the fact that $\dot{\hat{s}} = \vec{\omega} \times \hat{s}$. Now we multiply vectorially both sides of this last equation by $\hat{s}(t)$. If we take into account that $\vec{f}_1^{int}$ is parallel to $\hat{s}$, we get

$$\frac{L}{\sqrt{6}}\hat{s} \times (\vec{\omega} \times \hat{s}) = \hat{s} \times \vec{v}_1^R \tag{3.50}$$

This equation reduces to

$$\vec{\omega} = \frac{\sqrt{6}}{L}\hat{s} \times \vec{v}_1^R \tag{3.51}$$

provided that we have neglected rotations around $\hat{s}$-axis. Therefore, the angular velocity correlation function is

$$< \vec{\omega}(t)\vec{\omega}(t') >= \frac{6k_B T}{\pi\eta L^3}\left[\ln\frac{L}{2a} + 2\ln 2 - \frac{11}{6}\right]\delta(t-t')\,(\vec{\vec{1}} - \hat{s}\hat{s}) \tag{3.52}$$

where use has been made of eq. (3.40) and eq. (3.44). The rotational diffusion coefficient then follows from the corresponding Green-Kubo formula, yielding

$$D_r = \frac{1}{2}\int_0^\infty dt < \vec{\omega}(t) \cdot \vec{\omega}(0) >= \frac{3k_B T}{\pi\eta L^3}\left[\ln\frac{L}{2a} + 2\ln 2 - \frac{11}{6}\right] \tag{3.53}$$

also in agreement with Yamakawa's results [35,36].

4 Conclusions

In this paper we have presented a new formalism, which includes retardation effects in the fluid, to study the dynamics of polymers in solution. Our approach is based on the method of induced forces [18,19]. The method accounts for the perturbation introduced by the presence of the chain by considering induced force fields in the Navier-Stokes equation. We have been able to derive equations of motion of the polymer in a rather general situation: presence of time-dependent hydrodynamic interactions and of excluded volume effects. Our theory permits, in a systematic way, to obtain the mobility matrix and to prove the validity of the fluctuation-dissipation theorem for the Langevin forces. Moreover, its applicability ranges from flexible to rigid chains.

The dynamics of the polymers then follows from the equation of motion we have obtained. In fact, for flexible chains we have analyzed the motion of the centre of mass and the conformational changes of the polymer. We have found that the centre of mass velocity correlation function exhibits a $t^{-3/2}$ long-time tail identical to the one encountered for rigid spheres. Furthermore, the mean-square displacement shows corrections to the diffusion regime described by the Zimm model. Concerning conformational changes, the velocity correlation function for the different modes contributing to these changes decays as $t^{-5/2}$, showing a close analogy with the decay of the angular velocity correlation function of a Brownian particle. The retardation effects in the fluid on the dynamics of the normal coordinates are not very important. The dynamics is dominated by the elastic forces. For rigid rods, our formalism enables us to obtain the mobility kernel and from it, the diffusion coefficients given by Green-Kubo formulae without explicitly constructing the velocity field. This makes our analysis somewhat simpler than the analysis of Yamakawa who obtains the same formulae.

Aknowledgements

This work has been supported by the European Economic Community under Contract No. ST2*0246. J.B.A. and J.M.R. also whish to thank CICyT, Grant No. PB87-0782.

References

[1] J. Happel and H. Brenner, *Low Reynolds Number Hydrodynamics*, Noordhoff, Leiden (1973).

[2] Th.G.M. van de Ven, *Colloidal Hydrodynamics*, Academic Press, London (1989)

[3] W.B. Russell, D.A. Saville and W.R. Schowalter, *Colloidal Dispersions*, Cambridge University Press, Cambridge (1989)

[4] H.L. Goldsmith and S.G. Mason, "The microrheology of dispersions" in *Rheology: Theory and Applications*, ed. F.R. Eirich; New York: Academic (1967).

[5] Kirkwood and Riseman, J. Chem. Phys, **16**, (1948), 565

[6] Debije, Bueche, J. Chem. Phys, **16**, (1948), 573

[7] W. Kuhn, Zeits. f. Phys. Chem. **161**, (1932), 1-32; Kolloid-Zeitschrift, **62**, (1933), 269-285.

[8] W. Kuhn and H. Kuhn, Helv. Chim. Acta, **28**, (1945), 97-127

[9] J.J. Hermans, Physica, **10**, (1943), 777-789.

[10] G.K. Fraenkel, J. Chem. Phys, **20**, (1952), 642-647.

[11] R.B. Bird, R.C. Armstrong, O. Hassager and C.F. Curtiss *Dynamics of polymeric liquids*, John Wiley & Sons. (1987)

[12] H. Yamakawa, Ann. Rew. Phys. Chem. **25**, (1974), 179-200

[13] H. Yamakawa, Ann. Rew. Phys. Chem. **35**, (1984), 23-47

[14] R. Zwanzig, J. of Research Nat. Bur. Stand. -B. **68B** (1964), 143

[15] P.E. Rouse, J. Chem. Phys, **21**, (1953), 1272

[16] B.H. Zimm, J. Chem. Phys, **24**, (1956), 269

[17] J. Bonet Avalos, J.M. Rubí and D. Bedeaux (submitted for publication in Macromolecules)

[18] P. Mazur and D. Bedeaux, Physica **76A**, (1974), 235-246

[19] D. Bedeaux and P. Mazur, Physica **76A**, (1974), 247-258

[20] L.D. Landau and E.M. Lifshitz, *Fluid Mechanics*, Pergamon Press, Oxford (1982)

[21] W. van Saarloos and P. Mazur, Physica A, **1983**, *120*, 77-102.

[22] Y. Oono, Adv. Chem. Phys. **61**, (1985), 301; Y Shiwa, Y. Oono and P. Baldwin, Macromol. **21**, (1988), 208

[23] R. Zwanzig, "Linear response theory" in *Molecular Fluids*, eds. R. Balian and G. Weill; Gordon & Breach, London (1976)

[24] W.H. Stockmayer, Macromol. Chem. **35**, (1960), 54

[25] S.F. Edwards, Proc. Phys. Soc. (London), **85**, (1965), 613

[26] S.F. Edwards and K. Freed, J. Chem. Phys. **61**, (1974),1189

[27] M. Doi and S.F. Edwards, *The Theory of Polymer Dynamics*, Claredon Press, Oxford (1986)

[28] Hauge and Martin-Löf, J. Stat. Phys. **7**, (1973), 259

[29] W. Stockmayer and B. Hammouda, Pure and Appl. Chem. **56**, (1984), 1373

[30] Fixman, M. and Mansfield, M.L. Macromol. **17**, (1984), 522

[31] P.G. De Gennes, *Scaling Concepts in Polymer Physics*, Cornell University Press, Ithaca (1979)

[32] P. Debije, J. Phys. Colloid Chem **51**, (1947), 18

[33] S.F. Edwards, Proc. Phys. Soc. (London), **93**, (1965), 605

[34] H. Yamakawa, Macromol. **16**, (1983), 1928

[35] H. Yamakawa, Macromol. **8**, (1975), 339

[36] H. Yamakawa, T. Yoshizaki and M. Fujii, Macromol. **10**, (1977), 934

MESOSCOPIC DYNAMICS AND THERMODYNAMICS: APPLICATIONS TO POLYMERIC FLUIDS

Miroslav Grmela
École Polytechnique de Montréal
Montreal, H3C 3A7, Canada

1 Introduction

In the course of polymer processing operations (e.g. injection molding) a polymer melt (complex viscoelastic fluid) flows and simultaneously undergoes phase transitions. Modeling of the processing operations requires thus understanding of both thermodynamic and flow (rheological) properties. The interplay of these two types of properties is put into focus in this lecture. Thermodynamics and rheology have arisen independently of each other, their traditional formulations are therefore not well suited for such discussions. A novel and unified formulation of thermodynamics and dynamics is introduced in this lecture.

The lecture is organized as follows. In Section 2 we note that understanding of the flow and the thermodynamic behaviour of complex fluids requires several levels of description. Thermodynamics on any level of description is introduced in Section 3 as a geometry of the state space. The physical foundation of thermodynamics is provided by an analysis of dynamics introduced on a more microscopic level of description (Section 4). Section 5 is devoted to applications in the context of polymeric fluids. The theory is adapted to driven systems in Section 6.

2 Hierarchy of descriptions

A rheological investigation begins with the identification of physical systems under consideration, our interests in them, objectives and intended applications. These considerations lead then to the choice of observations (i.e. to the choice of our interactions with the systems). The second stage of the investigations consists of a search for a model that would summarize and explain the experience collected in the course of making the observations. Modeling starts with the choice of the state space. Elements of the state space (called state variables) are the quantities that are chosen to characterize states of the systems. It is useful to keep the state space as small as possible because in general, smaller is the state space simpler is the model. The simplicity is then essential for satisfying our expectations placed on the model. Two complementary situations can arise. Either there exists a single simple model (i.e. the model whose state space is composed of the state variables that are closely related to results of the chosen observations) or no such single model exists. The first situation arises for example if the chosen physical systems are simple fluids (e.g. water) and the interests are those arising in engineering practice. The single simple model that satisfies all our needs is in this case hydrodynamics. The second situation is more typical. It arises for example if either the class of system is enlarged to include also complex fluids (e.g. egg white or industrial polymer melts) and/or our curiosity increases (e.g. some more microscopic characteristics like for example results of the slow neutron scattering are observed). If the second situation arises a single model has to be replaced by a family of mutually interrelated models introduced on different levels of description. Every model in the family is relatively simple but it satisfies our needs only partially. In this section we begin a systematic study of a multilevel family of models by introducing their state spaces.

2.1 Mathematical formulation

Let the state space introduced on the i-th level of description be denoted by the symbol M_i, its elements by u_i (i.e. $u_i \in M_i$). Let us consider now two state spaces: M_i introduced on the i-th level of description and M_j introduced on the j-th level of description. If M_i and M_j can be related by constructing a bundle (M_i, M_j, Π^i_j), where M_i is the total space, M_j is the base space and $\Pi^i_j : M_i \to M_j$ is the bundle projection, then we say that M_i is more microscopic (less macroscopic) than M_j. From the physical point of view, the fact that M_i projects on M_j signifies that $u_i \in M_i$ represents a more detailed information about the system under consideration than $u_j \in M_j$. The space M_j and the projection mapping Π^i_j can be regarded as an additional structure (playing the similar role as an introduction of coordinates) introduced in M_i. By presenting M_i as the bundle (M_i, M_j, Π^i_j) the space M_i is divided into the base space M_j and the associated fibres $\{(\Pi^i_j)^{-1}u_j | u_j \in M_j\}$. We shall use the notation: $u_i \equiv (u_j, v_j)$, where $u_j \in M_j$ and $v_j \in (\Pi^i_j)^{-1}u_j$. The notation convention that will be used in the rest of this lecture will be that if $i > j$ then M_i is more microscopic than M_j (i.e. there exists a bundle (M_i, M_j, Π^i_j)). All the state spaces considered in this lecture will be linear spaces, the pairing in these linear spaces will be denoted by the symbol $<,>$. The organization of the state spaces into bundles will be found to be very useful in the subsequent analysis. In the rest of this section we shall illustrate the bundles on a few examples. For the introduction of bundles in the geometrical context see Refs. [1], [2].

2.2 Examples

The most macroscopic state space introduced in classical equilibrium thermodynamics (we shall denote it by the symbol M_{th1}) is a subset of $\mathbb{R}^2$. Its elements u_{th1} are $u_{th1} \equiv (e, n)$, where e is the energy per unit volume and n is the number of moles also per unit volume (see Ref. [3]). An example of a more microscopic point of view in classical equilibrium thermodynamics is to regard the system under consideration as an N-component system. This means that in this case we have in our disposition more detailed measurements that are able to distinguish in the system N different components. The state space $M_{thN} \subset \mathbb{R}^{N+1}; u_{thN} \equiv (e, n_1, \ldots, n_N)$, where n_i is the number of moles per unit volume of i-th component. We can construct a bundle $(M_{thN}, M_{th1}, \Pi^{thN}_{th1})$ with $\Pi^{thN}_{th1} : \mathbb{R}^{N+1} \to \mathbb{R}^2$, $(e, n_1, \ldots, n_N) \mapsto (e, n), n = n_1 + n_2 + \cdots + n_N$. Another example of a more microscopic state space that has been introduced in classical equilibrium thermodynamics is M_{ths}. The systems under consideration are in this case elastic solids (see Refs. [3], [4]), $u_{ths} \equiv (e, n, m)$, where m, the strain tensor, is a traceless symmetric tensor that has the physical meaning of the measure of deformation. We can again construct a bundle $(M_{ths}, M_{th1}, \Pi^{ths}_{th1})$ with Π^{ths}_{th1} given by $(e, n, m) \mapsto (e, n)$.

The state space introduced in the Boltzmann kinetic theory [5] will be denoted by the symbol M_{kt1}. Its elements are one particle distribution function $f_{kt1}(r, v)$, r is the position vector and v the velocity of one particle. All particles composing the system under consideration are assumed to be identical and without internal degrees of freedom. In the case of systems composed of macromolecules, the distribution function $f_{Nkt1}(r_1, v_1, \cdots, r_N, v_N)$ can be chosen to replace $f_{kt}(r, v)$; (r_i, v_i) denotes the position and velocity of i-th link in a chain representing one macromolecule. The projection operators Π^{kt1}_{th1} in the bundle $(M_{kt1}, M_{th1}, \Pi^{kt1}_{th1})$ can be specified for example as: $n = \frac{1}{V}\int d^3r \int d^3v f_{kt1}(r, v)$ and $e = \frac{1}{V}\int d^3r \int d^3v \frac{1}{2}mv^2 f_{kt1}(rv) + \frac{1}{V}\int d^3r \int d^3v \epsilon(f_{kt1}; r)$, where V is the volume of the space region in which the system under consideration is confined, m is the mass of one particle and ϵ is the intermolecular potential energy. Similarly, we can suggest the projection operation Π^{Nkt1}_{th1} in the bundle $(M_{Nkt1}, M_{th1}, \Pi^{Nkt1}_{th1})$ (see e.g. Refs. [6], [7], [8]).

In hydrodynamics, states of fluids are described by five hydrodynamic fields $\rho(r), e(r), u(r)$, denoting the mass density, the energy density and the momemtum density. The bundle $(M_{hyd}, M_{th1}, \Pi^{hyd}_{th1})$ is specified by

$$n = \frac{1}{mV}\int d^3r\rho(r), e = \frac{1}{V}\int d^3re(r),$$

the bundle $(M_{kt1}, M_{hyd}, \Pi^{kt1}_{hyd})$ by

$$\rho(r) = m\int d^3v f_{kt1}(r, v), u(r) = m\int d^3v v f_{kt1}(r, v)$$

and

$$e(r) = \int d^3\frac{1}{2}mv^2 f_{kt1}(r, v) + \int d^3v\epsilon(f_{kt1}, r).$$

We note that another candidate for the projection Π^{kt1}_{hyd} could be:

$$\rho(r) = m\int d^3r' \int d^3v\theta(r, r') f_{kt1}(r', v)$$

and similarly for $u(r)$ and $e(r)$, where $\theta(r, r')$ is a given function. If $\theta(r, r') \sim \delta(r - r')$ then the space localization in kinetic theory and hydrodynamics are identical (Π^{kt1}_{kyd} reduces in this case to the projection Π^{kt1}_{hyd} introduced above as the first candidate). If $\theta(r, r')$ is different from $\delta(r - r')$ then the space localization on the two levels of description, namely kinetic theory and hydrodynamics, are note the same.

Other examples of state spaces and bundles constructed out of them will be introduced in subsequent sections.

3 Thermodynamics – Geometry of the state space

We shall divide all physical systems into two groups. The first group is composed of the physical systems that are under the influence of external forces (these systems are called driven or externally forced). The physical systems in the second group are under no such external influence (they will be called nondriven or unforced). We note that we could always transform driven systems into nondriven by enlarging them, namely by bringing the external influence into the interior of the systems. This is however not useful since the division between external and internal is dictated by our interactions with the systems (that are in turn determined by our interests in the systems). The external influence is usually controled seperately from the system of interest. In Sections 3–5, only nondriven systems will be considered. In Section 6 we shall show how the analysis introduced in Sections 3–5 has to be modified in order to be applicable also to driven systems.

In this section we shall consider all nondriven systems but only after they have been specially prepared. The process of preparation consists of leaving the systems sufficiently long time without any disturbance. It will be only in Section 4 where we shall be in position to include the preparation process into the mathematical formulation. Our interests in the prepared nondriven systems will be determined by equilibrium thermodynamic measurements described in all textbooks of equilibrium thermodynamics. The experience acquired in the process of making the thermodynamic measurements shows that results of the measurements are not independent. Leaving the explanation of this experience to Section 4, we shall here only describe it in mathematical terms. The leading idea is that a relation can be conveniently described by specifying a submanifold.

3.1 Mathematical formulation

We begin by considering only one component fluids and thus by considering the state space M_{th1} (see Section 2.2). Following Gibbs, we introduce three dimensional space with coordinates (e, n, s), where s is the entropy per unit volume. Thermodynamics is specified by the fundamental thermodynamic relation $s = s(e, n)$ (see Refs. [3], [9]). This relation can be represented geometrically as a two dimensional surface imbedded in the three dimensional space with coordinates (e, n, s) (i.e. the image of $(e, n) \mapsto (e, n, s(e, n))$. Since the derivatives of $s(e, n)$ with respect to e and n are quantities directly accessible by thermodynamic measurements (recall that $\frac{\partial s}{\partial e} = \frac{1}{T}$ and $\frac{\partial s}{\partial n} = -\frac{\mu}{T}$, where T and μ are the temperature and the chemical potential respectively) a more complete geometrical representation of thermodynamics is obtained by making 1-jet extension of the graph of the function $s = s(e, n)$ (see Refs. [2], [10]). By this we mean that we introduce a five dimensional space (we shall call it an augmented thermodynamic state space) with coordinates $\left(e, n, \frac{1}{T}, -\frac{\mu}{T}, s\right)$. Thermodynamics is then represented by the image of the mapping $(e, n) \mapsto \left(e, n, \frac{\partial s}{\partial e}(e, n), \frac{\partial s}{\partial n}(e, n), s(e, n)\right)$. This two dimensional surface imbedded in the five dimensional extended thermodynamic state space will be called a Gibbs surface. The experience acquired while making thermodynamic measurements is now expressed as the statement that the results of these measurements lie on the Gibbs surface. The individual features of the physical systems are expressed in classical equilibrium thermodynamics in the Gibbs surface. The association between the Gibbs surfaces and physical systems can be obtained, in the context of equilibrium thermodynamics, only by collecting results of thermodynamic measurements. A theoretical study of this association requires more microscopic levels of description. For this reason and also because our interest is focussed on polymeric liquids that can be effectively studied only on several levels of description (see Section 2) we turn now our attention to the formulation of thermodynamics on more microscopic levels of description.

Before introducing thermodynamics in a general state space M, we shall cast the formulation of equilibrium thermodynamics introduced in the preceeding paragraph into the geometrical language (for an exposition of the geometrical concepts see for example [1], [2]).

We note that the augmented thermodynamic state space is in fact $T^*M_{th1} \times \mathbb{R}$ (by T^*M_{th1} we denote the cotangent bundle of M_{th}; $T^*_{(e,n)}M_{th1}$ is the cotangent space attached to $(e,n) \in M_{th1}$, its coordinates are $\left(\frac{1}{T}, -\frac{\mu}{T}\right)$; $s \in \mathbb{R}$). This space has the natural contact structure defined by the one form $\omega \equiv ds - \frac{1}{T}de + \frac{\mu}{T}dn$. The Gibbs surface is a Legendre submanifold of $T^*M_{th1} \times \mathbb{R}$ (i.e. the surface on which the one form ω equals zero). We shall denote it by the symbol $\mathcal{L}_{th1}$. Next, we note that the coordinates e and n of M_{th1} have a distinct physical meaning. This means that the thermodynamic state space M_{th1} has an additional structure that we can express by $M_{th1} \equiv (M_{th1}, M_e, \Pi_e^{th1})$, where $M_e \equiv \{e|(e,n) \in M_{th1}\} \subset \mathbb{R}$. We can introduce now a Legendre transformation (i.e. the transformation preserving the contact structure of $T^*M_{th1} \times \mathbb{R}$) for example by $\left(e, n, \frac{1}{T}, -\frac{\mu}{T}, s\right) \mapsto \left(\frac{1}{T}, -\frac{\mu}{T}, e, n, -\frac{P}{T}\right)$, where $-\frac{P}{T} = -s + \frac{1}{T}e - \frac{\mu}{T}n$, P is the pressure. This Legendre transformation transforms the fundamental thermodynamic relation $s = s(e,n)$ into its dual (Gibbs-Duhem) form $P = P(\mu, T)$. We can also formulate this Legendre transformation as follows: first we introduce $\Phi^{th1}_{th1} : T^*M_{th1} \to \mathbb{R}$ (we shall call this function a thermodynamic potential; the reason for choosing the particular notation will become clear later in this section) by

$$\Phi^{th1}_{th1}\left(e, n, \frac{1}{T}, -\frac{\mu}{T}\right) = -s(e,n) + \frac{1}{T}e - \frac{\mu}{T}n. \tag{1}$$

Second, we look for solutions of

$$\frac{\partial \Phi^{th1}_{th1}}{\partial e} = 0, \quad \frac{\partial \Phi^{th1}_{th1}}{\partial n} = 0. \tag{2}$$

Solution of eq. (2) will be denoted by $[u_{th1}]_{th1}$ and called thermodynamic states. Third, we evaluate the thermodynamic potential Φ^{th1}_{th1} at thermodynamic states and equate it with $-\frac{P}{T}$, i.e.

$$\Phi^{th1}_{th1}\left([(e,n)]_{th1}, \frac{1}{T}, -\frac{\mu}{T}\right) = -\frac{P}{T}. \tag{3}$$

Equation (3) is the Gibbs-Duhem form $P = P(\mu, T)$ of the fundamental thermodynamic relation $s = s(e,n)$. Finally, we recall that a Legendre transformation transforms one Legendre submanifold into another Legendre submanifold.

Now we introduce thermodynamics in a general state space M. Thermodynamics in M is specified by specifying a Legendre submanifold in $T^*M \times \mathbb{R}$. The coordinates in $T^*M \times \mathbb{R}$ will be (u, p, s), $u \in M$, $p \in T^*_uM$, $s \in \mathbb{R}$. The fundamental one form ω that defines the contact structure of $T^*M \times \mathbb{R}$ is $\omega \equiv ds - pdu$. Let now $M \equiv M_i \equiv (M_i, M_{th1}, \Pi^i_{th1})$. In accordance with the notation introduced in Section 2.1, the coordinates in $T^*M_i \times \mathbb{R}$ will be denoted by $u_i \equiv (u_{th1}, v_{th1})$, $u_{th1} \in M_{th1}$, (i.e. $u_{th1} \equiv (e,n)$), $v_{th1} \in (\Pi^i_{th1})^{-1}u_{th1}, p_i \equiv (p_{th1}, q_{th1})$, $\left(\frac{1}{T}, -\frac{\mu}{T}\right) \equiv p_{th1} \in T^*_{u_{th1}}M_{th1}$, $q_{th1} \in T^*_{u_{th1}}(\Pi^i_{th1})^{-1}u_{th1}$, $s \in \mathbb{R}$. The Legendre transformation corresponding to the Legendre transformation introduced in the preceeding paragraph is: $(u_{th1}, v_{th1}, p_{th1}, q_{th1}, s) \mapsto (p_{th1}, q_{th1}, u_{th1}, v_{th1}, -s + \langle p_{th1}, u_{th1} \rangle + \langle q_{th1}, v_{th1} \rangle)$. We shall again introduce the thermodynamic potential $\Phi^i = T^*M_i \to \mathbb{R}$ by

$$\begin{aligned} \Phi^i(u_i, p_i) &= -s(u_i) + \langle p_i, u_i \rangle \\ &= -s(u_i) + \langle p_{th1}, \Pi^i_{th1}u_i \rangle \\ &\quad + \langle q_{th1}, (\Pi^i_{th1})^{-1}u_{th1} \rangle . \end{aligned} \tag{4}$$

Moreover, we introduce

$$\begin{aligned}\Phi^i_{th1}(u_i, p_{th1}) &\overset{def}{=} \Phi^i(u_i, p_i)|_{q_{th1}=0} \\ &= -s(u_i) + < p_{th1}, \Pi^i_{th1} u_i > .\end{aligned} \tag{5}$$

Solutions of

$$\frac{\partial \Phi^i_{th1}}{\partial u_i} = 0 \tag{6}$$

will be called again thermodynamic states and they will be denoted by the symbol $[u_i]_{th1}$. If u_i is a function (for example a distribution function) then the derivative in (6) is the Volterra functional derivative. We shall not use any special symbol to distinguish the standard partial derivative and the Volterra functional derivative. Finally, we introduce

$$\Phi^i_{th1}([u_i]_{th}, p_{th1}) \overset{def}{=} \varphi_{th1} = -\frac{P}{T}, \tag{7}$$

that corresponds to eq. (3), and introduces the fundamental thermodynamic relation $P = P(\mu, T)$ that is implied in M_{th1} by the fundamental thermodynamic relation $s = s(u_i)$ introduced in M_i.

3.2 Examples

The Boltzmann entropy [5] $s = -\frac{1}{V}\int d^3r \int d^3v f_{kt1}(r, v) \ln f_{kt1}(r, v)$ provides a classical example of the fundamental thermodynamic relation in M_{kt1}. It is easy to verify that if we follow the analysis described in eqs. (4)–(7) with $(M_{kt1}, M_{th1}, \Pi^{kt1}_{th1})$ given in Section 2.2 (we consider only the particular case in which the interparticle potential energy $\epsilon(f_{kt1}; r) \equiv 0$) then eq. (7) becomes the thermodynamic relation $P = P(\mu, T)$ characterizing in classical equilibrium thermodynamics the ideal gas.

An example of the fundamental thermodynamic relation in M_{hyd} is: $s = \frac{1}{V}\int d^3r s(e(r), \rho(r))$, where s is the same function, pointwise for all r, as $s(e, n)$ introduced in M_{th1}. This thermodynamic relation arises when we assume that the system under consideration is locally at equilibrium (the hypothesis of local equilibrium). It is easy to verify that eq. (7) is in this case the Gibbs-Duhem form of $s = s(e, n)$.

Gibbs equilibrium statistical mechanics [11]–[14] can also be cast into the form of eqs. (4)–(7). In this case the elements of the state space are

$$u \equiv (u_0, u_1(x_1), u_2(x_1, x_2), \ldots, u_n(x_1, \ldots, x_n), \ldots), x_i \equiv (r_i, v_i),$$

u_m is the probability distribution of m isolated particles, $u_0, u_1, \ldots, u_n, \ldots$ are normalized, i.e. $u_0 + \sum_{n=1}^{\infty} \frac{1}{n!}\int u_n dx_1, \ldots dx_n = 1$. The projection on M_{th1} is introduced by

$$n = \sum_{n=1}^{\infty} \frac{1}{n!} n \int u_n dx_1, \ldots, dx_n, e = \sum_{n=1}^{\infty} \frac{1}{n!}\int h_n(x_1, \ldots, x_n) u_n dx_1 \ldots dx_n,$$

h_n is the n-particle Hamiltonian. The fundamental thermodynamic relation is:

$$s = -\sum_{n=1}^{\infty} \frac{1}{n!}\int u_n \ln u_n dx_1 \ldots dx_n.$$

Other examples will be introduced in Section 5.

3.3 Remarks, references

We have already mentioned that the association between physical systems and fundamental thermodynamic relations in M_{th1} (or equivalently Gibbs surfaces $\mathcal{L}_{th1}$) can be found, in the context of classical equilibrium thermodynamics, only by fitting results of observations. We have seen however that we can obtain fundamental thermodynamic relations in M_{th1} also from fundamental thermodynamic relations formulated on a more microscopic level M_i (see eqs. (4)–(7)). The physical insight is expressed in M_i in the function $s = s(u_i)$ and in the projection mapping Π^i_{th1}. On the most microscopic level of description, namely in the context of the Gibbs equilibrium statistical mechanics – see Section 3.2 – , the fundamental thermodynamic relation $s = s(u_i)$ is universal for all systems. The individual features of the systems are expressed only in $e = e(u_i)$ (i.e. in the specification of the forces among the particles composing the system under consideration). On a mesoscopic level of description, the relation $s = s(u_i)$ can either be suggested by interpreting s as a measure of an order (information) [15] or as an approximation of the entropy arising in the Gibbs theory.

The geometrical formulation of classical equilibrium thermodynamics presented in Section 3.1 has been suggested in Refs. [16]–[18], [10]. The extension of this point of view to general state spaces is new [19]. Implicitely however, this point of view has been routinely used in equilibrium statistical mechanics (see Refs. [12]–[14]). The unification of the structure of equilibrium thermodynamics and equilibrium statistical mechanics achieved in Section 3.1 is essential for introducing the multilevel point of view. The usefulness of the geometrical formulation will also be demonstrated in Section 6 where thermodynamics of unforced systems is extended to thermodynamics of driven systems.

4 Dynamics – Physical foundation of thermodynamics

Now we turn our attention to the time evolution in the state space M_i (in other words to the introduction of a vector field on M_i). We begin by listing some examples. If $M_i \equiv M_{th1}$ then the time evolution equations are particularly simple,

$$(u_{th1})_t = 0, \tag{8}$$

the symbol $(\ )_t$ denotes the time derivative. Equation(8) represents the conservation of the total mass and the total energy. If the elements of the state space M are position vectors (denoted by q) and momentum (denoted by p) of all particles composing the system under consideration then the time evolution is governed by Hamilton's equations

$$\begin{pmatrix} q \\ p \end{pmatrix}_t = L^C \begin{pmatrix} \dfrac{\partial h}{\partial q} \\ \dfrac{\partial h}{\partial p} \end{pmatrix}, \tag{9}$$

where $L^C \equiv \begin{pmatrix} 0 & 1 \\ -1 & 0 \end{pmatrix}$. If $M_i \equiv M_{kt1}$ then the time evolution equation could be for example the Boltzmann kinetic equation. An example of the time evolution equations in M_{hyd} are the Navier-Stokes-Fourier hydrodynamic equations.

Our objective is to introduce a vector field on M_i such that the trajectories (obtained by solving the governing time evolution equation) will agree with the following results of observations:

(i) the existence of the approach to thermodynamic equilibrium states (the process of preparation of physical systems – see Section 3),

(ii) equilibrium thermodynamics describes the behaviour of the systems at thermodynamic equilibrium states.

These observations are made on all levels of description. We can therefore expect that there exists a universal structure of the vector fields that guarantees the properties (i) and (ii) of the trajectories.

In order to identify the structure we recall first some well known and well tested results. We begin with the result of Onsager and Casimir [20], [21], according to which the linear equation governing the time evolution in a small neighborhood of a thermodynamic equilibrium state (defined as a solution of eq. (6)) is

$$(u_i)_t = L^C \frac{\partial \Phi^i_{th1}}{\partial u_i} - L^0 \frac{\partial \Phi^i_{th1}}{\partial u_i}, \tag{10}$$

where Φ^i_{th1} is the thermodynamic potential introduced in Section 3, L^C is a skewadjoint linear operator (called hereafter a linear Casimir operator) and L^0 is a linear selfadjoint and nonnegative operator (called hereafter a linear Onsager operator). Equation (10) is a linear time evolution equation if $\frac{\partial \Phi^i_{th1}}{\partial u_i}$ is linearized about equilibrium states (i.e. about states for which $\frac{\partial \Phi^i_{th1}}{\partial u_i} = 0$).

Now we are looking for a nonlinear extension of eq. (10). First, we focus our attention on the second term on the right hand side of eq. (10). In order to guarantee that the time evolution equation reproduces the experience (i) and (ii) mentioned above, the thermodynamic potential Φ^i_{th1} has to play the role of the Lyapunov function. This means in particular that

$$(\Phi^i_{th1})_t \leq 0. \tag{11}$$

If $L^0 \equiv 0$ then eq. (10) implies that $(\Phi^i_{th1})_t = 0$. We shall see later that this property will be kept also if the linear operator L^C is replaced by its nonlinear extension. We thus want to construct L^0 in such a way that the inequality (11) holds. It has been observed in Refs. [22]–[24] that (11) is satisfied if

$$L^0(p_i) = \frac{\partial^2 \Lambda}{\partial p_i \partial p_i}, \tag{12}$$

where $M_i^* \ni p_i = \frac{\partial \Phi^i_{th1}}{\partial u_i}$ and $\Lambda : M_i^* \to \mathbb{R}$, called dissipative potential, satisfies the following properties: $\Lambda(0) = 0$, Λ reaches a minimum at 0 and Λ is concave in a neighborhood of 0.

Now we focus our attention on the nonlinear extension of the first term on the right hand side of eq. (10). We recall that the nondissipative part of the Boltzmann kinetic equation [25], the Navier-Stokes-Fourier hydrodynamic equation [26] as well as the time evolution equation (9) introduced on the most microscopic level of description prossesses the Hamiltonian structure. Following the general (noncanonical) formulation of Hamiltonian systems (see e.g. Refs [27], [1], [2]), that is most convenient in our multilevel analysis, a time evolution equation

$$(u_i)_t = L^C(u_i)\frac{\partial \Phi^i_{thi}}{\partial u_i} \tag{13}$$

represents a Hamiltonian system if L^C is a Poisson operator. This means that if we introduce a bracket $\{A,B\}$ by

$$\{A,B\} = \left\langle \frac{\partial A}{\partial u_i}, L^C(u_i)\frac{\partial B}{\partial u_i}\right\rangle, \tag{14}$$

where A, B are sufficiently regular functions $M_i \to \mathbb{R}$, then this bracket is a Poisson bracket, i.e. $\{A,B\} = -\{B,A\}$ and $\{A,B\}$ satisfies the Jacobi identity $\{A,\{B,C\}\} + \{B,\{C,A\}\} + \{C,\{A,B\}\} = 0$. We note that if L^C is independent of u_i (as it is in eq. (10)) then the skewadjointness of L^C itself guarantees that (14) is a Poisson bracket. We also note that if the time evolution of u_i is governed by eq. (13) then $(\Phi^i_{th1})_t = 0$.

On the basis of the above considerations, we propose (see Refs. [27]–[30]) the following time evolution equation in the state space M_i:

$$(u_i)_t = L^C(u_i)\frac{\partial \Phi^i_{th1}}{\partial u_i} - L^0\left(\frac{\partial \Phi^i_{th1}}{\partial u_i}\right)\frac{\partial \Phi^i_{th1}}{\partial u_i}, \tag{15}$$

where L^0 is introduced in eq. (12), L^C in eq. (13). In addition, we require that the operator L^C and L^0 are degenerate so that no time evolution takes place in the base space, i.e.

$$(\Pi^i_{th1} u_i)_t = 0. \tag{16}$$

It has been shown that many well known and well tested time evolution equations on many different levels of description (e.g. all well know kinetic equations and hydrodynamic equations) can indeed be cast into the form of eq. (15) (see Refs. [28]–[38]). If eq. (15) is linearized about an equilibrium state then the linear Onsager-Casimir equation (10) is recovered. We suggest therefore to call eq. (15) a nonlinear Onsager-Casimir time evolution equation [19].

We end this subsection by a few remarks.

Remark 1

Equation (15) together with the properties required from the operators L^C and L^0 implies

$$(\Phi^i_{th1})_t \leq 0. \tag{17}$$

This inequality is called a dissipation inequality. It guarantees that the thermodynamic potential Φ^i_{th1} plays the role of the Lyapunov function that is associated with the approach (as $t \to \infty$) to thermodynamic equilibrium states. At the equilibrium states the behaviour of the system under consideration is then well described by the thermodynamic potential Φ^i_{th1}. This means that if we accept eq. (15) as a formulation of an experience arising in observing the time evolution on the level of description that uses M_i as the state space then the equilibrium thermodynamic experience formulated in equilibrium thermodynamics arises as a consequence. The nonlinear Onsager-Casimir time evolution equation provides thus a physical foundation of equilibrium thermodynamics. We recall that equilibrium thermodynamics is usually based on the experience with thermal machines (i.e. on the experience that belongs and is expressed entirely on the equilibrium thermodynamics level of description) or on the experience serving as a basis of Gibbs equilibrium statistical mechanics (i.e. the experience that belongs to the level of description on which classical or quantum mechanics is formulated)

Remark 2

In order to manifest even more clearly the relationship of the vector field (15) with the geometrical formulation of thermodynamics (see Section 3) we extend the vector field (15) introduced in M_i into a vector field in $T^*M_i \times \mathbb{R}$ (see Ref. [19]):

$$\begin{aligned}(u_i)_t &= -\frac{\partial K}{\partial p_i}\\ (p_i)_t &= \frac{\partial K}{\partial u_i} - p_i\frac{\partial K}{\partial s}\\ (s)_t &= K - \left\langle p_i, \frac{\partial K}{\partial p_i}\right\rangle,\end{aligned} \tag{18}$$

where K, called a contact Hamiltonian, is given by

$$K = \left\langle p_i, \frac{\partial \Lambda}{\partial p_i}\right\rangle - \left\langle p_i, \left[\frac{\partial \Lambda}{\partial p_i}\right]_{p_i=\frac{\partial \Phi^i_{th1}}{\partial u_i}}\right\rangle - \left\langle p_i, L^C\frac{\partial \Phi^i_{th1}}{\partial u_i}\right\rangle. \tag{19}$$

We make the following observations:

(i) The vector field (18) is a cononical formulation of a contact vector field [1],[2]. This means that the flow generated by the vector field (18) preserves the contact structure of $T^*M_i \times \mathbb{R}$.

(ii) The Legendre submanifold $\mathcal{L}_{th}$ of $T^*M_i \times \mathbb{R}$ defined as the image of the imersion $u_i \mapsto \left(u_i, \frac{\partial \Phi^i_{th1}}{\partial u_i}(u_i), \Phi^i_{th1}(u_i)\right)$ is an invariant submanifold. This means that the vector field (18) is tangent to the Legendre submanifold $\mathcal{L}_{th}$. Moreover, the vector field (18) restricted to $\mathcal{L}_{th}$ is identical with the vector field (15). We also note that $K|_{\mathcal{L}_{th}} = 0$. All statements made in this observation can be readily verified (we verify that eq. (18.1) $|_{\mathcal{L}_{th}} \equiv$ eq. (15); eq. (18.2) $|_{\mathcal{L}_{th}} \equiv \left(\frac{\partial \Phi^i_{th1}}{\partial u_i}\right)_t$; and eq. 18.3 $|_{\mathcal{L}_{th}} = (\Phi^i_{th1})_t$). This observation permits us to regard the vector field (18) as a lift of the nonlinear Onsager-Casimir vector field (15). We see that the flow generated by the vector field (18) preserves the contact structure (i.e. the fundamental structure of thermodynamics) and the manifold $\mathcal{L}_{th}$ (that specifies thermodynamics) provides the state space for the time evolution.

(iii) If $L^0 \equiv 0$ (i.e. if there is no dissipation) then the lift (18) from M_i to $T^*M_i \times \mathbb{R}$ is equivalent to the complete lift introduced in [39].

In order to give a simple illustration of the lift of the vector field (15) into the vector field (18), we consider the one dimensional damped harmonic oscillator. We have thus

$M_i \equiv \mathbb{R}^2, (x,y) \in M_i,\ (x,y,p,q) \in T^*M_i,\ \Phi^i_{th1} = \frac{1}{2}(x^2+y^2),\ \Lambda = \frac{1}{2}kp^2, k > 0$. The vector field (15) is now $\binom{x}{y}_t = \begin{pmatrix} 0 & 1 \\ -1 & 0 \end{pmatrix}\binom{x}{y} - \begin{pmatrix} 0 & 0 \\ 0 & k \end{pmatrix}\binom{x}{y}$. The corresponding lift (18) is thus:

$$\begin{array}{rcl} K & = & kp^2 - kpx - py + qx \quad ; \quad (x)_t = -2kp + kx + y; \\ (y)_t & = & -x \quad ; (p)_t = -kp + q \quad ; \quad (q)_t = -p \quad ; \quad (s)_t = -kp^2. \end{array}$$

Remark 3

The nonlinear Onsager-Casimir time evolution equation (15) provides a unified formulation of dynamics of unforced systems on all levels of description. It serves us also as a formulation of the physical foundation of thermodynamics. In addition, eq. (15) can be used with advantage in two other ways. First, the mathematical structure of eq. (15) expressed in the properties of L^C, L^0, Φ^i_{th1}, allows to make some conclusions about its solutions. By recasting a known time evolution equation into the form (15) we reveal some properties of its solutions. This type of applications of eq. (15) has been introduced for the first time in the context of hydrodynamics by Arnold [40] (see also [27]). In the second type of application the time evolution equation of systems under consideration is not known. Equation (15) is then used to introduce it. This application of eq. (15) will be considered in detail in Section 5.

4.1 Examples

We shall introduce here only one illustrative example that is taken from kinetic theory. Some other examples will be shown in Section 5. Still many other examples have been worked out in [31], [25]–[37].

We shall introduce M_{kt1} and thermodynamics in M_{kt1} as in Section 2.2 and 3.2. The Poisson bracket in M_{kt1} has been introduced in [25], [27],

$$\begin{aligned} \{A,B\} = & \int d^3r \int d^3v f_{kt1}(\mathbf{r},\mathbf{v}) \left[\frac{\partial}{\partial r_\alpha}\left(\frac{\partial A}{\partial f_{kt1}(\mathbf{r},\mathbf{v})}\right)\frac{\partial}{\partial v_\alpha}\left(\frac{\partial B}{\partial f_{kt1}(\mathbf{r},\mathbf{v})}\right)\right. \\ & \left. - \frac{\partial}{\partial r_\alpha}\left(\frac{\partial B}{\partial f_{kt1}(\mathbf{r},\mathbf{v})}\right)\frac{\partial}{\partial v_\alpha}\left(\frac{\partial A}{\partial f_{kt1}(\mathbf{r},\mathbf{v})}\right)\right]. \end{aligned}$$

The dissipative potential generating the dissipative part of the Fokker-Planck equation is:

$$\Lambda = \Lambda_0 \frac{\partial}{\partial v_\alpha}, \left(\frac{\partial \Phi^{kt1}_{th1}}{\partial f_{kt1}(\mathbf{r},\mathbf{v})}\right)\frac{\partial}{\partial v_\alpha}\left(\frac{\partial \Phi^{kt1}_{th1}}{\partial f_{kt1}(\mathbf{r},\mathbf{v})}\right)$$

where $\Lambda_0 \in \mathbb{R}$, $\Lambda_0 > 0$, and the summation convention is used. It is now easy to verify that with these specifications eq. (15) becomes the Fokker-Planck kinetic equation. Other kinetic equations (e.g. the Boltzmann and the Enskog kinetic equations) have been cast into the form of eq. (15) in [31].

5 Rheological modeling

The principal objective of rheological modeling is to provide a setting that allows to organize and understand results of the chosen observations. By understanding we usually mean establishing a bridge between microscopic (molecular) and macroscopic (flow and thermodynamic) properties of polymeric fluids. We shall describe briefly three modeling techniques. First, it is a technique that will be called an ideal modeling technique, second, it is the most widely used modeling technique (we shall call it Kirkwood's modeling technique) and third, it is a technique based on the nonlinear Onsager-Casimir time evolution equation (15) (we shall call it a thermodynamic modeling technique).

5.1 Ideal modeling technique

It is useful to begin the analysis of modeling by realizing what is the best that could be done while leaving the question of feasibility aside. The actual modeling techniques can be then judged by comparing them with this ideal.

Step 1

The starting point that we choose is quantum mechanics of all molecules composing the polymeric fluid under consideration.

Step 2

We solve the governing equations of quantum mechanics for all possible initial conditions and for a large class of boundary conditions corresponding to our actual macroscopic control over the polymeric fluid. We shall refer to the result as a phase portrait.

Step 3

We extract from the phase portrait a pattern that is pertinent to our macroscopic observations. This step is thus essentially a pattern recognition process.

Step 4

Following now only the pertinent pattern in the phase portrait (i.e. forgetting all impertinent details) we obtain a simple macroscopic rheological model whose consequences agree with results of our observations.

5.2 Kirkwood's modeling technique

Now we describe the basic steps in the modeling technique introduced by Kirkwood [6] (see also Ref. [7]).

Step 1

The starting microscopic (molecular) point of view is chosen. We shall denote the chosen state space by the symbol $M_1(u_1 \in M_1)$ and the time evolution equation $(u_1)_t = R_1(u_1)$. The symbol R_1 stands for the right hand side of the time evolution equation on the level of description denoted by the index 1.

Step 2

The level of description on which the model is formulated is chosen. We shall denote the state space corresponding to this level of description by the symbol M_0.

Step 3

The projection $\Pi_0^1 : M_1 \to M_0$ is suggested. In other words, the bundle (M_1, M_0, Π_0^1) is introduced.

Step 4

The projection operator Π_0^1 is applied on the time evolution equation $(u_1)_t = R_1(u_1)$. One obtains $(u_0)_t = \Pi_0^1 R_1(u_1)$. In order to make this equation a time evolution equation in M_0, the equation has to be closed. This means that we have to express u_1 that remains on the right hand side as a function of u_0. We introduce thus an operator $I_1^0 : M_0 \to M_1$ that we shall call a closure operator. It specifies an imbedding of the space of M_0 in the larger space M_1. It is obvious that we shall require that $\Pi_0^1 \circ I_1^0 =$ identity mapping in M_0.

By comparing this modeling technique with the ideal modeling technique we see that both operators Π_0^1 and I_1^0 should arise as a result of the pattern recognition process in the phase portrait. The choice of Π_0^1 can usually be justified, at least partially, by plausible physical arguments. There is however no clue offered in Kirkwood's modeling technique for choosing the closure mapping I_1^0. It is in fact only the a posteriori analysis of the closed equation (comparison of its consequences with results of observations) that brings a justification for choosing I_1^0. However succesfull has been the Kirkwood modeling technique in introducing many useful rheological models (see Ref. [27]) it would be certainly useful to introduce another completely different modeling technique that would complement the Kirkwood technique. Such modeling technique is introduced in the next subsection.

5.3 Thermodynamic modeling technique

Step 1

The level of description on which the model is formulated is chosen. We shall use again the symbol M_0 to denote the state space introduced on the chosen level of description. Note that this step is exactly the same as Step 2 in Kirkwood's modeling technique.

Step 2

We assume that the time evolution equation in M_0 is the nonlinear Onsager-Casimir time evolution equation (15). In order to specify this equation we need to specify three quantities, namely the thermodynamic potential Φ^0_{th1}, the Casimir operator L^C and the Onsager operator L^0. These are the three basic building blocks of the rheological model. It is in these blocks where we express (in the next steps) our physical insight into the polymeric fluid under consideration.

Step 3

Specification of the thermodynamic potential Φ^0_{th1}. This task has been discussed in Section 3.

Step 4

Specification of the Casimir operator L^C. From the physical point of view, this problem is the problem of identifying kinematics in the state space M_0 (see more in the last paragraph in Section 5.4).

Step 5

Specification of the Onsager operator L^0 (or equivalently – see eq. (12) – the dissipative potential Λ). There are three types of arguments that we can advance to approach this problem. First, it is the argument of simplicity. The quadratic dissipative potential Λ can be suggested as the first candidate. Second, we follow the Kirkwood technique, look at the projected equation $(u_0)_t = \Pi^1_0 R_1(u_1)$ and try to suggest the operator L^0 by trying to cast it into the form of eq. (15) with Φ^0_{th1} and L^C already know from the two previous steps. The third type of arguments that can be used to specify L^0 will be mentioned in Section 6.

We end this subsection by listing advantages of the thermodynamic modeling technique.

1. During the experimental investigation, experimentalists acquire an intuitive understanding of the physics of the polymeric fluid under investigation. Such understanding

can be then easily expressed, by following the steps of the thermodynamic technique, in a rheological model. A confrontation of the consequences of the model with results of observations represents a test of the understanding on the basis of which the model was formulated. This test can be then used to improve the understanding. It should be emphasized in particular that the formulas exprssing the extra stress tensor as a function of the state variable u_0 is obtained as a result if the thermodynamic modeling technique is followed. The ambiguities that arise in the specification of these formulas in Kirkwood's modeling (reflecting the ambiguity in choosing the closure mapping – see an example in Ref. [41]) are thus avoided.

2. We desire, of course, that consequences of the rheological models agree with results of observations. We recall that already the fact that the thermodynamic modeling technique is used to introduced the models guarantees agreement with one particular experience, namely the equilibrium thermodynamic experience.

3. Rheological model constructed by following the thermodynamic modeling technique stay on their own. By this we mean that the arguments on which the models are based belong to the same level of description on which the model is formulated. There is no need of more microscopic levels of description to give the models a clear physical meaning. The advantage of this feature of modeling is well illustrated on the following example. It is well known that hydrodynamics can be founded physically in the state space M_{hyd} (local expression of global conservation laws). Let us suppose for a moment that this macroscopic foundation is not known to us and that we are familiar only with the Boltzmann kinetic equation describing the time evolution of a rarefied gas on a more microscopic level of description, namely on the level of descirption that uses M_{kt1} as its state space. From the Boltzmann equation, by following the Kirkwood modeling technique (in this case it is the Chapman-Enskog method that can be used to suggest the closure mapping), we arrive at hydrodynamic equations. We then conclude that hydrodynamic equations describe well the asymptotic time evolution of rarefied gases. In reality, of course, hydrodynamic equations describe well the asymptotic time evolution of a much larger class of physical systems. We have no problem with this finding since we can provide an alternative physical foundation of hydrodynamic equations that is based only on the physical insight associated with the hydrodynamic level of description and thus applicable to larger class of physical systems than rarefied gases. A similar situation arises in rheological modeling. Many useful rheological models that use M_c as its state space (see Section 5.4) can be introduced, by following Kirkwood's modeling technique, from the kinetic equations describing the time evolution of noninteracting dumbbells [7]. The resulting models formulated in M_c are then found to be good modeles also for a much larger class of polymeric fluids. An independent introduction of rheological models in the state space M_c, provided by the thermodynamic modeling technique, is thus very useful.

4. If we place the thermodynamic modeling technique in the context of the Kirkwood modeling technique, we observe that the thermodynamic modeling technique offers in fact a way to suggest the closure mapping. First, we may use the requirement that after closing the time evolution equation has the structure of the nonlinear Onsager-Casimir time evolution equation (15). The second agument leading to the closure mapping will be introduced in Section 6.

5.4 Examples

Since many examples of the application of the thermodynamic modeling technique to formulate rheological models have already been published, we shall limit ourselves only to making a few remarks and listing the references.

The choice of the state space M_0 in Step 1 is always a compromise between completeness (the ability to express molecular details) and simplicity (the easiness of solving the governing equations of the model). In general, we have followed the experience collected in [6], [7] and used three state spaces M_ψ, M_c and M_{nw}. Elements of M_ψ are $(\rho(\boldsymbol{r}), e(\boldsymbol{r}), \boldsymbol{u}(\boldsymbol{r}), \psi(\boldsymbol{r}, \boldsymbol{R}))$, where $\rho, e, \boldsymbol{u}$ are elements of M_{hyd} (see Section 3.2) and $\psi(\boldsymbol{r}, \boldsymbol{R})$, the so called configuration space distribution function ($\boldsymbol{R}$ is the end-to-end vector of a macromolecule), satisfies the normalization condition $\int d^3\boldsymbol{r} \int d^3\boldsymbol{v}\psi(\boldsymbol{r}, \boldsymbol{R}) = 1$; $\psi(\boldsymbol{r}, \boldsymbol{R})d^3\boldsymbol{r}d^3\boldsymbol{R}$ is physically interpreted as the probability that a macromolecule with the end-to-end vector in $(\boldsymbol{R}, \boldsymbol{R}+d\boldsymbol{R})$ is found in the position $(\boldsymbol{r}, \boldsymbol{r}+d\boldsymbol{r})$. Elements of M_c are $(\rho(\boldsymbol{r}), e(\boldsymbol{r}), \boldsymbol{u}(\boldsymbol{r}), \boldsymbol{c}(\boldsymbol{r}))$, where again $\rho, e, \boldsymbol{u}$ are elements of M_{hyd} and $\boldsymbol{c}(\boldsymbol{r})$, called a conformation tensor, is a symmetric positive definite tensor that is physically interpreted as a measure of macromolecular deformations (a macromolecular strain tensor). By M_{nw} we denote the state space used in the Ericksen-Leslie theory of liquid crystals (see Refs. [47], [48], [35], [37]).

The list of references is presented in the following table (the references with star, [ ]*, include experimental results and comparison of the experimental results with model predictions)

Polymeric Fluid	State Space	Reference
compressible, non-isothermal	M_ψ M_c	[33] [41]
polyelectrolytes	M_c	[42]*
semiflexible macromolecules	M_ψ M_c	[43], [37] [43], [44], [45]*
liquid crystals	M_ψ M_c M_{nw}	[43], [33], [37] [34], [35], [37] [35]
network of macromolecules	M_c	[46]*

Finally, we shall make a remark concerning the statement that we made in Step 4 (i.e. the choice of the Casimir operator L^C or equivalently the Poisson bracket). We have stated that L^C represents a mathematical formulation of kinematics in M_0. Let G be a Lie group of the fundamental kinematic transformations in M_0 and g its corresponding Lie algebra. In hydrodynamics, G is the group of mappings $\mathbb{R}^3 \to \mathbb{R}^3$, in kinetic theory, G is the group of canonical transformations $\mathbb{R}^6 \to \mathbb{R}^6$, $(\boldsymbol{r}, \boldsymbol{v}) \mapsto (\boldsymbol{r}', \boldsymbol{v}')$). If we succeed to present the state space M_0 as dual of the semidirect product of the Lie algebra g and a direct sum of vector spaces then the Poisson bracket arises naturally (see e.g. [49], [27]). For example, the Lie algebra corresponding to the group of canonical transformations is the algebra of the generating functions of the canonical transformations. Its dual is the space whose elements are one particle distribution functions, thus the space M_{kt1}.

6 Thermodynamics and dynamics of driven systems

Until now, we have considered only unforced systems. In this section, our attention will be turned to driven systems. We can think for example about the Bénard system (a horizontal layer of fluid heated from below – the driving force in this example is the temperature gradient and the gravitational force) or a polymeric fluid subjected to a flow (the driving force in this example is the imposed flow). We begin by recalling some experience that has been collected about driven systems.

First, we note that driven systems do not approach thermodynamic equilibrium states. In fact, they, in general, do not even approach stationary states (recall the experience collected about the Bénard system). Since the approach to thermodynamic equilibrium states was taken as a basis for introducing thermodynamics of unforced systems (Section 4), we have to look, in the context of driven systems, for other properties on the basis of which thermodynamics could be introduced. Another observation that points against the existence of thermodynamics of driven systems is that the organization in driven systems in many cases increases as the time evolution proceeds and not decreases as it is in the case of all unforced systems. To see this we can recall again the familiar behaviour of the Bénard system. There are however also observations indicating that a simple modification of the free energy can be used to explain certain behaviour of driven systems. For example the experimental observations and the theoretical analysis of solubility of polymer solutions subjected to a flow (see Refs. [50]–[53]) are of this nature.

The observation on which we shall base the analysis of driven systems presented below is that the time evolution of all (or at least many) driven systems is well described on a level of description that is not the most microscopic one. We shall denote the state space used on this level of description by the symbol M_0. From this observation we conclude that if we describe the time evolution of the driven system under consideration in a state space M_1 that is more microscopic than M_0 then this more microscopic description will approach (as $t \to \infty$) the description in M_0. For example, it is well known that the behaviour of the Bénard system is well described by hydrodynamic equations (Boussinesq equations). If we thus describe the time evolution of this system on, say, kinetic theory level, i.e. in M_{kt1}, then this description will approach, as $t \to \infty$, the hydrodynamic description. Similarly, in the context of polymeric fluids subjected to a flow, observations seem to indicate that their behaviour is well described in M_c. If we thus formulate the time evolution in M_ψ, we can expect that this description will approach (as $t \to \infty$) the description in M_c. Thermodynamics of unforced systems has been based (Section 4) on the observed existence of the approach to thermodynamic equilibrium states or equivalently on the approach to equilibrium thermodynamics level of description. We suggest to base thermodynamics of driven systems on the observed existence of the approach to a more macroscopic (but still involving time evolution) level of description. In order to prepare the analysis of driven systems (Section 6.2), we shall consider first the approach to a more macroscopic (but still involving the time evolution) level of description in the context of unforced systems (Section 6.1).

6.1 Approach of the time evolution in M_1 to the time evolution in a more macroscopic state space M_0: nondriven systems

In this subsection, we shall still consider nondriven systems. In Sections 3 and 4, we have based our analysis on the comparison of descriptions in M_i and M_{th1}. In this subsection, we shall compare three levels of description, namely M_i, M_j $j < i$ and M_{th1}. For example: $M_i \equiv M_{th1}$, $M_j \equiv M_{hyd}$ or $M_i \equiv M_\psi$ and $M_j \equiv M_c$. We shall assume that the three bundles $(M_i, M_{th1}, \Pi^i_{th1})$, $(M_j, M_{th1}, \Pi^j_{th1})$ and (M_i, M_j, Π^i_j) are know. By following Sections 3 and 4, the first two bundles can serve us to introduce the time evolution expressing the approach to the equilibrium thermodynamic description (we shall denote it symbolically by $M_i \overset{t\to\infty}{\longrightarrow} M_{th1}$ and $M_j \overset{t\to\infty}{\longrightarrow} M_{th1}$) and the corresponding thermodynamics. We shall now try to extend the analysis introduced in Sections 3 and 4 so that it could also be applied to the approach $M_i \overset{t\to\infty}{\longrightarrow} M_j$. In this way we then also expect to introduce the corresponding thermodynamics. The generality of the mathematical formulation introduced in Sections 3 and 4 will be particularly useful.

First, we turn our attention to the extension of thermodynamics by replacing in Section 3 the bundle $(M_i, M_{th1}, \Pi^i_{th1})$ by the bundle (M_i, M_j, Π^i_j). The notation introduced in Section 2.1 will be used: $u_i \equiv (u_j, v_j)$, $v_j \in (\Pi^i_j)^{-1} u_j$. In addition, $p_i \in T^*_{u_i} M_i, p_j \in T^*_{u_j} M_j, p_i \equiv (p_j, q_j)$, $q_j \in T^*_{u_j} v_j$. The thermodynamic potential (4) is now

$$\begin{aligned} \Phi^i(u_i, p_i) &= -s(u_i) + \langle p_i, u_i \rangle \\ &= -s(u_i) + \langle p_j, \Pi^i_j u_i \rangle \\ &\quad + \langle q_j, (\Pi^i_j)^{-1} u_j \rangle . \end{aligned} \tag{20}$$

We introduce also (see eq. (5))

$$\begin{aligned} \Phi^i_j(u_i, p_j) &\overset{def}{=} \Phi^i(u_i, p_i)|_{q_j \equiv 0} \\ &= -s(u_i) + \langle p_j, \Pi^i_j u_i \rangle . \end{aligned} \tag{21}$$

Solutions of

$$\frac{\partial \Phi^i_j}{\partial u_j} = 0 \tag{22}$$

will be denoted $[u_i]_j$ and called thermodynamic states. Finally, we introduce

$$\Phi^i_j([u_i]_j, p_j) \overset{def}{=} \Sigma. \tag{23}$$

The equation $\Sigma = \Sigma(p_j)$ is the fundamental thermodynamic relation in M_j implied by the fundamental thermodynamic relation $s = s(u_i)$ in M_i.

Now we turn our attention to the time evolution. Let

$$(u_i)_t = R_i(u_i) \tag{24}$$

denote symbolically the equation that governs the time evolution in M_i, and

$$(u_j)_t = R_j(u_j) \tag{25}$$

the equation that governs the time evolution in M_j. In Section 4, we have considered the case $M_j \equiv M_{th1}$, eq. (25) $\equiv$ eq. (8) and eq. (15) has been introduced as a reformulation of eq. (24) in which the approach to $[u_i]_{th}$ is manifestly displayed. Now, if $M_j \not\equiv M_{th1}$, we look for a reformulation of eq. (24) which will manifestly display the approach to the time evolution governed by eq. (25). In other words, we want to formulate the time evolution in M_i (governed by eq. (24)) as a two step evolution. The first step is the "fast" approach to the time evolution in M_j (governed by eq. (25)) and then the "slow" evolution governed by eq. (25). In the case of $M_j \equiv M_{th1}$ the "slow" dynamics is the slowest possible which means, of course, that the dynamics in M_{th1} is absent. It seems natural (see [54], [55]) to suggest that the straighforward modification of eq. (15) obtained by replacing M_{th1} by M_j, namely

$$(u_i)_t = L^C(u_i)\frac{\partial \Phi^i_j}{\partial u_i} - L^0\left(\frac{\partial \Phi^i_j}{\partial u_i}\right)\frac{\partial \Phi^i_j}{\partial u_i} \tag{26}$$

$$(\Pi^i_j u_i)_t = 0, \tag{27}$$

will govern the first step, i.e. the "fast", time evolution. Equations (26), (27) describe the approach of u_i to $[u_i]_j$ that is then governed by eq. (25). Note that the thermodynamic potential Φ^i_j plays now the role of the Lyapunov functional associated with this approach.

As in Section 4, we can apply eqs. (25)–(27) in two ways. First, we take eq. (24) as known. Equations (25)–(27) represent then in fact a recasting eq. (24) into a form revealing the properties of its solutions that are of our particular interest. We recall that in Section 4 this application has been illustrated by the example: eq. (24) $\equiv$ Fokker-Planck kinetic equation (for other illustrations see [31]). In the context of eq. (24) versus eqs. (25)–(27), we have no example in which eqs. (25)–(27) would be strictly equivalent to a given equation (24). Some examples in which solutions of eqs. (25)–(27) can be regarded as a good approximation of eq. (24) will be introduced later. The second application of eqs. (25)–(27) follows the application of eq. (15) in Section 5. We assume that the time evolution in M_j (i.e. eq. (25)) is known. By using a physical insight (similar to the one used in Section 5) we then introduce L^c, L^0 and Φ^i_j appearing in eq. (26). Equation (25)–(27) are then regarded as a model of the time evolution in M_i. We shall discuss here only this second type of the application of eqs. (24)–(26).

In the rest of this subsection, we shall discuss the process of expressing our physical understanding of the system under consideration in the thermodynamic potential Φ^i_j. If we assume that the bundle projection Π^i_j is known, we need only the fundamental thermodynamic relation $s = s(u_i)$ (see eqs. (20), (21)) to specify Φ^i_j. We note that the relation $s = s(u_i)$ introduced in the context of the study of the approach $M_i \overset{t\to\infty}{\longrightarrow} M_j$ will, in general, be different from the relation $s = s(u_i)$ introduced in the context of the study of the approach $M_i \overset{t\to\infty}{\longrightarrow} M_{th1}$. This is because these two time evolutions are different and thus their corresponding Lyapunov functions are different. While studying the approach $M_i \overset{t\to\infty}{\longrightarrow} M_{th1}$ (Section 3–5) we could use the Gibbs equilibrium statistical mechanics to suggest the fundamental thermodynamic relation $s = s(u_i)$. How shall we proceed now when we consider the approach $M_i \overset{t\to\infty}{\longrightarrow} M_j$? We suggest two types of arguments:

(i) Let eqs. (24) and (25) as well as the bundle projection Π^i_j be known. If we apply Π^i_j on eq. (24) (compare with Step 4 in Kirkwood's modeling technique), we obtain

$$(u_j)_t = \Pi^i_j R_i(u_i) \tag{28}$$

We shall now require that if we choose the closure by

$$u_i = [u_i]_j \tag{29}$$

(i.e. we insert (29) into the right hand side of (28)), then eq. (28) will become equivalent to the known eq. (25). We thus look for Φ^i_j such that solutions of eq. (22) (i.e. the thermodynamic state $[u_i]_j$) are those that make eq. (28) equivalent to eq. (25). An illustration of this argument is introduced below.

(ii) The second argument is inspired by Muschik's [56] analysis. The entropy that we are looking for is the entropy of the system that is constrained by $\Pi^i_j u_i = \text{const}$. We shall denote it by the symbol $[\text{entropy}]_{[\text{constraint}]}$. If we now remove the constraint the system will reach the thermodynamic equilibrium state with entropy that we denote $[\text{entropy}]_{[\text{no constraint}]}$. Following Muschik's arguments, we suggest

$$[\text{entropy}]_{\text{no constraint}} = [\text{entropy}]_{\text{constraint}} + \begin{bmatrix} \text{entropy created} \\ \text{when constraints} \\ \text{are removed} \end{bmatrix} \tag{30}$$

We expect, of course, that the relations $s = s(u_i)$ obtained by following the argument (i) and (ii) will be the same.

Now, we mention some examples. Let eq. (24) be the Boltzmann kinetic equation (i.e. $M_1 \equiv M_{kt1}$) and eq. (25) the Navier-Stokes-Fourier hydrodynamic equations (i.e. $M_0 \equiv M_{hyd}$). If we choose $s = s(u_i)$ as in Section 3.2 (i.e. the Boltzmann entropy) and construct Φ^{kt1}_{hyd} by following eqs. (20), (21) (the projection Π^{kt1}_{hyd} is introduced in Section 2.2), then solution of eq. (22) (i.e. the thermodynamic state $[u_{kt1}]_{hyd}$) the well known local Maxwellian distribution function. It is also well known that if we introduce the local Maxwellian distribution into eq. (28) we obtain the Euler hydrodynamic equations and not the Navier-Stokes-Fourier hydrodynamic equations. But we know what is in this case the distribution function that, if inserted into eq. (28), implies the Navier-Stokes-Fourier equations. It is, of course, the distribution function obtained as the first approximation in the Chapman-Enskog solution of the Boltzmann kinetic equation (Ref. [57]). We thus look for a modification of the Boltzmann entropy for which the solution of eq. (22) is this distribution function (see Ref. [55]). It can be shown that in this case the Muschik argument leads to the same result.

As another example, we consider $M_1 \equiv M_\psi, M_0 \equiv M_c$, eqs. (24) and (25) are some of the equation introduced in Section 5. In this case, the distribution function that plays the role that is analogous to the role of the local Maxwellian distribution function in Boltzmann kinetic theory is the distribution function introduced in Refs. [58],[59]:

$$\psi(\mathbf{r}, \mathbf{R}) = (2\pi)^{-3/2}(\det p(\mathbf{r},t))^{1/2} \exp(-R_\alpha p_{\alpha\beta}(\mathbf{r},t) R_\beta), \tag{31}$$

where is thermodynamic dual of the tensor $\mathbf{c}$. We shall mention this example also in Section 6.2.

6.2 Approach of the time evolution in M_1 to the time evolution in a more macroscopic state space: driven systems

The system under consideration is now a driven system. We follow Section 6.1 and consider a description in M_i (we shall denote it M_1) and M_j (we shall denote it M_0). We cannot describe our system in M_{th1} since driven systems do not approach thermodynamic equilibrium states. The time evolution describing the approach $M_1 \overset{t\to\infty}{\longrightarrow} M_0$ will be considered in the same way as in Section 6.1. The fact that the system is now driven enter our consideration in the bundle projection Π^i_j (e.g. in the example of the Bénard system the gravitational energy will be a part of the total energy) and in the second term on the right hand side of eq. (30). We shall illustrate the construction of the thermodynamic potential of driven systems on the following example.

Let the system under consideration be a polymer solution (studied in Refs. [50]–[53]) subjected to a flow. Experimental results show that solubility depends on the imposed flow. Our objective is to use the discussion of thermodynamics of driven systems presented in this section to understand these experimental results. We shall only sketch this application. Details will be published elsewhere.

We shall assume that elements of the most macroscopic state space in which the system can be described are: $M_0 \ni u_0 \equiv (n_s(\boldsymbol{r}), n_p(\boldsymbol{r}), e(\boldsymbol{r}), c_p(\boldsymbol{r}))$, where n_s is the number of moles per unit volume of the solvent, n_p is the number of moles per unit volume of the polymer, e is the internal energy per unit volume and c_p is the conformation tensor characterizing states of the polymer macromolecules. Elements of the more microscopic state space M_1 will be chosen to be: $M_1 \ni u_1 \equiv (n_s(\boldsymbol{r}), n_p(\boldsymbol{r}), e(\boldsymbol{r}), \psi_p(\boldsymbol{r}, \boldsymbol{R}))$, where n_s, n_p, e have the same meaning as in u_0 and ψ_p is configuration space distribution function describing states of the polymer macromolecules. The projection mapping Π^1_0 is introduced by: $(n_s, n_p, e, \psi_p) \mapsto (n_s, n_p, e, \int d^3\boldsymbol{R}\boldsymbol{R}\boldsymbol{R}\psi_p(\boldsymbol{r}, \boldsymbol{R}))$. Following eqs. (20), (21), (30), the thermodynamic potential Φ^1_0 is introduced as:

$$
\begin{aligned}
\Phi^1_0\left(u_1, \frac{1}{T(\boldsymbol{r},t)}, -\frac{\mu_s(\boldsymbol{r},t)}{T(\boldsymbol{r},t)}, -\frac{\mu_p(\boldsymbol{r},t)}{T(\boldsymbol{r},t)}, \frac{p_p(\boldsymbol{r},t)}{T(\boldsymbol{r},t)}\right) = & \\
& -s_{FH}(u_1) + \frac{1}{V}\int d^3r \frac{1}{T(\boldsymbol{r},t)}(e(\boldsymbol{r}) + \epsilon(u_1, \boldsymbol{r})) \\
& -\frac{1}{V}\int d^3r \frac{\mu_s(\boldsymbol{r},t)}{T(\boldsymbol{r},t)} - \frac{1}{V}\int d^3r \frac{\mu_p(\boldsymbol{r},t)}{T(\boldsymbol{r},t)} \\
& +\frac{1}{V}\int d^3r \frac{(p_p)_{\alpha\beta}(\boldsymbol{r},t)}{T(\boldsymbol{r},t)} \int d^3\boldsymbol{R} R_\alpha R_\beta \psi_p(\boldsymbol{r}, \boldsymbol{R}) \\
& -s_{flow}(u_1),
\end{aligned}
\tag{32}
$$

where p_p denotes the tensor that is the thermodynamic dual of the conformation tensor c_p. The quantities that remain unspecified in eq. (32) are ϵ, s_{FH} and s_{flow}. We shall discuss them one after the other.

The quantity ϵ is the intra and inter molecular energy. It depends on the state of the macromolecules (described by the state variable u_1). Many examples of ϵ can be found in the references listed in Section 5.

The quantity s_{FH} is the Flory-Higgins entropy formulated in the state space M_1. What we require is that Φ^1_0 corresponding to no imposed flow (i.e. $s_{flow} \equiv 0$) and $p_p \equiv 0$ (we shall denote it $(\Phi^1_0)_0$) and evaluated at solutions of $\frac{(\Phi^1_0)_o}{\partial\psi_p} = 0$ becomes the well known Flory-Higgins free energy (see Refs. [50]–[53]). The explicit form of $s_{FH}(u_1)$ will be published elsewhere.

The quantity s_{flow} is the new term in the entropy (see the second term on the right hand side of eq. (30)) arising due to the imposed flow. In the particular case when T = const., we suggest that $s_{flow}\,(u_1) = -\tau(\Pi_0^1(u_1))\frac{\partial u_\alpha}{\partial r_\beta}\sigma_{\alpha\beta}(\Pi_0^1(u_1))$, where $\frac{\partial u_\alpha}{\partial r_\beta}$ is the imposed velocity gradient, τ and σ are respectively the relaxation time and the extra stress tensor introduced in the time evolution in M_0. Explicit expression for these quantities can again be found in the references listed in Section 5.

Following the procedure described in Section 6.1 we, first, look for solutions of $\frac{\partial \Phi_0^1}{\partial u_1} = 0$. Solutions of this equation are denotes $[u_1]_0$. Let us assume, for the sake of simplicity, that μ_s, μ_p and T are constants independent of r. Now, we insert $[u_1]_0$, that is a function of μ_s, μ_p, p_p and $\frac{\partial u}{\partial r}$ (assumed to be a given function), into Φ_0^1 and obtain

$$-\frac{P}{T} = \Phi_0^1\left([u_1]_0, -\frac{\mu_s}{T}, -\frac{\mu_p}{T}, p_p; \frac{\partial u}{\partial r}\right) \tag{33}$$

that we write finally as

$$P = P\left(T, \mu_s, \mu_p, p_p; \frac{\partial u}{\partial r}\right). \tag{34}$$

In addition to eq. (34) we have still the equation governing the time evolution of c_p (i.e. eq. (25)) that we shall write formally as

$$(c_p)_t = C_p\left(u_0; \frac{\partial u}{\partial r}\right). \tag{35}$$

We note that eq. (35) can be obviously transformed into equation governing the time evolution of the thermodynamic dual p_p of c_p. Equation (34) is now the fundamental thermodynamic relation of polymer solutions submitted to a flow (specified by $\frac{\partial u}{\partial r}$) that replaces the fundamental thermodynamic relation

$$P = P(T, \mu_s, \mu_p) \tag{36}$$

of polymer solutions subjected to no external flow.

Even at this state, i.e. before the quantities $\epsilon, s_{PH}, s_{flow}$ and C_p (see eqs. (32) and (35)) have been specified, we can draw from eqs. (32)–(36) interesting conclusions.

1. If a polymer solution (a two component system) is subjected to no flow (no driving force) then

   (a) states are described in thermodynamics by $(e, n_p, n_s) \equiv u_{th2} \in M_{th2} \subset \mathbb{R}^3$, their time evolution is governed by

   $$(u_{th2})_t = 0$$

   and

   (b) $s = s(e, n_p, n_s)$ or eq. (36) is the fundamental thermodynamic relation.

   If the flow is switched on then the above thermodynamic description generalizes as follows:

(a) states are described in thermodynamics by $(e, n_p, n_s e_p) \equiv u_0 \in M_0$, their time evolution is governed by

$$(u_0)_t = C_p(u_0; flow)$$

and

(b) $s = s(e, n_p, n_s, e_p; flow)$ is the fundamental thermodynamic relation.

The generalization of thermodynamics consists thus in, first, generalizing the thermodynamics state space, and, second, in generalizing the fundamental thermodynamic relation. We note that the generalized thermodynamic state space M_0 is in fact the thermodynamic state space used in thermodynamics of elastic solids (in Section 2.2 we have denote it by the symbol $M_{ths} \ni u_{ths}$, e_p is in this context replaced by the elastic deformation tensor that we have denoted by the symbol m). The fundamental thermodynamic relation depends now on the imposed flow and consists of two parts. First, it is the thermodynamic relation of the same type as it arises for example in thermodynamics of elastic bodies (the specific relation between the entropy and the rest of the variables will not, of course, be the same as in the context of elastic bodies). Second, it is an equation governing the time evolution of the thermodynamic variables. In the context of unforces systems, this equation reduces to the statement that the thermodynamic state variables do not evolve in time. The physical origin of the fundamental thermodynamic relation in both unforced and forced systems lies in the time evolution on a more microscopic level of description.

2. The fundamental thermodynamic relation $s = s(e, n_p, n_s, e_p; flow)$; $(u_0)_t = C_p(u_0; flow)$ involves kinetic coefficients (like e.g. the viscosity coefficient, normal stress coefficients, etc.). This is because s introduced in M_1 involves the term s_{flow}. The setting introduced above implies thus relations among the flow induced changes of thermodynamic properties (like solubility or specific heat) and temperature, pression, concentration dependence of kinetic coefficients. These relations become now a part of the usual Maxwell relations.

3. It can be easily shown that if the imposed flow is considered as a small parameter then, in the lowest approximation, the type of results presented in Refs. [50]–[53], [60] is recovered. A detailed analysis will be published elsewhere.

7 Concluding remarks

The objective that we have pursued in this lecture is to increase the understanding of dynamics and thermodynamics so that the thermodynamic analysis could be applied to polymeric fluids in conditions that arise in polymer processing operations. Different routes leading to the same or similar objectives have been explored by other authors. We shall end this lecture by making a comment about the relationship between the approach presented in this lecture and the approaches known as irreversible thermodynamics (IT) (see Refs. [61], [62]) and extended irreversible thermodynamics (EIT) (see Refs. [63]–[66]).

While there is no limitation placed on the choice of the state space in the approach presented in this lecture, the choice of the state spaces in IT and EIT (denoted M_{IT} and M_{EIT}) is limited. The state variable $u_{IT} \in M_{IT}$ is either u_{hyd} (see Section 2.2) or

$u_{IT} \equiv (u_1(r), \ldots, u_n(r))$, where $u_i(r)$, $i = 1, \ldots, n$ are fields (i.e. quantities depending on the position vector r). Elements of $u_{EIT} \in M_{EIT}$ are $u_{EIT} \equiv (u_{IT}, \dot{u}_{IT})$, where $\dot{u}_{IT}$ are time derivatives of the fields u_{IT}. From the physical point of view, the extension from u_{IT} to u_{EIT} is the extension that allows to include inertia into the time evolution (recall the extension from the position vector r to $(r, \dot{r})$ in classical mechanics). Instead of $\dot{u}_{IT}$ we can also introduce the fluxes (or a part of the fluxes – e.g. the dissipative part) associated with the field u_{IT}. The particularity of M_{IT} and M_{EIT} limits the applicability of *IT* and *EIT* (for example in the context of polymeric fluids other state spaces as for example M_ψ and M_{kin} are very useful – see Section 5) but simplifies the problem of specifying the time evolution equations. If $M_{IT} \equiv M_{hyd}$ then Φ_{th1}^{hyd} is introduced usually be using the local equilibrium hypothesis (see Section 3.2) and the time evolution equations by requiring that the inequality $(\Phi_{th1}^{hyd})_1 \leq 0$ holds and that the linearized time evolution equations have the structure of the linear Onsager-Casimir equation (10). In *EIT*, the equations governing the time evolution of u_{IT} are known (since the extra state variables $\dot{u}_{IT}$ have been chosen to be in a one-to-one relation with the right hand side of the time evolution equations for u_{IT}). The equations governing the time evolution of the extra state variables $\dot{u}_{IT}$ are chosen so that $(\Phi_{IT}^{EIT})_t \leq 0$. The approach developed in *EIT* represents thus a special case of the approach that we have presented in Section 6. It has been shown in some particular case (see Refs. [36], [67]) that the nonlinear Onsager-Casimir time evolution equation (26) is a nonlinear extention of the time evolution equations introduced in *EIT*.

The approach presented in this lecture can be thus regarded as an attempt to extend *EIT* to general state spaces and to fully nonlinear time evolution equations. The extension enlarges the domain of applicability of the theory and illuminates the underlying physics (e.g. the assumption that *EIT* can also be applied to driven systems is shown to be equivalent to the assumption that even if the systems under consideration are under the influence of an external force the time evolution in M_{EIT} approach as $t \to \infty$ the time evolution in M_{IT}).

Acknowledgement

The author acknowledges the financial support received from the Natural Sciences and Engineering Research Council of Canada and from the Province of Québec. The author has enjoyed the hospitality of the Laboratoire de modélisation en mécanique, Université Paris VI during the summer of 1990 where a part of this lecture was written. Discussions with David Jou, Daniel Lhuillier and Gérard Maugin are gratefully acknowledged.

References

1. P. Liberman and C.-M. Marle, *Sympletic Geometry and Analytical Mechanics*, D. Reided Publ. Comp. (1987).

2. V.I. Arnold, *Les méthodes mathématiques de la mécanique classique*, Edition Mir, Moscou (1976).

3. H.B. Callen, *Thermodynamics*, Wiley, N.Y. (1963).

4. J.H. Weiner, *Statistical Mechanics of Elasticity*, J. Wiley (1983).

5. L. Boltzmann, in *Wissenchaftlichen Abhandlungen von Ludwig Boltzmann* Vol. 2, N.Y. Chelsea (1968).

6. J.G. Kirkwood, in *Documents in Modern Physics*, edited by P.L. Auer, Gordon and Breach, N.Y. (1967).

7. R.B. Bird, O. Hassager, R.C. Armstrong and C.F. Curtiss, *Dynamics of Polymeric Fluids*, Vol. 2, Wiley, N.Y. (1987).

8. M. Doi and S.F. Ewards, *The Theory of Polymer Dynamics*, Oxford Calderon Press (1986).

9. J.W. Gibbs, *Collected Works*, Longmans, Green Comp. N.Y. (1978).

10. V.I. Arnold, Lecture given at the Gibbs Symposium, Yale Univ. May (1989).

11. J.W. Gibbs, *Elementary Principles of Statistical Mechanics*, Dover (1960).

12. T. Morita and K.Hiroike, Prog. Theor. Phys. *25*, 537 (1964).

13. C. de Dominics, J. Math. Phys. *3* 983 (1962), C. deDominics and P. Martin, J. Math. Phys *15*, 14 (1964).

14. M.S. Green, in *Cargèse Lectures in Theoretical Physics*, edited G. B. Jancovici, Gordon and Breach (1966).

15. R.D. Levin and M. Tribus, editors, *Maximum Entropy Formalism*, MIT Cambridge (1979).

16. R. Hermann, *Geometry, Physics and Systems*, Marcel Dekker N.Y. (1973).

17. F. Weinhold, J. Chem. Phys. *63*, 2479 (1975), *65*,559 (1976).

18. P. Salamon, E. Ihring and R.S. Berry, J. Math. Phys. *24*, 2515 (1983).

19. M. Grmela, in *Hamiltonian Systems, Transformation Groups and Spectral Transform Methods*, eds. J. Harnad and J.E. Marsden, Les Publications CRM, Université de Montréal(1990).

20. L. Onsager, Phys. Rev. *37*, 405 (1931), *38*, 2265 (1931).

21. H.G. Casimir, Rev. Mod. Phys. *17*, 343 (1945).

22. J.J. Moreau, C.R. Acad. Sci. Paris *271*, 608 (1970).

23. D.G.B. Edelen, Int. J. Eng. Sci. *10*, 481 (1972).

24. N.G. van Kampen, Physica A *67*, 1 (1973).

25. P.J. Morrison and J.M. Greene, Phys. Rev. Lett. *45*, 790 (1980).

26. A. Clebsh, J. Reine Angew. Math *56*, 1 (1895).

27. D.D. Holm, J.E. Marsden, T. Ratiu and A. Weinstein, Phys. Reports, *123*, 1 (1985).

28. M. Grmela, Contemp. Math. *28*, 125.

29. A.N. Kaufman, Phys. Lett. A. *100*, 419 (1984).

30. P.J. Morrison, Phys. Lett. A *100*, 423 (1984), Physica D, *18*, 410 (1986).

31. M. Grmela, Physica D, *21*, 179 (1986).

32. M. Grmela, Phys. Lett. A *130*, 81 (1988).

33. M. Grmela, J. Phys. A: Math. & Gen. *22*, 4375 (1989).

34. B.J. Edwards, A.N. Beris and M. Grmela, J. Non Newtonian Fluid Mech. *35*, 51 (1990).

35. B.J. Edwards, A.N. Beris and M. Grmela, Mol. Cryst. Liquid Cryst. (to appear). B.J. Edwards, A.N. Beris, M. Grmela and R. Larson, J. NonNewtonian Fluid Mech. *36*, 243 (1990).

36. M. Grmela and G. Lebon, J. Phys. A: Math & Gen. *23*, 3341 (1990).

37. M. Grmela, in *Polymer Rheology and Processing*, eds A.A. Collyer and L.A. Utracki, Elsevier (to appear).

38. R. Salmon, Ann. Rev. Fluid Mech. *20*, 225 (1988).

39. K. Yano and S. Ishilara, *Tangent and Cotangent Bundles*, Marcel Dekker (1973).

40. V.I. Arnold, Dokl. Acad. Nauk Math. *165*, 773 (1965).

41. M. Grmela, J. Rheol. *33*, 207 (1989).

42. A. Ait-Kadi, P.J. Carreau and M. Grmela, Rheol. Acta *27*, 241 (1988).

43. M. Grmela and Chhon Ly, Phys Letters A. *120*, 281 (1987); Spatial Nonuniformities in Lyotropic Liquid Crystals, Prepint, École Polytechnique de Montreal, January (1988).

44. P.J. Carreau and M. Grmela, Contribution in this Proceedings.

45. T. Aubry, P. Navard and M. Grmela, Rheol. Acta (submitted).

46. A. Ajji, P.J. Carreau, M. Grmela and H.P. Schreiber, J. Rheol. *33*, 401 (1989).

47. J.L. Ericksen, Trans. Soc. Rheol. *5*, 23 (1961).

48. F.M. Leslie, Proc. R. Soc.London Ser. A *307*, 359 (1968).

49. J.E. Mardsen and A. Weinstein, Physica D *4*, 394 (1982).

50. A. Silberger and W.Kuhn, J. Polym. Sci. *13*, 21 (1954).

51. C. Rangel-Nafaile, A.B. Metzner and K.F. Wissburn, Macromol. *17*, 1187 (1984).

52. B.A. Wolf, Macromol. *17*, 615 (1984).

53. D. Jou, *Thermodynamics under Flow* (Critical review), École Polytechnique de Montreal, July 1990.

54. M. Grmela, J. Chem. Phys. *85*, 5689 (1986).

55. M. Grmela, in *Trends in Applications of Mathematics to Mechanics*, eds. J.F. Besseling and W. Eckhaus, Springer (1988) (pp. 329-342).

56. W. Muschik, J. Noneq. Thermod. *4*, 277 (1979).

57. S. Chapman and J.G. Cowling, *Mathematical Theory of Non-Uniform Gases*, Cambridge (1952).

58. P.H. van Wiechen and H. C. Booij, J. Eng.Math. *5*, 89 (1971).

59. A.S. Lodge and Y.J. Wu, Rheol. Acta *10*, 539 (1971).

60. G. Marrucci, Trans. Soc. Rheol. *16*, 321 (1972).

61. I. Prigogine, *Étude thermodynamique des phénomènes irréversible*, Deover, France (1974).

62. S.R. de Groot and P. Mazur, *Non Equilibrium Thermodynamics*, Amsterdam: North Holland, (1962).

63. I. Müller, *Thermodynamics*, Pitman, London (1985).

64. D. Jou, J. Casas-Vazquez and G. Lebon, Rep. Prog. Phys. *51*, 1105 (1988).

65. L.S. Garcia-Colin, Rev. Mex. Fisica *34*, 3244 (1989).

66. R.E. Nettleton and E.S. Freidkin, Physica A, *158*, 672 (1989).

67. M. Grmela and D.Jou, J.Phys. A: Math & General (submitted).

CONFORMATION TENSOR RHEOLOGICAL MODELS

Pierre J. Carreau and Miroslav Grmela
Centre de recherche appliquée sur les polymères (CRASP)
École Polytechnique de Montréal
C.P. 6079, Succc. A
Montréal (Québec), Canada, H3C 3A7

1. INTRODUCTION

Most polymeric systems such as high polymer weight melts, blends, composites and solutions are known to exhibit complex rheological behaviour. Constitutive equations or rheological models are needed for understanding and clarifying the strange flow phenomena encountered in polymer processing: extrudate swell, vortices, instabilities, etc.. Parameters contained in rheological models can be advantageously used to correlate data and obtain master curves incorporating such effects as molecular weight, molecular weight distribution, polymer concentration, chain flexibility, solvent properties, etc. Finally, a ultimate objective in developing constitutive equations is to obtain a correct expression for the stress tensor in terms of the thermo-mechanical history of any given polymeric system.

Three basic approaches can be followed in writing down a constitutive relation. One can use the principles of continuum mechanics to propose admissible forms and assess these forms with appropriate sets of experimental data. A more promising and rewarding route is, however, through the so-called molecular theories. These theories should lead to a number of meaningful parameters that can be used not only for correlating data but for extrapolation and formulating new products. Molecular theories leading to constitutive equations applicable to large deformation flows can be subdivided into four main categories: i) the phase space and configuration space kinetic theories such as those proposed by Kirkwood [1] and refined and extended by Bird and co-workers [2], ii) the network theories derived for polymeric liquids from solid rubber theories by Lodge [3] and by Yamamoto [4] and extended by Carreau [5], Macdonald [6] as well as many others, iii) reptation theories, based on the original concept of Edwards [7] or de Gennes [8], proposed by Doi and Edwards [9], extended to large deformation flows by Marrucci [10] and put into the phase space kinetic framework by Curtiss and Bird [11] and iv) the rheological models based on the conformation tensor, as those proposed by Han [12] and Giesekus [13]. Useful reviews of the first three categories of molecular theories have been presented by Bird [14] and a review by Carreau and Grmela [15] includes the conformation tensor models. The third route, combining the previous ones, has been proposed by Grmela [16] and by Grmela and Carreau [17]. It improves the first route by allowing to use molecular state variables as for example distribution functions and the second route by guaranteeing the compatibility of the governing equations of the model with thermodynamics and by providing a general formula for the extra stress tensor.

This lecture summarizes our work on conformation tensor models based on the third route. Ait Kadi et al. [18] introduced a FENE-Charged potential to account for the "coil-stretch" transformation of macromolecules and explain rheological data obtained for polyacrylamide solutions in aqueous solvents containing various quantities of salt. The case of rigid polymeric chains has been discussed by Grmela and Carreau [17] and Grmela [16] and the modification needed to account for the semi-flexibility of chains has been introduced by Grmela and Chhon Ly [19], and discussed by Carreau et al. [20]. Ajji et al. [21] used two conformation tensors to describe polymer melts as a blend of free macromolecular chains and entangled chains in a network. With that model, they could account for changes in the entanglement structure due to a controlled solution precipitation treatment.

Trying to describe mathematically the time evolution of the complex fluids we have to, first, choose the quantity characterizing the internal structure (we shall call it an internal state variable) and, secondly introduce equations that govern the time evolution of the hydrodynamic fields and the chosen internal state variable. The choice of the internal variables is always a compromise between completeness (the ability to describe molecular details) and simplicity (the easiness with which the governing equations are solved). Since the focus of this lecture is on solutions, we shall emphasize the latter requirement and choose the conformation tensor $\underline{\underline{c}}$ as the internal state variable [12,13]. The tensor $\underline{\underline{c}}$ is assumed to be symmetric and positive definite. From the physical point of view, we can regard it for example as a molecular deformation tensor. Following Han [12] and Giesekus

[13], the conformation tensor can be viewed in reference to a distribution function $\psi(\underline{r},t)$ as the second moment of the end-to-end vector $\underline{R}$ (see Figure 1), i.e.

$$\underline{\underline{c}}(\underline{r},t) = \langle \underline{R}\,\underline{R} \rangle = \int \psi(\underline{r},\underline{R},t)\ \underline{R}\,\underline{R}\ d\,r \tag{1}$$

Here it represents the average conformation of coil structure under any flow conditions. Other interpretations will be introduced in Section 3.

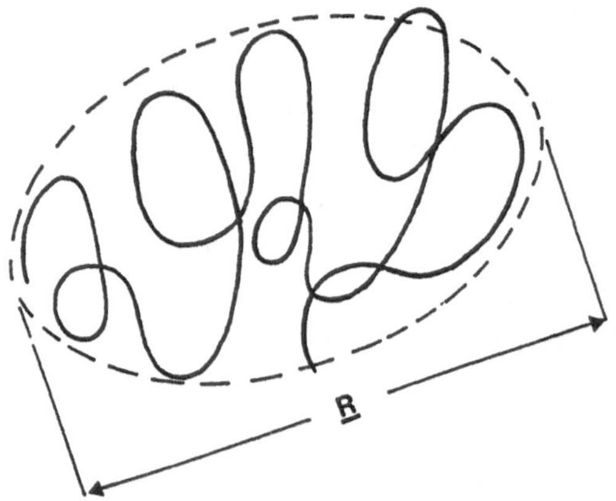

Figure 1 Polymeric chain in a coil conformation

We shall assume, however, that the internal structure of the fluids that we shall discuss is spatially homogeneous (i.e. $\underline{\underline{c}}$ is independent of the position vector $\underline{r}$) and that the fluids are isothermal and incompressible. This means that the only state variables that we shall consider are the velocity field $\underline{v}(\underline{r},t)$ and the conformation tensor $\underline{\underline{c}}(t)$.

In formulating the time evolution equations we are led by the following requirements:

(i) The governing equations should have a clear physical meaning. This means that we should be able to express clearly the molecular nature of the fluids in the governing equations and thus consider the governing equations as a bridge between the molecular properties and the thermodynamic and flow properties that are observed in thermodynamic and rheological measurements. What we want to avoid in particular is the introduction of physically unjustified approximations, like for example the closure approximations, that are necessary if the governing equations are searched as a reduced (simplified) form of equations that have been formulated originally with the internal state variables that can describe more molecular details than the chosen internal state variable.

(ii) We naturally require that solutions of the governing equations agree with results of observations. Let us consider first a general observation on which equilibrium thermodynamics is based (i.e. isolated fluids reach, after some time, a state at which results of thermodynamic measurements do not change in time and the behaviour of the fluids is well described by equilibrium thermodynamics). It has been shown [16,22] that the agreement with this observation is guaranteed if the governing equations possess some general structure. The structure plays moreover the role of a pivotal point about which the process of expressing the physical (molecular) nature of the fluids in the governing equations can be organized.

In Section 2, we introduce a general form of the governing equations for the state variables $\underline{v}(\underline{r},t)$, $\underline{c}(t)$. Solutions of the equations and their comparison with rheological observations are discussed in Section 3.

1.1 Material functions

As we are mostly concerned here with rheological behaviour of polymers in relation to processing, we introduce only the rheological or material functions commonly used to describe the non-linear behaviour. The steady and transient simple shear flow is defined by the following velocity profile:

$$v_1(x_2,t) = \dot{\gamma}(t)x_2 \tag{2}$$

$$v_2 = v_3 = 0$$

where $\dot{\gamma}(t) = (\partial v_1/\partial x_2)$ = shear rate.

For <u>steady shear flow</u>, $\dot{\gamma}$ is constant and the material functions are

$$\eta(\dot{\gamma}) = \sigma_{12}/\dot{\gamma} \tag{3}$$

$$\psi_1(\dot{\gamma}) = (\sigma_{11} - \sigma_{22})/\dot{\gamma}^2 \tag{4}$$

$$\psi_2(\dot{\gamma}) = (\sigma_{22} - \sigma_{33})/\dot{\gamma}^2 \tag{5}$$

For <u>stress relaxation</u> after cessation of steady simple shear

$$\dot{\gamma}(t) = \dot{\gamma}_o\ [1 - h(t)] \tag{6}$$

where h(t) is the unit step function: h(t) = 0 for t < 0 and h(t) = 1 for t > 0. The constant $\dot{\gamma}_o$ is the initial constant shear rate. The shear stress and normal stress relaxation functions are introduced by

$$\eta^-(t;\dot{\gamma}_o) = \sigma_{12}/\dot{\gamma}_o \tag{7}$$

$$\psi_1^-(t;\dot{\gamma}_o) = (\sigma_{11} - \sigma_{22})/\dot{\gamma}_o^2 \tag{8}$$

For <u>stress growth</u> after onset of steady simple shear,

$$\dot{\gamma}(t) = \dot{\gamma}_\infty\ h(t) \tag{9}$$

$$\text{and}\quad \eta^+(t;\dot{\gamma}_\infty) = \sigma_{12}/\dot{\gamma}_\infty, \tag{10}$$

$$\psi_1^+(t;\dot{\gamma}_\infty) = (\sigma_{11} - \sigma_{22})/\dot{\gamma}_\infty^2 \tag{11}$$

where $\dot{\gamma}_\infty$ is the constant applied shear rate.

In the case of uniaxial <u>elongation</u> at constant volume, the flow is non-viscometric and the velocity profile is given by

$$v_1(x_1,t) = \dot{\varepsilon} x_1$$

$$v_2(x_2,t) = -\frac{\dot{\varepsilon}}{2}x_2,\quad v_3(x_3,t) = -\frac{\dot{\varepsilon}}{2}x_3 \tag{12}$$

where $\dot{\epsilon}$ is the elongational rate (constant for steady state). The elongation viscosity is defined by

$$\eta_e = \frac{\sigma_{11} - \sigma_{22}}{\dot{\epsilon}} \tag{13}$$

2. GOVERNING EQUATIONS

Following [16,22], the equations that govern the time evolution of the velocity field $\underline{v}$ $(\underline{r},t)$ and the conformation tensor $\underline{\underline{c}}(t)$ are:

$$\frac{\partial \underline{v}}{\partial t} = - \underline{v} \cdot \nabla \underline{v} - \nabla p + \nabla \cdot \underline{\underline{\sigma}} \tag{14}$$

$$\nabla \cdot \underline{v} = 0 \tag{15}$$

$$\frac{D\underline{\underline{c}}(t)}{Dt} = (\nabla \underline{v}^{+} \cdot \underline{\underline{c}}) + (\underline{\underline{c}} \cdot \nabla \underline{v}) - \underline{\underline{\Lambda}} (\underline{\underline{c}}) \cdot \underline{\underline{c}} \cdot \frac{dA(\underline{\underline{c}})}{d\underline{\underline{c}}} \tag{16}$$

where ∇ denote the spatial gradient, p is the scalar pressure (determined by the incompressibility condition (15)), $\underline{\underline{\sigma}}$ is the extra stress tensor, $\underline{\underline{\Lambda}}$ is the mobility tensor (specified below), A is the free energy (also specified below). In order that eqs. (14)-(16) be compatible with equilibrium thermodynamics [16,22], $\underline{\underline{\Lambda}}$ is such that

$$\underline{\underline{\Lambda}} \cdot \underline{\underline{c}} \text{ is positive definite} \tag{17}$$

and

$$\underline{\underline{\sigma}} = -2n \, \underline{\underline{c}} \cdot \frac{dA(\underline{\underline{c}})}{d\underline{\underline{c}}} \tag{18}$$

where n is the number density of the macromolecules. Equations (17) and (18) guarantee that eqs. (14)-(16) represent a particular realization of the general nonlinear Onsager-Casimir time evolution equation [16,22]. This means in particular that eqs. (14)-(16) are compatible with equilibrium thermodynamics. Now we shall discuss the quantities A and $\underline{\underline{\Lambda}}$ arising in eq. (16).

The free energy A is given by

$$A(\underline{\underline{c}}) = E(\underline{\underline{c}}) - T\, S(\underline{\underline{c}}) \tag{19}$$

where E represents the inter and intra molecular energy, T is the temperature (assumed to be a constant) and S is the entropy. Led by equilibrium statistical mechanics, we can express our insight into the molecular nature of the polymeric fluid under consideration in the quantities E ($\underline{c}$) and S ($\underline{c}$). Several examples will be discussed in the next section.

The mobility tensor $\underline{\Lambda}$ characterizes the frictional forces. The choice of this tensor is determined by phenomenological considerations (see the next section) and by the condition (17).

It we consider E, S and $\underline{\Lambda}$ as undetermined quantities (parameters), equations (14) to (19) represent a large family of rheological equations. It is important to note that the compatibility with equilibrium thermodynamics of all rheological equations in the family is guaranteed and that the quantities E, S and $\underline{\Lambda}$ have a clear physical meaning. This then implies that we know how these quantities reflect the molecular nature of the polymeric fluids and we can thus suggest their form for a given polymeric fluid. Many well known rheological models lie inside the family (14) to (19). The relation between the models introduced previously and the choice of E, S and $\underline{\Lambda}$ is discussed in detail in Ref. [23].

3. PREDICTIONS AND COMPARISON WITH EXPERIMENTS

3.1 Rheological Invariants

The question that we can ask is what are the predictions (i.e. properties of solutions of eqs. (14) to (19)) that are common for all the family (i.e. that are independent of the choice of the quantities E, S and $\underline{\Lambda}$)? We shall call these properties rheological invariants of the family of the rheological models (14) to (19). It is obviously very interesting to find the invariants since if we find that any invariant does not agree with experimental data, then we have to reject the whole family.

We know already one such rheological invariant, namely the properties of solutions that imply the compatibility with equilibrium thermodynamics. We are, however, interested in particular in flow properties observed in rheological measurements. We shall look for such properties in a somewhat smaller subfamily. The subfamily that we shall discussed first will be called a family of isotropic rheological models [17]. The choice of E, S, $\underline{\Lambda}$ is restricted by:

$$S(\underline{c}) = \frac{1}{2} k_B \ln \det \underline{c} \tag{20}$$

$$E(\underline{c}) = E(c) \tag{21}$$

$$\underline{\Lambda}(\underline{c}) = \Lambda(c)\underline{\delta} \tag{22}$$

where $c = \mathrm{tr}\,\underline{c}$ (trace of $\underline{c}$), k_B is the Boltzmann constant, $\underline{\delta}$ is the unit tensor, E(c) and $\Lambda(c)$ remain undetermined. The entropy (20) is the entropy of the noninteracting macromolecules (recall that the entropy for an ideal gas is proportional to ln(volume) and that $(\det \underline{c})^{\frac{1}{2}}$ is proportional to the volume occupied by a molecule). The subfamily of isotropic rheological models is thus parametrized by two scalar functions E and Λ of $c = \mathrm{tr}\,\underline{c}$. Particular examples of models lying in this subfamily are the Maxwell upper convected model ($E(c)$ = const. X $\mathrm{tr}\,\underline{c}$, $\Lambda(c)$ = const.) and the FENE-P model [24] ($E(c)$ = const. X $1/(1-\mathrm{tr}\,\underline{c}/R_o)$, R_o is constant).

The rheological invariants of the subfamily of isotropic rheological models are the following:

1. $\{\sigma_{33}(\dot{\gamma})-\sigma_{22}(\dot{\gamma})\}_{shear} = 0$ (23)

2. $n\, k_B\, T\, \{\sigma_{22}(\dot{\gamma})-\sigma_{11}(\dot{\gamma})\}_{shear} = 2\{\sigma_{12}(\dot{\gamma})\}^2_{shear}$ (24)

3. $\{\sigma_{12}(\dot{\gamma})\}_{shear}$ is related to $\{\sigma_{11}(\dot{\epsilon})-\sigma_{22}(\dot{\epsilon})\}_{elong}$ by

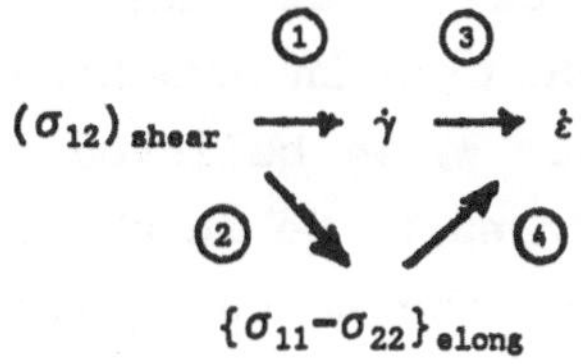

where 1,2,3 are transformations specified as follows:

1 is known if $(\sigma_{12})_{shear}$ is known as a function of $\dot{\gamma}$;

2: $(\sigma_{11}-\sigma_{22})_{elong} = - \frac{(\sigma_{12})^2_{shear}}{2n\ k_B\ T} + (\sigma_{12})_{shear} \left(\frac{9}{4} \frac{(\sigma_{12})^2_{shear}}{(n\ k_B\ T)^2} + 3\right)^{1/2}$;

3: $\dot{\varepsilon} = \dot{\gamma} \frac{(\sigma_{12})_{shear}}{(\sigma_{11}-\sigma_{22})_{elong}}$;

4: is known if $(\sigma_{11}-\sigma_{22})_{elong}$ is known as a function of $\dot{\varepsilon}$.

We shall now explain the notation and discuss the above three rheological invariants one after the other. We solve eqs. (14)-(22) for shear flow:

$$(\nabla \underline{v}) = \gamma \begin{bmatrix} 0 & 0 & 0 \\ 1 & 0 & 0 \\ 0 & 0 & 0 \end{bmatrix} \tag{26}$$

and unaxial elongational flow:

$$(\nabla \underline{v}) = \dot{\varepsilon} \begin{bmatrix} 1 & 0 & 0 \\ 0 & -\frac{1}{2} & 0 \\ 0 & 0 & -\frac{1}{2} \end{bmatrix} \tag{27}$$

The extra stress tensor that arises as a solution of eqs. (14)-(22) is denoted as $(\underline{\sigma})_{shear}$ if $\nabla\underline{v}$ is given by (26) and $(\underline{\sigma})_{elong}$ if $\nabla\underline{v}$ is given by (27). The extra stress tensor $\underline{\sigma}$ appearing in eqs. (23) to (25) is a stationary solution of eqs. (14) to (22), (26), or (14) to (22), (27).

The first invariant (eq. (23)) implies that the secondary normal stress difference or ψ_2 is equal to zero. This is also known as the Weissenberg hypothesis. Very few ψ_2 data have been reported in the literature, due mainly to difficulties in measuring with accuracy this material function. Nevertheless, ψ_2 is believed to be small compared to ψ_1 and negative: ψ_2 has been found to be in the range of 10 to 13% of $-\psi_1$ [2,25]. The only way to make the second normal stress difference different from zero (in the context of conformation tensor models) is to introduce a modified generalized time derivative (e.g. a slip is introduced) or a tensorial mobility (see [13]). We note that the use of a slip parameter in eq. (16) less than unity will lead to negative secondary normal stress differences [17,18].

The second invariant (eq. (24)) implies that the primary normal stress coefficient is proportional to the shear viscosity. This is shown in Figure 2 to represent well the behaviour of five different polymer solutions. The slope of the curves representing ψ_1 as a function of η in logarithmic scales equals two for five very different polymeric solutions. The contribution of the solvent to the viscosity has been subtracted, as the governing equations (14) to (22) describe the contribution of the polymeric chains only. We conclude therefore that the isotropic rheological models (21)-(22) are good candidates for describing all five solutions. The choice of the conformational dependence of the mobility Λ and the modulus H will be, of course, different for every one of the five solutions. Eq. (24) is an excellent test of the applicability of the family of models (eqs. (14) to (22)). Further assessment of this relation is in progress.

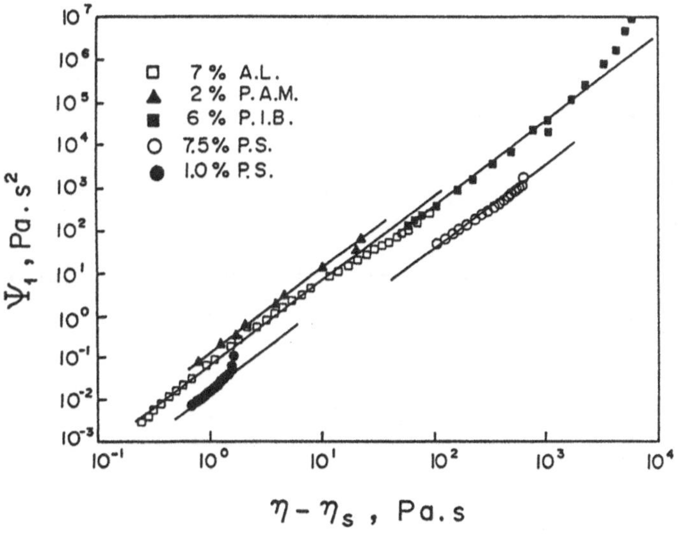

Figure 2 Primary normal stress coefficient as a function of the reduced viscosity for five different polymer solutions. The solutions are described in [17]. The straight lines have a slope of 2.

The third invariant (eq. (25)) could be called a generalized Trouton relation. It derived as follows [26]. If we write eqs. (16), (18) with is $D\underline{c}/Dt = 0$ and S, Λ, E, $\nabla\underline{v}$ given by (20) to (22), (26) we obtain a nonlinear algebraic equation that we write finally as

$$\underline{\underline{F}}_{shear} \{(\underline{\underline{g}})_{shear}, (\underline{\underline{c}})_{shear}, \dot{\gamma}\} = 0 \tag{28}$$

The same equations obtained for $\nabla\underline{v}$ given by (27) will be denoted formally by

$$\underline{\underline{F}}_{elong} \{(\underline{\underline{g}})_{elong}, (\underline{\underline{c}})_{elong}, \dot{\varepsilon}\} = 0 \tag{29}$$

We now look for a transformation T

$$\{(\underline{\underline{g}})_{shear}, (\underline{\underline{c}})_{shear}, \dot{\gamma}\} \overset{T}{\rightarrow} \{(\underline{\underline{g}})_{elong}, (\underline{\underline{c}})_{elong}, \dot{\varepsilon}\} \tag{30}$$

such that $\underline{\underline{F}}_{shear} \{T [(\underline{\underline{g}})_{shear}, (\underline{\underline{c}})_{shear}, \dot{\gamma}]\} = \underline{\underline{F}}_{elong} \{(\underline{\underline{g}})_{elong}, (\underline{\underline{c}})_{elong}, \dot{\varepsilon}\}$. The transformation described in (25) is such a transformation. We note that if we consider $\dot{\gamma} \rightarrow 0$ then (25) implies the Trouton relation:

$$\eta_{eo} = 3\eta_o \tag{31}$$

and if $\dot{\gamma} \rightarrow \infty$

$$m = \frac{2n}{1-n} \tag{32}$$

where m and n are the power-law indices at high deformation rates for η_e and η respectively i.e.

$$\eta_e (\dot{\varepsilon}) \sim | \dot{\varepsilon} |^{m-1}$$

as $\dot{\varepsilon} \rightarrow \infty$

and $\dot{\varepsilon} (\dot{\gamma}) \sim | \dot{\gamma} |^{n-1}$

as $\dot{\gamma} \rightarrow \infty$

We note that for polymeric solutions, the solvent contribution has to be considered ($\eta_{es} = 3\eta_s$).

Unfortunately, there exist very few reliable elongational data at high elongational rates and under steady-state conditions to test relation (32). If we take a value of 0.5 for n (typical value for polyolefine melts), then m is found to be equal to 2. Hence the elongational viscosity is expected to increase with increasing elongational rate. This has been reported for low density polyethylenes [27], but decreasing elongational viscosity with increasing rate has been observed for linear polyethylenes. A value of m smaller than 1 implies a power-law index, n, smaller than 1/3. This, in light of the very few available data, appears to be too restrictive.

Finally, we note that the invariants 2 and 3 refer to relationships between two different (independent) rheological measurements. To check these relationships is very important, more than the verification of isolated measurements. In the following sections, we give three examples of models which describe quite different physics.

3.2 FENE - Charged Macromolecules

Charged polymers known as polyelectrolytes are widely used in various industrial processes encountered in the food, pharmaceutical, paint, pulp and paper industries and are also used for mobility control of fluids in porous media. A rheological model for polyelectrolytes has been proposed by Dunlap and Leal [28] who used a Coulombic potential to account for electrostatic repulsive forces. The concept of isotropic but conformation-dependent friction coefficient [8,29,30] has also been introduced into the setting of the FENE-P model [18] to represent the coil-stretch transition experienced by macromolecules during flow.

As shown in Ait Kadi et al. [18], we used a FENE potential and a Coulombic potential to write the Helmholtz free energy as:

$$A(\underline{\underline{c}}) = - H R_o^2 \ln (1-c/R_o^2) + E c^{-\frac{1}{2}} - \frac{1}{2} k_B T \ln \det \underline{\underline{c}} \qquad (33)$$

where H is the coil (spring) modulus, R_o its maximum extension, E is a constant and the second term of the right-hand-side implies that the electrostatic forces derived from the electrostatic potential U_c decrease as the inverse of the square of the extension (c) of the macromolecule [28]. The electrostatic charges are regarded as being concentrated at the ends of the polymer chain. This is equivalent to assuming that the distance between chain sub-elements is proportional to $\sqrt{c}$. The parameter E is related to the amount of available electrostatic charges. The last term is due to the Brownian motion (entropic contribution). It has been justified in [16,22].

Following de Gennes [8], Hinch [29] and Tanner [30], we use the following relation for the mobility

$$\Lambda(c) = \Lambda_o / (1 + \beta \xi) \tag{34}$$

This relation expresses the physics that as the coil is deformed and stretched under flow, its mobility is reduced. Λ_o and β are two parameters and ξ $(=\sqrt{c/R_o^2})$ is the reduced extension.

For any flow kinematics, the set of governing equations can be solved to obtain material functions in terms of the following four parameters:

$$b = 2HR_o^2 / k_B T, \text{ extensibility parameter} \tag{35}$$

$$e = E / k_B T, \text{ electrostatic parameter} \tag{36}$$

$$\lambda_e = \frac{2\, \xi_e^2\, R_o^2}{3k_B T \Lambda_o} (1 + \beta\, \xi_e), \text{ time constant} \tag{37}$$

and β, is the conformation or friction parameter.

In (37) ξ_e is the equilibrium (no flow) value of ξ, the reduced end-to-end distance or extension of the coil. It is given by the positive root of the following equation [18]:

$$\xi_e^3 + \frac{e}{3+b} \xi_e^2 - \frac{3}{3+b} \xi_e - \frac{e}{3+b} = 0 \tag{38}$$

and λ_e is related to the zero-shear viscosity by

$$\lambda_e = [\frac{\eta_o - \eta_s}{c}] \frac{M}{RT} \tag{39}$$

where c is the polymer concentration, M its molecular weight and R, the gas constant.

Model Predictions

Many examples of model predictions for steady shear and elongational flows have been presented by Aid Kadi et al. [18]. We summarize the effects of the electrostatic and friction parameters. We first note that if e and β are taken to be equal to zero, then the model is identical to the FENE-P model [24]. Figure 3 shows the influence of the parameter e. Part a) of the figure reports the steady shear and elongational viscosity as a function of a dimensionless deformation rate that is the product of λ_e times the square root of the second invariant of the rate-of-deformation tensor (this is equal to $\lambda_e\dot{\gamma}$ for shear and $\sqrt{3}\lambda_e\dot{\epsilon}$ for elongational flow). Part b) of the figure reports the reduced extension as a function of the dimensionless deformation rate.

Model predictions are reported for b = 1000, β = 0, and e ranging from 0 to 1000. As the electrostatic parameter increases, an intermediary region appears between the low rate of deformation zone (constant shear and elongational viscosities) and the high rate of deformation zone (shear-thinning viscosity with a constant power-law index equal to - 2/3 and a constant elongational viscosity). The electrostatic parameter affects the onset of the decrease of viscosity in shear flow and shifts the onset of the increase of the elongational viscosity towards lower values of deformation rate. Moreover, as e increases, the zero-shear viscosity increases and the high-rate-of-deformation elongational viscosity decreases.

Figure 3b shows that the equilibrium extension increases with increasing e. The equilibrium hydrodynamic volume is then larger, due to the repulsive forces acting on the macromolecule chain, but the deformation due to the flow field is considerably restricted and the behaviour is that of a rigid chain. We also note that e has no apparent effect on the transition from the equilibrium conformation to the fully extended conformation in elongational flow. In shear flow large values of the electrostatic parameter result in a longer transition from the equilibrium to the stretched conformation. At the same time the onset of the transition is shifted towards high values of the velocity gradient. The results described above are comparable to those obtained by Dunlap and Leal [28].

3a

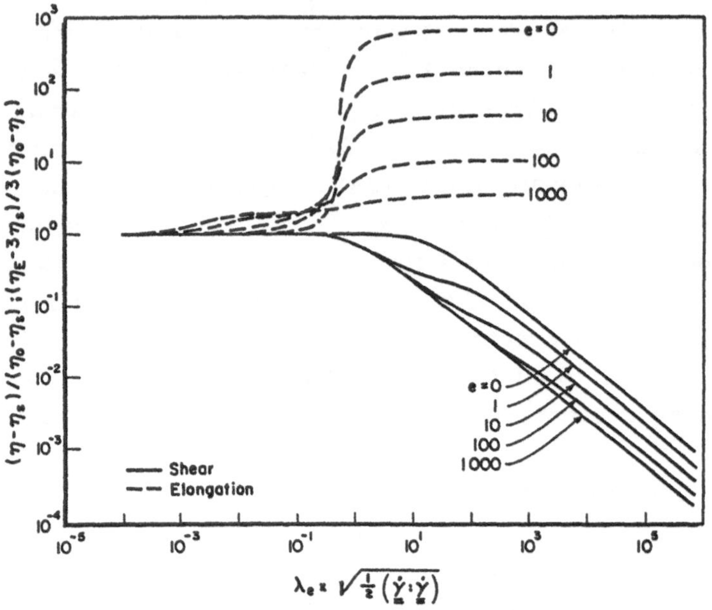

3b

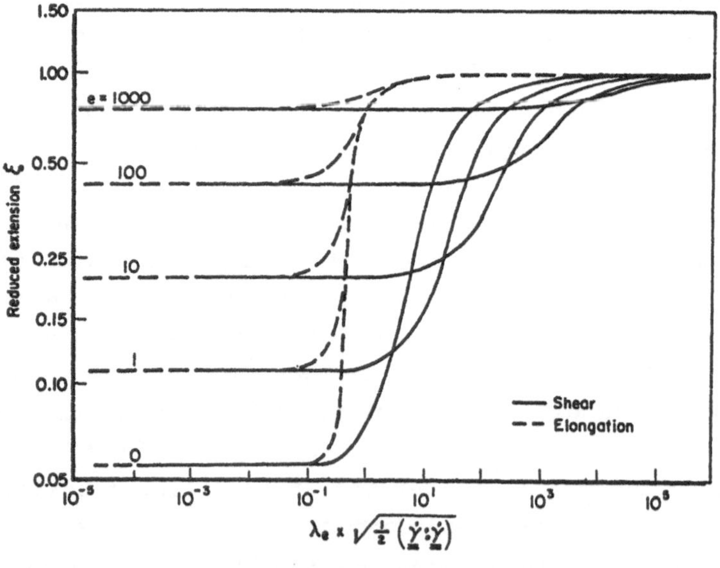

Figure 3 Effect of the electrostatic parameter e on (a) the steady shear and elongation reduced viscosity and (b) on the reduced extension in shear and elongational flows; b = 1000, β = 0.

The influence of the friction parameter β is shown in Figure 4. In shear flow, β has little influence on the onset of non-linear effects. However, the transition from equilibrium to full extension is more rapid as β increases. At the onset of the deformation state, for $\beta > 0$, the shear viscosity starts to increase with increasing rate of deformation (shear-thickening), up to a maximum corresponding to 50% of the full extension. Then the effect of the FENE connector becomes important, and the viscosity starts to decrease, following a power-law model with a constant exponent equal to - 2/3. The amplitude of the shear-thickening phenomenon increases with increasing values of β, but tends towards an asymptotic value for high values of the friction parameter. Tanner [30], using a Dirac-delta function to approximate the configuration distribution function, has obtained identical results. Equivalent results were also obtained by Fuller and Leal [31] and Dunlap and Leal [28] using the "Peterlin pre-averaging approximation".

The effect of β is more pronounced in the case of elongational flow. The onset of the transition from undeformed to extended conformation is dramatically shifted to lower values as β increases. This corresponds to an increase in the elongational viscosity at lower rates of deformation. Moreover, the value of the second elongational viscosity plateau increases with β up to a limiting value reached for $\beta > 100$. The model also predicts a "S-shaped" curve, for both the elongational viscosity and the molecular extension for non-zero values of β. This phenomenon has been predicted by several authors (de Gennes [8], Tanner [30], Fuller and Leal [31], Dunlap and Leal [28] and others.

One objective in developing the model was to describe shear-thickening behaviour observed for polyacrylamide solutions in simple shear experiments. Figure 5 compares the viscosity data of partially hydrolyzed HPAAm (Pusher 700 from Dow Chemicals) solutions in a solvent mixture contained 20 percent (by weight) distilled water and 80 percent glycerol. The solvent also contains 20g/L of added sodium chloride [32]. The shear-thickening behaviour observed for dimensionless shear rates, $\lambda_e\dot{\gamma}$, larger than 10 is fairly well predicted by the model using a friction parameter β equal to 5. We stress than none of these polymer solutions except may-be the 170 ppm are dilute solutions and λ_e is an increasing function of the polymer concentration. Hence, intermolecular interaction and possibly entanglements may play an important role on the observed phenomenon. Further discussion can be found in [18].

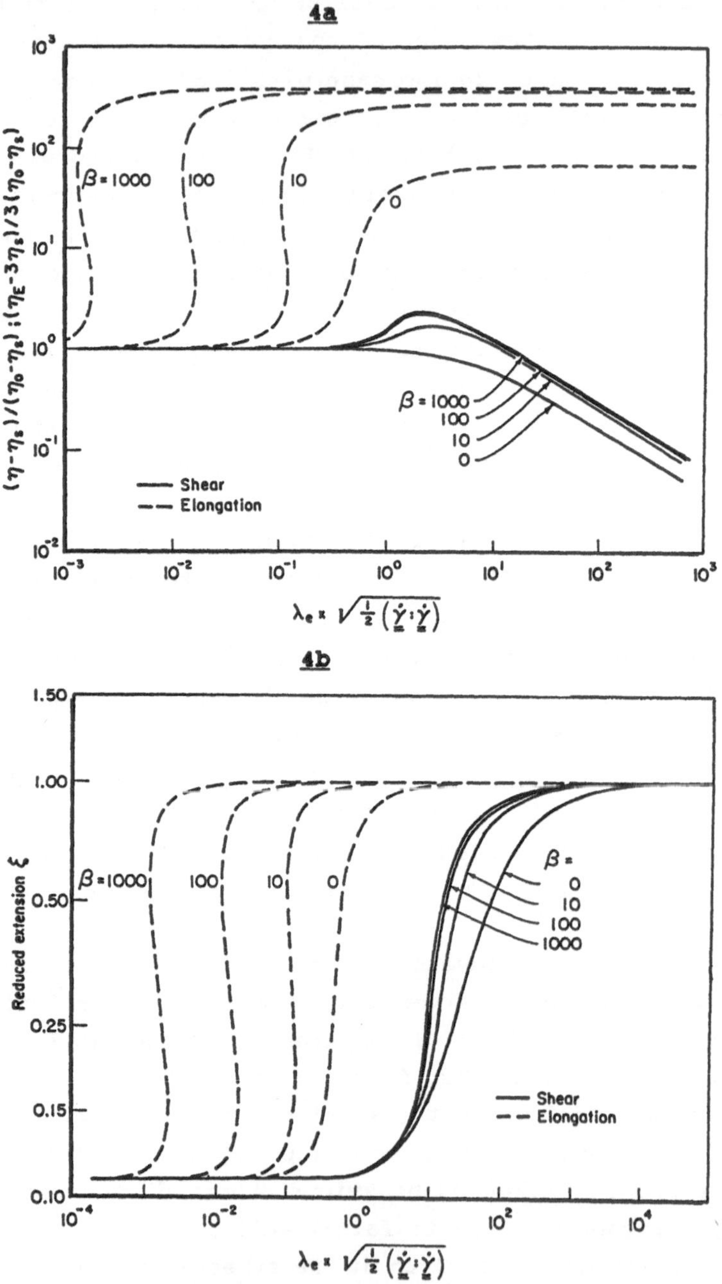

Figure 4 Effect of the friction parameter β on (a) the steady shear and elongation reduced viscosity and (b) on the reduced extension in shear and elongational flows; $b = 100$, $e = 0$.

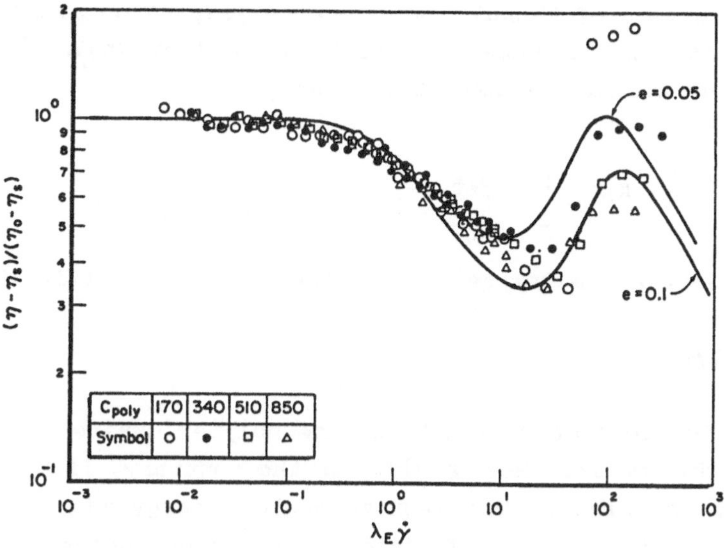

Figure 5 Reduced shear viscosity data for HPAAm Pusher-700 solutions. C_{NaCl} = 20g/L. Polymer concentrations are in ppm. The model parameters are: b = 10^5, β = 5 and e = 0.01.

3.3 Wormlike Macromolecules

In this section, we focus on rod-like or semi-flexible wormlike macromolecules. We consider the polymer chains to be rigid in average, but we allow for some flexibility in the chains. We choose the wormlike model with Kuhn segments of length ℓ, as shown in Figure 6.

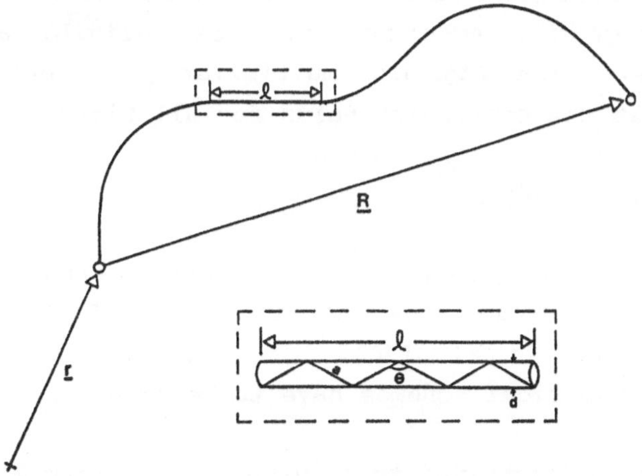

Figure 6 Sketch of a wormlike molecule

The results of Khokhlov and Semenov [33,34] obtained for entropy of chains made of sub-segments vibrating about an angle θ close to 2π can be written within a constant term as:

$$S = \frac{1}{2} k_B T \left[\ln \det \left(\frac{\underline{\underline{c}}}{R_o^2}\right) - b \operatorname{tr} \left(\left(\frac{\underline{\underline{c}}}{R_o^2}\right)\right)^{-1} \right] \tag{40}$$

where b is a parameter related to the chain flexibility that we take as

$$b = L/\ell \tag{41}$$

where L is the contour length of the polymer chains. We note that if $b \ll 1$, the chains are rigid. On the contrary, if $b \gg 1$, the chains are quite flexible. The intramolecular energy will be replaced by a constraint $F(\underline{\underline{c}}) = \operatorname{tr} \underline{\underline{c}} = R_o^2$ (rigid-in-average chains).

Equation (19) for the free energy becomes then:

$$A(t) = \frac{1}{2} k_B T \left[\beta F(\underline{\underline{c}}) - a \ln \det \left(\frac{\underline{\underline{c}}}{R_o^2}\right) + b \operatorname{tr} \left(\left(\frac{\underline{\underline{c}}}{R_o^2}\right)^{-1}\right) \right] \tag{42}$$

In Eq. (42) we have introduced also the parameter a, which weighs the influence of the standard contribution of the Brownian motion with respect to the effect of the chain flexibility. The physical justification is the following. Brownian motion is of importance for small or molecular size particles. For larger particles such as crystallites, fillers, fibers, the contribution of the Brownian motion to the free energy will become more and more negligible as the particles size increases. The Lagrange multiplier β is obtained from the requirement that the constraint holds for all times, i.e.

$$\frac{dF}{dt} = \frac{d}{dt} (c - R_o^2) = 0 \tag{43}$$

To obtain the components for the stress tensor and material functions under various flow situations, we follow the procedure outlined by Carreau et al. [20]. The solutions are in general not analytical and numerical schemes have to be used.

If we restrict ourselves to a constant mobility $\Lambda = \Lambda_o$, then the material functions can be described in terms of three parameters: b as defined by eq. (41) (it plays a similar role as in the FENE-type models); the parameter a which weighs the Brownian motion and a time constant defined by:

$$\lambda = 2R_o^2/k_BT\Lambda_o \tag{44}$$

and it is related to the zero-shear viscosity by:

$$\eta_o = \lim_{\dot{\gamma} \to 0} \sigma_{12}/\dot{\gamma} = 1/3\ nk_BT\lambda \tag{45}$$

On the other hand, the zero-shear primary normal coefficient is expressed by:

$$\psi_{10} = \frac{2nk_BT\lambda^2}{9(a+12b)} = \frac{2\eta_o\lambda}{3(a+12b)} \tag{46}$$

As expected, both contributions to the Brownian motion affects the material's elasticity as well as the shear-thinning properties.

Figure 7 shows steady-shear viscosity master curves for different values of the parameters a and b. It is interesting to note that with increasing chain flexibility (increasing value of b), the onset of shear thinning appears at higher reduced shear rate, $\lambda\dot{\gamma}$. Also, with increasing flexibility, the viscosity becomes less shear-thinning: the slope in the power-law region increases from -2/3 for rigid chains to approximately -1/2 for semi-flexible chains.

We notice that this result for b = 0 is significantly different from that obtained by Bird et al. [2] for rigid dumbbells in the context of the configuration phase space kinetic theory, which does not admit a closed form solution. The power-law exponents for the viscosity and the primary normal stress coefficient were found to be respectively -1/3 and -2/3, i.e. half of the values obtained for the rigid-in-average model. It is not clear to us why such differences in the high-shear rate behaviour are observed. We finally note that this case is identical to the FENE-P model for $b \to 0$, $e = \beta = 0$, discussed in section 3.2.

The primary normal stress coefficient describes similar master curves as shown in Figure 8.

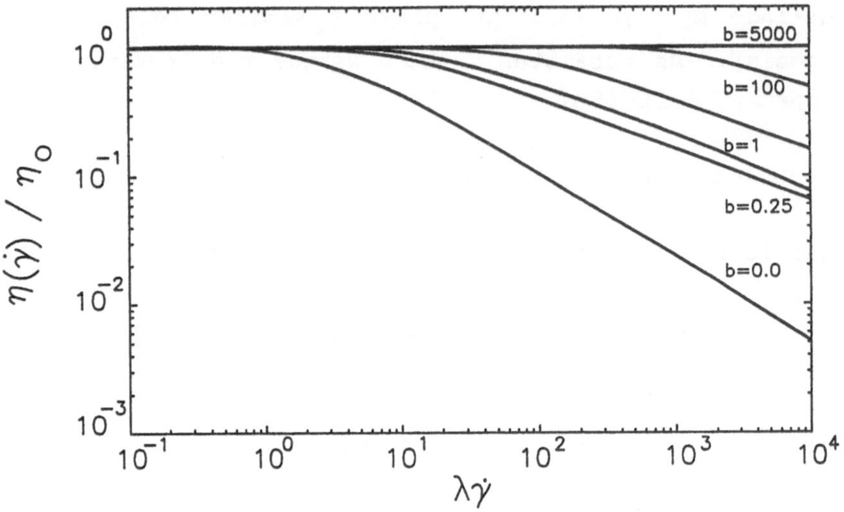

Figure 7 Effect of chain flexibility on the reduced shear viscosity; a = 1.0.

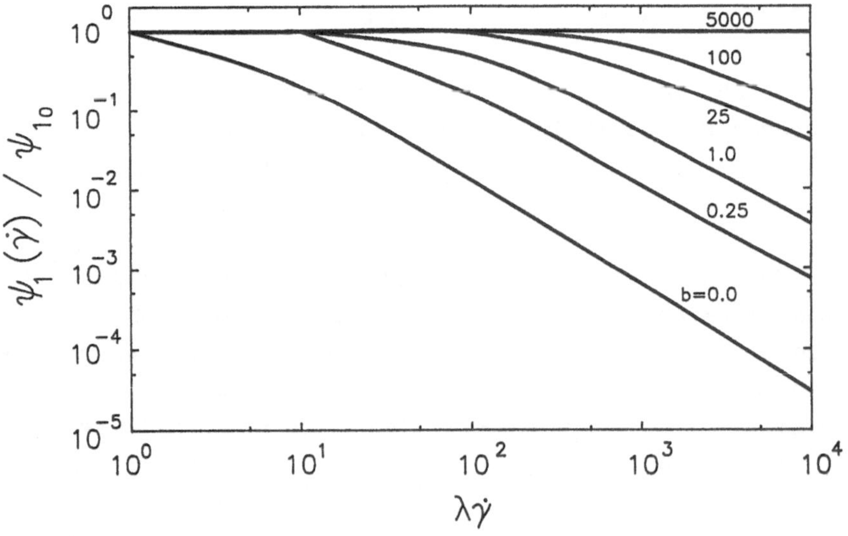

Figure 8 Effect of chain flexibility on the reduced primary normal stress coefficient; a = 1.0.

The results obtained numerically when flexibility is considered appear to verify the relationship (24) between the primary normal stress coefficient and the shear viscosity. The power-law slope for ψ_1 varies from -4/3 for rigid chains to -1 for semi-flexible chains. We notice, however, that the proportionality factor is no longer the same, due to the parameter a affecting the primary normal stress (see eq. (46)).

Figure 9 illustrates the conformation tensor for simple shear flow, using the principal values to generate ellipses. In Figure 9a, the conformation in the case of rigid rods (b=0) is shown to be more oriented in the flow direction as the dimensionless shear rate is increased. Figure 9b shows the influence of flexibility on the orientation of the conformation tensor. As expected the more flexible chains are less oriented by the flow at the same dimensionless flow rate.

The transient behaviour in shear flow and the elongational properties predicted by the model are of considerable interest. Figure 10 compares the model predictions of the shear stress growth and relaxation functions for the rigid and semi-flexible chains. In Figure 10a, we show the influence of the dimensionless rate on the stress growth and stress relaxation functions for rigid rods. With increasing shear rate, the model predicts overshoots in the growth function (η^+) that increase with increasing shear rate and the maximum of the overshoot occurs at shorter time with increasing $\lambda\dot{\gamma}$. This is in agreement with experiments with typical viscoelastic fluids. The relaxation curve, however, is a unique function of the dimensionless time t/λ. This is not supported by experimental observations which show faster relaxation at higher shear rates.

Figure 10b shows how the chain flexibility affects the transient behaviour. With increasing value of b, the behaviour becomes closer to that of a Newtonian fluid, i.e. the shear stress grows more rapidly to the steady-state value but depicts no overshoot; the stress relaxes more rapidly after cessation of steady shear flow. This is in line with the predictions of the onset of shear-thinning at higher shear rates (see Figures 7 and 8).

9a

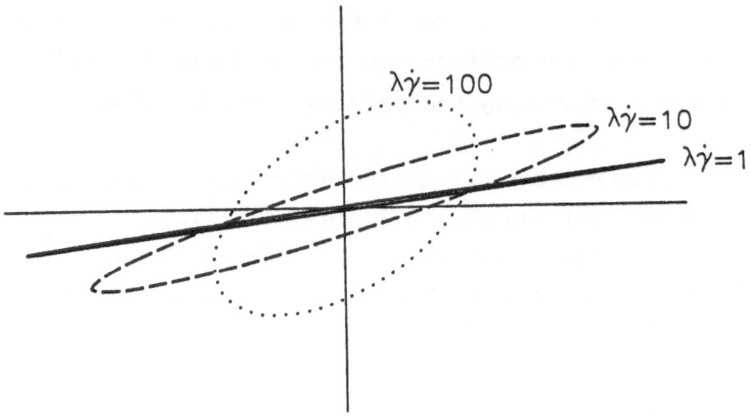

9b

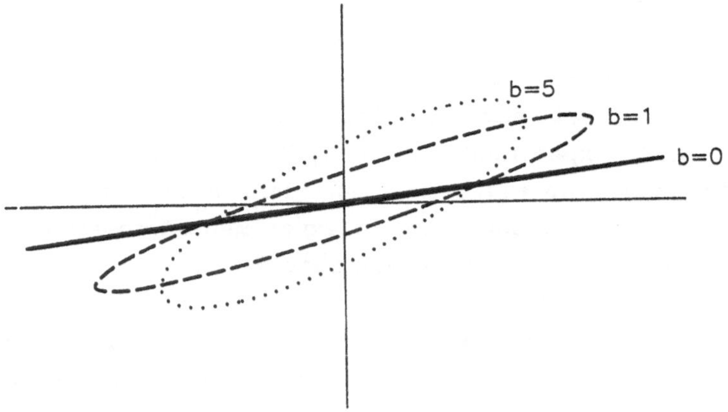

Figure 9 Conformation tensor in simple shear flow
a) Rigid rods a = 1 and b = 0
b) Effect of flexibility, a = 1, $\lambda\dot{\gamma}$ = 100

10a

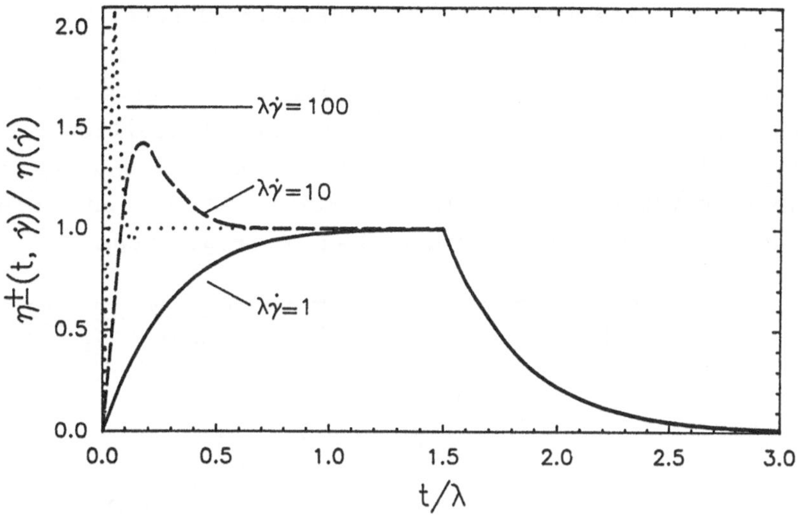

10b

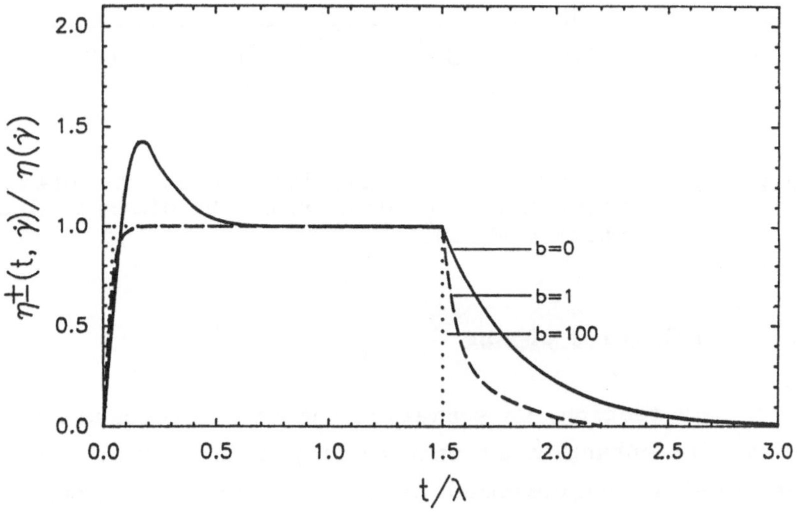

Figure 10 Stress growth and stress relaxation functions as functions of dimensionless time. The parameter a is taken to be 1.0
a) rigid-rod case
b) effect of chain flexibility

As a last example, we show in Figure 11 the effect of chain flexibility on the elongational viscosity. The elongational viscosity increases with the dimensionless elongational rate, as it was predicted for coil type chains (section 3.2), however, the high strain rate plateau is equal to 2 times the low deformation rate elongational viscosity, independently of the chain flexibility. We observe, moreover, that strain-hardening occurs at much smaller dimensionless strain rate for rigid rods (b=0). For b=100, the onset is observed at a value of $\lambda\dot{\epsilon}$ 1000 times larger than for rigid rods.

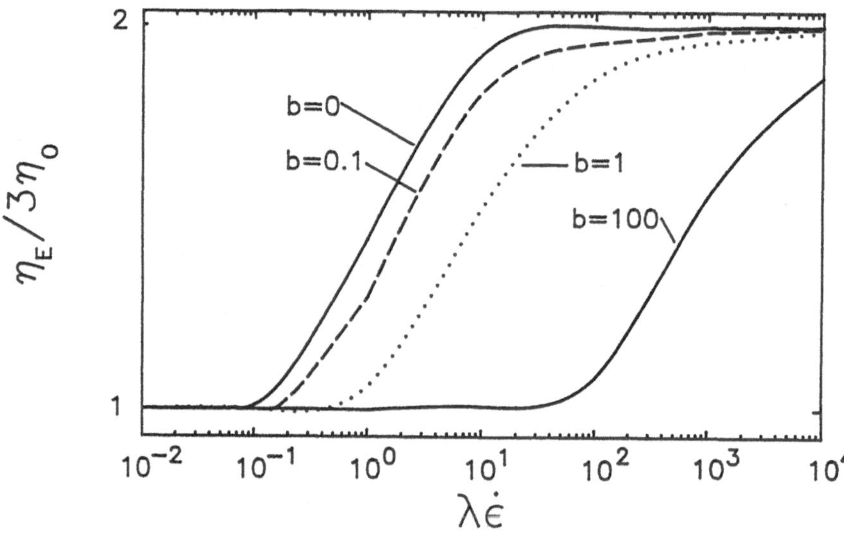

Figure 11 Effect of chain flexibility on the steady elongational viscosity as a function of the dimensionless elongational rate; a = 1.0.

3.4 Entangled Chains

In this section, we summarize the ideas proposed by Ajji et al. [20] to describe entanglement control in polymer melts. The polymer melt is considered to be composed of free chains and chains in an entangled network as illustrated in Figure 12.

The change in the morphological structure and the properties are represented by changes in two conformation tensors and by a change in the relative fraction (m) of chains in the network and free chains. This model can be viewed as an extension of the Jongschaap model [35].

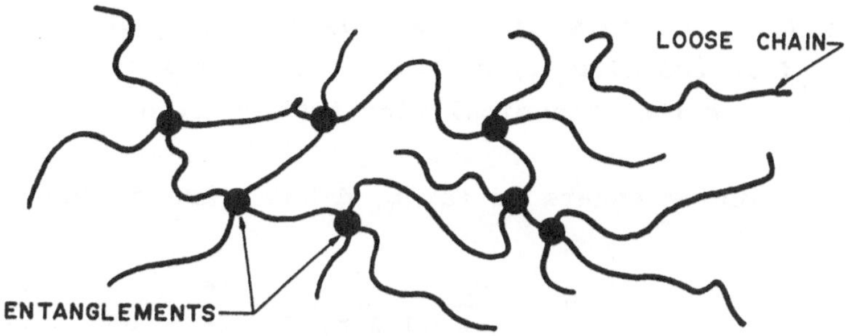

Figure 12 Polymer melt in an entangled network

We use two conformation tensors, $\underline{c}$ to describe the free chains and $\underline{d}$ for the network. The free energy A is taken as

$$A(\underline{c},\underline{d},T) = \underbrace{mA_n(\underline{d},T)}_{\text{network contribution}} + \underbrace{(1-m)A_f(\underline{c},T)}_{\text{free chains contribution}} + \underbrace{m(1-m)A_i(\underline{c},\underline{d},T)}_{\text{interaction contribution}} \tag{47}$$

If m = 0, that is, if the polymer is composed of only free chains, then it follows from eq. (47) that $A = A_f$. We shall assume that the free chains are the FENE chains and thus eq. (33) for E = 0 and using the reduced extension ($\xi = (\text{tr } c/R_o^2)^{\frac{1}{2}}$ becomes:

$$A_f = - H R_o^2 \ln (1-\xi^2) - \tfrac{1}{2} k_B T \ln (\det \underline{c}) \tag{48}$$

If m = 1, that is, if the polymeric medium is composed of only the network then if follows from eq. (47) that $A = A_n$. Following Leonov [36], we take

$$A_n = \alpha(\text{tr } (\underline{d} - \underline{\delta}))^2 + \beta \text{ tr}((\underline{d} - \underline{\delta})^2) \tag{49}$$

where α and β are constants.

The term A_i has the physical meaning of chemical affinity for the entanglement-disentanglement process regarded as a chemical reaction. It is possible to use similar considerations as those that led to A_f and A_n (essentially based on equilibrium statistical mechanics) and suggest A_i. At this stage we prefer, however, another approach. Instead of introducing A_i, we suggest an equation that governs the time evolution of m (the "chemical kinetics" equation).

The evolution equations for $\underline{\underline{c}}$, $\underline{\underline{d}}$ and m are chosen as:

$$\frac{D\underline{\underline{c}}}{Dt} = \nabla\underline{v}^{+} \cdot \underline{\underline{c}} + \underline{\underline{c}} \cdot \nabla\underline{v} - \frac{2}{\lambda_f k_B T(1-m)} \underline{\underline{c}} \cdot \frac{\partial A}{\partial \underline{\underline{c}}} - \frac{2}{k_B T \lambda_i} \{\frac{\partial A}{\partial \underline{\underline{c}}} - \frac{\partial A}{\partial \underline{\underline{d}}}\}_0 \tag{50}$$

$$\frac{D\underline{\underline{d}}}{Dt} = \nabla\underline{v}^{+} \cdot \underline{\underline{d}} + \underline{\underline{d}} \cdot \nabla\underline{v} - \frac{2}{\lambda_f k_B T m} \{\underline{\underline{d}} \cdot \frac{\partial A}{\partial \underline{\underline{d}}}\}_0 - \frac{2}{k_B T \lambda_i} \{\frac{\partial A}{\partial \underline{\underline{c}}} - \frac{\partial A}{\partial \underline{\underline{d}}}\}_0 \tag{51}$$

$$\frac{dm}{dt} = - \frac{1}{k_B T \lambda} \frac{\partial A_i}{\partial m} \tag{52}$$

We shall now explain the meaning of the symbols and the origin of the terms arising in these equations. The operator $\{\ \}_0$ makes the tensor inside the bracket traceless [i.e., $\{\underline{\underline{T}}\}_0 = \underline{\underline{T}} - (\mathrm{tr}\underline{\underline{T}})\underline{\underline{\delta}}/3$ for any tensor $\underline{\underline{T}}$]. The first two terms on the right-hand side of eqs. (50) and (51) together with the substantial time derivative D/Dt form the Oldroyd upper convective time derivative. This means that we consider only affine deformations. The remaining terms on the right-hand side of eqs. (50) and (51) and the term on the right-hand side of eq. (52) are chosen so that the following requirements are met:

i) If m = 0 (i.e., if the polymer is composed of only free chains) then eqs. (50 to 52) reduce to the governing equation (16).

ii) If m = 1 (i.e., if the polymer is composed of the network only) then eqs. (50 to 52) reduce to the governing equations of the Leonov model [36].

iii) Equation (52) is the so-called Marrucci equation [37].

$$\frac{dm}{dt} = \frac{2}{\lambda} (m_{e} - m(1 - k_0|\mathrm{tr}(\underline{\underline{g}})|\mathrm{tr}(\underline{\underline{g}})^{*})) \tag{53}$$

where m plays the role of the structural parameter x_i eq. (19) of

Acierno et al.[37], λ, m_e, and k_0 are parameters. In this general case, the stress tensor is given by [16]

$$\underline{\underline{\sigma}} = -2n_f \left[\underline{\underline{c}} \cdot \frac{\partial A}{\partial \underline{\underline{c}}} + \left\{ \underline{\underline{d}} \cdot \frac{\partial A}{\partial \underline{\underline{d}}} \right\}_0 \right] \tag{54}$$

where n_f is the number density of free chains defined by $n_f k_B = \rho\ R/M$, ρ is the polymer density, and M its molecular weight.

Details on the solution of this set of governing equations can be found in Ajji et al [21]. We recall here the parameters of the model: the characteristic times associated with each phase (λ_f and λ_n), a characteristic time of the interactions (λ_i), a parameter b related to the flexibility of the free chains (identical to b in the FENE model of section 3.2), two parameters α and β related to the connectivity and rigidity of the network, finally, two parameters related to the kinetics of creation of entanglements: m_e, the equilibrium fraction of segments in the network, and k_0 a rate constant.

Figures 13 and 14 compare the model predictions with the steady-state shear viscosity and primary normal stress data for a polypropylene melt.

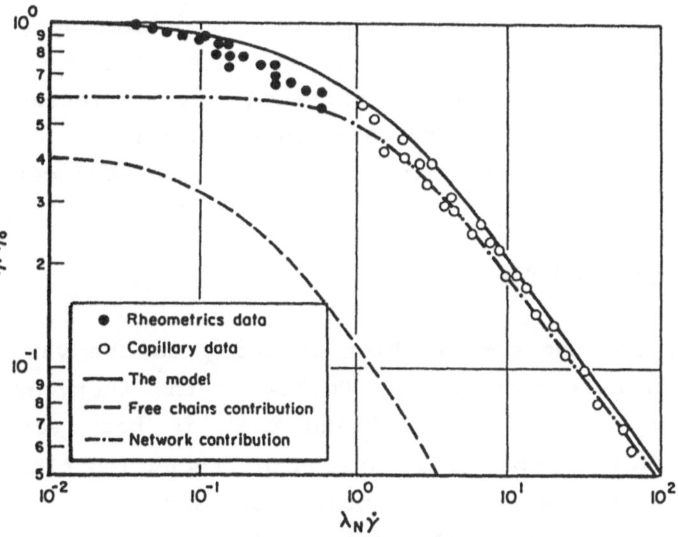

Figure 13 Model predictions and experimental results for the viscosity of PP at 200°C ($M_w = 2.3 \cdot 10^5$, $M_w/M_n = 5.90$); $\eta_0 = 47.7$ kPa.S., $\alpha = 0.4$, $\beta = 0.25$, $\lambda_n = 3.45$s, $\lambda_f = 13{,}7$s, b = 1, $k_o = 0.5\ Pa^{-1/2}$ and $m_e = 0.7$.

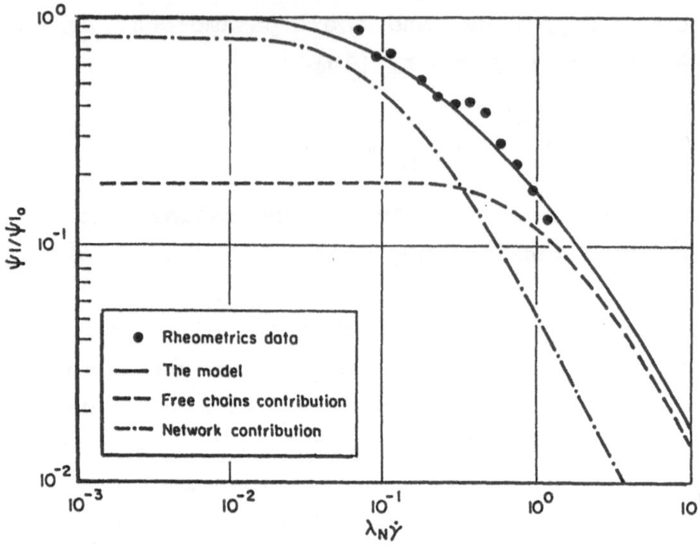

Figure 14 Model predictions and experimental results for the primary normal stress coefficient of PP at 200°C, ψ_{10} = 550 kPa.s^2, the model parameters are given in Fig. 13.

It is interesting to note that the model predicts a much smaller contribution of the free chains to the stresses compared to the network. This is to be expected from such a high molecular weight polymer melt. It is shown in Ajji et al. [21] that this model can describe entanglement modifications obtained by solution precipitation treatments. Simplified version of this model is now being assessed and the results will be reported in a forthcoming publication.

4. CONCLUDING REMARKS

By choosing the conformation tensor $\underline{\underline{c}}$ as the variable describing polymer macromolecules and by following the thermodynamic modelling technique we arrive at a setting that is both simple to use (in particular in numerical calculations of processing flows) and useful in providing a bridge between molecular and flow properties. The quantities (depending on the conformation tensor) that enter the models are the inter and intra molecular energies, the entropy and the mobility tensor. Considerations based on equilibrium statistical mechanics and molecular dynamics can be used to specify these quantities for the polymeric fluids under consideration. Once these quantities are specified the governing equations of the rheological model and the formula for the extra stress tensor arise as a result. There is no need for additional approximations (like for example the various closure approximations). Numerical calculations of processing flows that use some of the models presented above are in progress. We hope to be able to present the results in a near future.

In the chapter, we have presented three examples of rheological models that illustrate how flexible the conformation tensor approach is in the setting of diffusion-convection formulation. Completely different physics can be described and reasonable fits with experimental results can be obtained.

ACKNOWLEDGEMENTS

We wish to thank Mr. T. Gosh who has carried the numerical calculations for the section 3.3. The authors also acknowledge the financial support provided by the National Sciences and Engineering Council of Canada and by the programme FCAR of the Province of Quebec.

REFERENCES

1. J.G. Kirkwood in P. Auer (Ed.), Documents in Modern Physics, Gordon and Breach, N.Y. (1967).
2. R.B. Bird, C.F. Curtiss, R.C. Armstrong and J. Hassager, Dynamics of Polymeric Liquids, Vol. 2, Kinetic Theory -second edition, Wiley, N.Y. (1987).
3. A.S. Lodge, Rheol. Acta, 7, 379 (1968).
4. M. Yamamoto, J. Phys. Soc. Jpn, 11, 413 (1956).
5. P.J. Carreau, Trans. Soc. Rheol., 16, 99 (1972).
6. I.F. Macdonald, Rheol. Acta, 14, 801, 899, 906 (1975).
7. S.F. Edwards, Proc. Phys. Soc., London, 92, 9 (1967).
8. P.G. de Gennes, J. Chem. Phys., 60, 5030 (1974).
9. M. Doi and S.F. Edwards, J. Chem. Soc. Faraday, Trans. II, 74, 1789, 1802, 1818 (1978).
10. G. Marrucci, in Advances in Transport Processes, Vol. V, A.S. Majumdar and R.A. Mashelkar Eds., Wiley, N.Y. (1984).

11. C.F. Curtiss and R.B. Bird, J. Chem. Phys., 74, 2016, 2026 (1981).

12. Hand, C.L., J. Fluid Mech., 13, 33 (1964).

13. M. Giesekus, J. Non-Newt. Fluid Mech., 11, 69 (1982).

14. R.B. Bird, J. Rheol, 26, 277 (1982); Chem. Eng. Commun.,16, 175 (1982).

15. P.J. Carreau and M. Grmela, in "Recent Advances in Structured Continua", edited by D. De Kee and P.N. Kaloni, Pitman Publishing Company, Chap. 4, p. 105 (1986).

16. M. Grmela, Physics Letters, 111A, 36 and 41 (1985); Physica D, 21, 179 (1986).

17. M. Grmela and P.J. Carreau, J. Non-Newt. Fluid Mech., 23, 27 (1987).

18. A. Ait Kadi, M. Grmela and P.J. Carreau, Rheol. Acta, 27, 241 (1988).

19. M. Grmela and Chhon Ly, Phys. Lett. A, 120, 281 (1987).

20. P.J. Carreau, M. Grmela and A. Rollin, in "Recent Advances in Structured Continua", Volume II, edited by D. De Kee and P.N. Kaloni, Pitman Publishing Company, Chap. 4, p. 110 (1990).

21. A. Ajji, P.J. Carreau, M. Grmela and H.P. Schreiber, J. Rheol., 33, 401 (1989).

22. M. Grmela, in this proceedings.

23. A.N. Beris and B.J. Edwards, J. Rheol., 34, 55 (1990).

24. R.B. Bird, P.J. Dotson and N.L. Johnson, J. Non-Newtonian Fluid Mech., 7, 213 (1980).

25. E.B. Christiansen and W.R. Leppart, Trans. Soc. Rheol., 18, 65 (1974).

26. M. Gremla, J. Rheol., 30, 707 (1986).

27. J. Meissner, Chem. Eng. Commun., 33, 159 (1985).

28. P.N. Dunlap and L.G. Leal, Rheol. Acta, 23, 238 (1984).

29. E.J. Hinch, Colloques Internationaux du CNRS, No 233, p. 241.

30. R.I. Tanner, Trans. Soc. Rheol., 19, 557 (1975).

31. C.G. Fuller and L.G. Leal, Rheol. Acta, 19, 580 (1980).

32. A. Ait Kadi, P.J. Carreau and G. Chauveteau, J. Rheol., 31, 537 (1987).

33. A.R. Khokhlov and A.N. Semenov, Macromolecules, 17, 2678 (1984).

34. A.R. Khokhlov and A.N. Semenov, J. Stat. Phys., 38, 161 (1985).

35. R.J.J. Jongschaap, H. Ramphuis and D.K. Doeksen, Rheol. Acta, 22, 539 (1983).

36. A.I. Leonov, Rheol. Acta, 15, 85 (1976).

37. D. Acierno, F.P. La Mantia, G. Marrucci and G. Titomanlio, J. Non-Newtonian Fluid Mech., 1, 125 and 147 (1976).

BIOFLUIDS AS STRUCTURED MEDIA : RHEOLOGY AND FLOW PROPERTIES OF BLOOD

D. Quemada, LBHP, Tour 33/34, Univ. PARIS 7
2 Place Jussieu, 75005, Paris (France)

0. INTRODUCTION

Most of biofluids exhibit steady and unsteady rheological properties quite similar to those observed in many disperse systems. Such general properties must be interpreted as resulting from some general characteristics that these systems have in common, although they may appear very different (as blood or synovial fluid and a slurry, for ex.). The existence of an internal structure (usually called "micro"-structure, although it is observed at any scale) and shear induced changes in this structure are now recognized as responsible of these common properties. Therefore, "structural models" (i.e. models which involve such structural changes) are believed to be the most capable to allow us rheological characterisation of these materials.

Biofluids can be considered as either (more or less complex) polymer solutions or, more generally, dispersions of some inclusions in such solutions. We can crudely separate biofluids into two classes of materials:

(i) biofluids in which interactions between macromolecules are dominant, hence which behaves as polymer solutions, either "pure" (as synovial fluid, for ex.) or slightly modified by the presence of few inclusions (as bronchial secretion , cell protoplasm, for ex.).

(ii) biofluids containing a large amount of dispersed "particles" in a polymer solution, hence which behave as a dispersion, the properties of which are mainly governed by particle-particle and polymer-particle interactions (as blood, lymph,... for ex.).

Another type of differences comes from difficulties in both sampling and physico-chemical characterisation, which result into very different levels in improvement of biofluid modelling. From this point of view, models in blood rheology have reached the upper level in quantitative predictions of both blood rheological properties and blood flows, that made blood an exception among biofluids. This likely results from superimposition of several factors, likely (i) the "precise characterization" of the system, as a consequence of reaching a high level in self-regulation by physiological control of blood characteristics (e.g. protein content in plasma) and flow conditions (e.g. matching of cardiac frequency, vessel size,...), (ii) a very quantitative knowledge in basic physico-chemical properties of blood components (red and white cells, platelets, different species --especially ions and proteins-- in plasma...), (iii) cumulative efforts in research due to the important role blood circulation plays in human body (more especially the "microcirculation", in very small vessels) , in relation with heart diseases). On the other hand, as "simple" polymeric media, synovial fluid and differents kinds of mucus show basic properties which are also well-understood, however their sampling is often very hard to be reproducible and requires techniques which may, without caution, completly change the properties of the material. Other biofluids –as living cell content, secretions (saliva, sweet,...), fluid in cephalo-rachidian cavities, semen...– gather both kinds of difficuty. Therefore, in the following, we shall focus our attention on hemorheology and blood flow properties and modelling.

Basing us on the analogies we have just mentioned , the aim of this lecture is to discuss the possibility of applying to Blood and Red Cell suspensions some rheological models of dispersions, already developped from theoretical grounds.

1. MAIN CHARACTERISTICS OF BLOOD FLOWS.

1.1- Blood constituents (see e.g. Burton,1966; Lewis,1970; Shiga & al,1990)

Blood is a very highly concentrated suspension (about 40-45% in volume fraction) of different kinds of cells (suspended in a continuous phase called plasma): Red Cells (RBC),White Cells (WBC), Platelets, alternatively called erythrocytes, leucocytes and thrombocytes, respectively.

Plasma is an aqueous solution of electrolytes and organic substances, mainly proteins, in relative proportions as shown on Table 1. It behaves as a newtonian fluid.

TABLE 1. BLOOD CONSTITUANTS

CELLS ELEMENTS (5.106 particles/mm3)	*(Relative proportions)*
White cells (all kinds)	1
Platelets	30
Red Cells	600
PLASMA	*(Weight fraction)*
Water	0.91
Inorganic solutes	0.01
Proteins	0.07
Other organic substances	0.01

As RBCs represent, in normal conditions, about 97% of the total cell volume, mechanical properties of whole blood are dominated by the presence of RBCs and their properties. Of course, this is not true in some pathological cases as leukaemia, where the large number of WBC increases both blood viscosity and its shear dependence.

In order to quantify RBC concentration, Physicians use the volume fraction of packed Red Cells (plus the plasma volume trapped in between), called hematocrit H, which is obtained by centrifuge of a sample of whole blood prevented from clotting by addition of anticoagulant (Protocole are now well-defined that allows to observe blood rheological properties free from effects of drawing and sampling, especially those due to anticoagulants). Hematocrit strongly depends on RBC deformability and it is better to use the true volume fraction ϕ.

Red Cells. RBC dimensions are illustrated in Fig.1.1. The biconcave discoid rest shape of (mammals) RBC results from nucleus ejection on entering the circulation (from the bone marrow). This leads to a volume reduction of the cell, without changing its membrane area.

Hence, RBC can be considered as a partially inflated balloon having a very high deformability which decreases with cell aging (mean life time ~120 days).

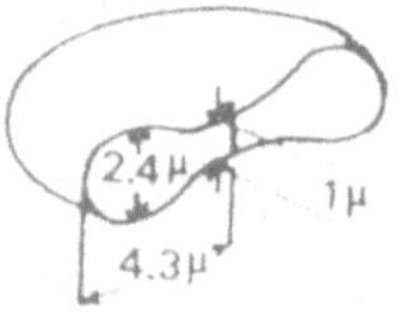

Fig.1.1- Human Red Blood Cell

Cell internal fluid is a water solution of hemoglobin (35 g/100 ml), the viscosity of which is about 6-8 cP. Mechanical study of motion of a single RBC suspended in a flowing fluid, demontrated that RBC rotates as a disc having an equivalent axis ratio, $r_e \cong .26$ (Goldsmith & Mason,1967). The RBC membrane consists of a bimolecular leaflet (a lipid bilayer with anchored proteins) rigidified by a skeletal structure (the actin-spectrin network, anchored on the internal lipid layer).
A negative electrical charge is found equivalent to about 6,000 electron charges.

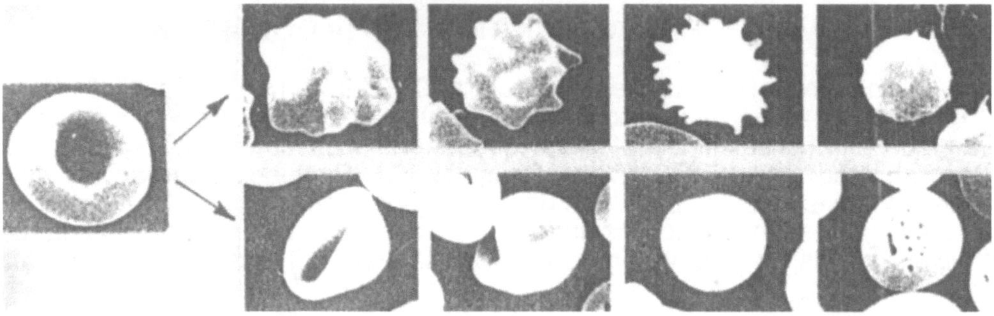

Fig.1.2- Morphological changes of erythrocytes. Stages of Transformation to echinocytes (upper array) and stomatocytes (lower array)
(from Kon & al,1983)

In steady conditions, blood settling occurs, however very dependent on blood internal structure (i.e. levels in RBC deformability and RBC aggregation, see hereafter) therefore allowing to distinguish pathological blood from normal one (the Erythrocyte Sedimentation Rate (ESR) is used in hospitals in this goal). Although negligible in almost all parts of the in vivo circulation, sedimentation effects becomes more and more important as shear rate is lowered (that requires to take precautions for rheometry). Important settling effects were found in horizontal narrow tube flows, in relation with RBC aggregation (Alonso & al,1989).

Change in osmotic conditions (normal ones correspond to those of 0.9% NaCl water solution) induced drastic changes in RBC shape (Fig.1.2). Indeed, in solutions of lower osmolarity than normal one, RBC becomes spherical after swelling (it is called spherocyte), with very negligible increase in membrane area. Further swelling leads to hemolysis of the cell, which eliminates hemoglobin towards the plasma, the resulting cell being a "ghost", i.e. a RBC reduced to its membrane only. On the contrary, above normal osmolarity, RBC volume is reduced, leading to formation of an echinocyte, the shape of which (reminiscent of sea-urchin one) is closely related to the actin-spectrin network. Further increase in salt concentration leads also to cell hemolysis.

Plasma proteins. Plasma proteins are mainly albumin and globulins which control water exchanges between blood and tissues by their action on osmotic balance. The remainders are lipoproteins and fibrinogen. The latter is known for a long time as playing a very important role in blood clotting. However, it was more recently recognized that, in normal conditions, fibrinogen (and in less extent, β-globulins) is strongly involved in reversible aggregation of RBC to form rouleaux (see Fig.1.3). We will see later that such a process is one of the fundamental determinants of blood rheology.

Clotting and Platelets. Whole blood reacts if it is put either in the presence of air oxygen or in contact with extracorporeal surfaces. In vitro, a clot is formed which mainly consists of a complex imbricated stucture of RBC and filaments of fibrin, which derives from fibrinogen by polymerization. After complete formation, a very slow clot contraction occurs, which ressembles syneresis, a similar retraction observed during gelation of colloids. On the other hand, in vivo clotting is governed by platelet activation, which promotes (irreversible) thrombus formation and/or platelet adhesion on (eventually injured) vessel wall.

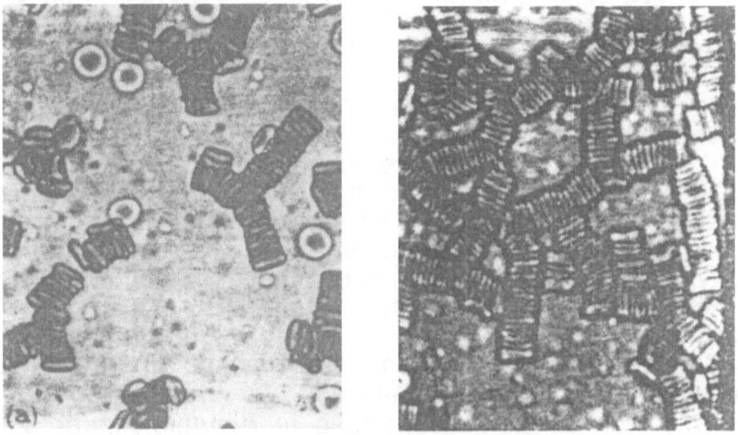

Fig.1.3 - Rouleaux of Human RBC

1.2- Blood Flows and Hydrodynamic description.

Three main classes in blood circulation can be distinguished, according to the value of the vessel diameter to particle size ratio, $\xi=2R/2a$.

a). Blood can be considered as a continuous medium only at the *macroscale* (i.e. in large arteries $\xi>\sim 50$, on the inside of which flow pulsatility promotes important deformations of vessel walls, hence leads to prior effects of (non-linear) visco-elasticity of vessel walls; idem, in less extent, in large veins, due to effects of external pressure and gravity). The effective (bulk) viscosity of blood is approximatively newtonian (however ϕ-dependent).

b). On the contrary, blood appears quite heterogeneous at the *microscale* (i.e. in small capillaries, with $\xi<\sim 1$, and the flow description reduces to solve the problem of transport of deformable capsules (as non-linear viscoelastic bodies) immersed in a newtonian fluid (plasma), in presence of "rigid" walls (Gaehtgens,1979; see a recent review by Skalak,1990).

c). The situation is more complex at the *mesoscale*, i.e. in small arteries and veins, arterioles and venules, $\sim 5<\xi<\sim 50$, where two kinds of structural effects appears: (i) the presence of RBC rouleaux, likely forming at rest a (labile) network due to rouleau branching and (ii) existence of plasma layer, close to the wall vessels, that rends the system macro-heterogeneous. Indeed, a kind of phase separation occurs, with formation of a particle rich axial core, surrounded by a particle depleted wall layer. Poiseuille (1835) was the first who observed such a separation, findings which were more or less considered as artefacts untill the quantitative studies of Farhaeus and Lindqvist (Farhaeus,1929; Farhaeus & Lindqvist,1931) who established the existence of a viscosity lowering as the vessel radius is reduced. This lowering mirrors directly lubricating effects resulting from the presence of the plasma layer. The presence of an endo-endothelial fibrin lining (Copley,1984) is believed to enhance the lubrication. A continuous description of this "diphasic annular flow" can be recovered if one considers each phase as homogeneous.

Such a description provides a framework for coherent interpretations of "anomaleous" features that blood flows through small vessels exhibit : in addition to the Farhaeus & Linqvist effect, many studies contain observations of (i) mean (tube averaged) hematocrit lesser than the (feed) reservoir one (the so-called Farhaeus' effect) (e.g. Cokelet,1976; Stadler & al,1990); (ii) blunted velocity profiles, near the vessel axis, that mirrors the non-newtonian properties of blood at this scale(e.g. Dufaux & al,1980).

1.3 - Blood Viscosity (steady shear viscosity).

Apparent shear viscosity, measured in Couette rheometers, is found non-newtonian. Under steady conditions, blood behaves as a shear-thinning fluid, in close relation with *levels* that *RBC Aggregation (RCA)* and *Deformation (RCD)* reach. Both processes are strongly dependent on physico-chemical properties of plasma, cell membrane and cell content. This was clearly demonstrated more than twenty years ago (Chien,1970) in comparing the following RBC suspensions at same hematocrit (H=.45), here refered as (NP), (NA) and (HA), that means respectively

(1) (NP)=Normal RBC suspended in Plasma, where rouleau formation is promoted by fibrinogen (and β-globulins)

(2) (NA)=Normal RBC suspended in Albumin-Ringer solution, without detectable RBC aggregation

(3) (HA)=Hardened RBC (by glutaraldehyde) suspended in the same Albumin-Ringer solution

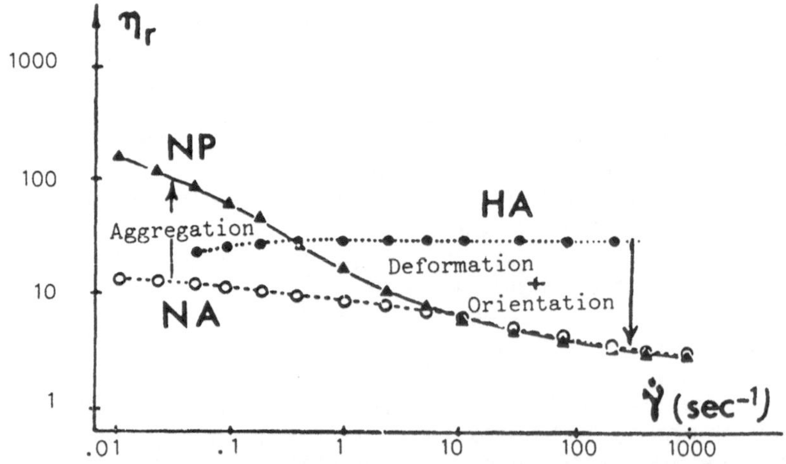

Fig.1.4 - Relative apparent viscosity vs. shear rate in (NP), (NA) and (HA) suspensions of RBC . -(from Chien,1967)

The albumin concentration in (NA) and (HA) was chosen in order to obtain a suspending fluid viscosity η_F equal to the plasma viscosity (here, η_P=1.2cP), hence getting the same shear stress σ acting on RBC at a given shear rate $\dot{\gamma}$. The corresponding rheograms $\eta = \eta(\dot{\gamma})$ are shown on Fig.1.3.

These results led to distinguish between two fundamental processes in blood rheology,

a) **reversible Red Cell Aggregation (RCA) at low shear rates**, which gives a large increase of NP viscosity as compared with the corresponding NA viscosity,

b) **Red Cell Deformation (RCD) and correlative orientation (RCO) at high shear rates** nevertheless small enough to avoid any hemolysis effect, protein degradation and/or hydrodynamic instabilities or turbulence onset. These processes lead to observe (i) same low values in (NP) and (NA) viscosities, as compared to the higher value of (HA) viscosity and (ii) identical high shear viscosities for (NP) and (NA), that confirms reaching a complete dispersion of rouleaux under shearing in (NP), with same RCD (and RCO) under the same shear stress (due to $\eta_P=\eta_F$).

Anticipating our goal in modelling blood rheology, we may underline main features from these findings: the viscosity vs. shear rate curve exhibits same general characteristics of concentrated dispersions of small particles (size ~ 1μm)

(i) two "viscosity plateaux" η_O and η_∞ both volume fraction dependent, $\eta=\eta(\phi)$,

(ii) a critical shear rate $\dot{\gamma}_C$ (e.g. the $\dot{\gamma}$-value at the inflexion point, $\dot{\gamma}_i$)

(iii) a characteristic slope s_C (e.g. $s=s_i$ at $\dot{\gamma}=\dot{\gamma}_i$)

showing that rheological modelling requires a minimum number of 4 parameters.

A large amount of work (of course, references given hereafter do not constitute an exhaustive list!) has been devoted to direct studies of RCA and RCD, mainly by optical measurements [as microscopy with imaging (Paulus & al,1986), light transmission (Schmid-Schonbein & al,1972; Thurston,1990), laser back-scattering, especially in the presence of shear (Mills & al,1980)], acoustic measurements [ultrasound back-scattering (Boynard & Hanss,1981; Boynard, 1986)] and mechanical measurements [micropore filtration (Hanss,1983; Bucherer & al,1988), micro-pipette apparatus (Evans & La Celle,1975)]. These methods provided quantitative evaluations of RCA and RCD, in both steady and unsteady conditions. In some studies, these evaluations were experimentally correlated to stress measurements (Snabre & al,1987). The main problem is to insert such a microscopic knowledge into a macroscopic model of the whole system. Of course, as the problem remains unsolved, even in the simplest case of concentrated suspensions of rigid spheres, the question is still open .

Factors governing RCD and RCA are briefly listed hereafter

++ RCD & RCO raises if we increase:

(i) shear stress (hence either suspending fluid viscosity or shear rate, or both), that results into cell alignment in the flow direction

(ii) hematocrit , due to enhancement of crowding effects and increasing of effective shear rate between adjacent cells

(iii) pH and ionic strength of plasma, giving changes in cell deformability (e.g. echinocytes less deformable than normal RBC)

(iv) intrinsic deformability of cell, which depends on membrane elasticity, viscosity and physico-chemical properties of internal fluid (e.g. drepanocytes, with abnormal hemoglobin molecules whose rigidity is strongly increased under low oxygen pressure, that constitute the prior cause of the "sickel-cell disease)

++ RCA mainly depends on

(i) shear rate, which promotes opposite effects (aggregate formation from "hydrodynamic" collisions and aggregate rupture by shear stresses)

(ii) physico-chemical properties of suspending fluid (essentially ionic and macromolecular content) and mechanical properties of the cell membrane (since RBCs in a rouleau are deformed).

Rouleau formation involves competitive interactions between adjacent cells, essentially (i) repulsion, either electrostatic one, thus leading to RCA level variations with ionic strength (see Fig.1.5) or steric one (between polymer chains adsorbed on the cell membrane, in the presence of a good solvant, and (ii) attraction from either macromolecular bridging (Chien & al,1971c; Chien & Jan,1973) or polymer depletion (as in colloids, e.g. Sperry & al,1981), the question is still open.

Number of studies on cell adhesion due to macromolecular bridging, from either a thermodynamic approach (e.g. Bell, 1978, Bell & al, 1984) or a model based on coupling elastic deformation of the cell membrane with chemical kinetics of the adhesion molecules (e.g. Evans, 1985, Dembo & al, 1988). A recent review (Skalak & Zhu, 1990) incorporates both the macroscopic viewpoint (which only introduces an adhesion energy, without details at the molecular level) and the microscopic one (including kinetics of bond formation and disaggregation, viscoelasticity of cross-bridges and their lateral mobility, resulting from either diffusion or peeling induced sliding.

Substances, as drugs, which modify the equilibrium between repulsive and adhesive forces, will act as either aggregative or anti-aggregative agents. For ex. Aspirin induces a strong lowering in low shear viscosity (Fig.1.6).

However, respective roles of RCA and RCD depend on the vessel size. In larger arterioles, RCA dominates and gives the above-described annular flow. As the size decreases, RCD becomes more and more important, giving in pre-capillaries (or in post-venules) an axial file of cells. Tese cells exibit the so-called "tank-treading" motion , i.e. a continuous rotation of cell membrane around its cytoplasm(Fischer & Schmid-Schonbein,1977; Fischer & al,1978). Characteristics of this motion allows estimating of intrinsic mechanical properties of the RBC membrane[Sutera & al, 1989])

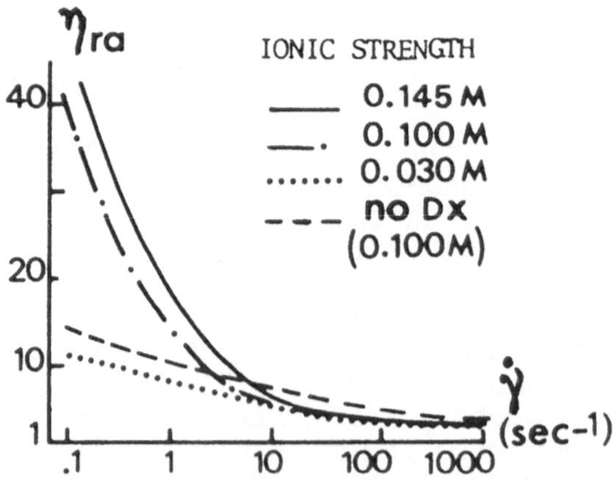

Fig.1.5- Relative viscosity vs. shear rate for washed human RBC suspended in saline at the ionic strengthes indicated. H=(50±.2)%, 3gr/100ml of Dextran70, T=25°C. -(from Brooks & al, 1974)

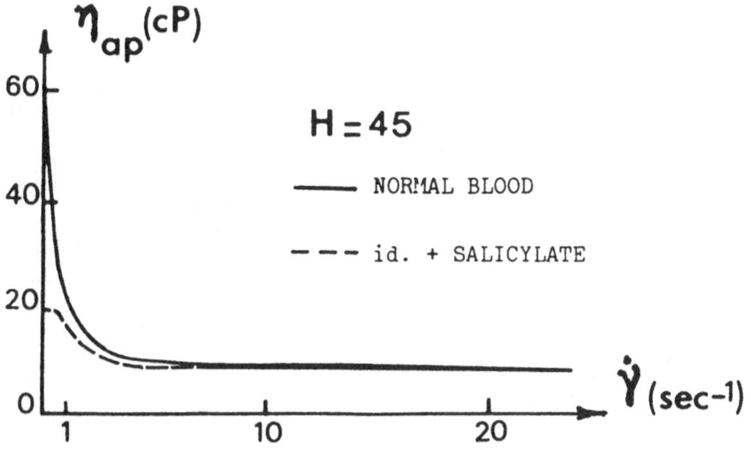

Fig.1.6- Apparent viscosity vs. shear rate of normal and disaggregated blood.
(from Healy & Joly, 1975)

Some evidence of normal stresses in blood exists . Moreover, elongational flows (hence elongational viscosity) are involved in blood circulation. Nevertheless, experiments are few and modelling almost absent.

1.4 - Time dependent effects (thixotropy and visco-elasticity).

As time dependent effects occur in blood flow in vivo (through pulsatility, vasomotricity,...) measurements of unsteady viscosity have been recently developped.

Under unsteady conditions, the rheological response (as stress relaxation in transient tests, or linear viscoelasticity in oscillatory experiments) not only mirrors RCA and RCD levels, but also RCA and RCD *kinetics*.

Stress Relaxation. Fig.1.7 displays the different flow curves (shear stress vs. time curves), $\sigma=\sigma(t)$, in response to increasing $\dot{\gamma}$-steps in coaxial Couette viscometer and effects of adding a disaggregative agent (aspirin), that leads to lowering in RCA, on these curves.

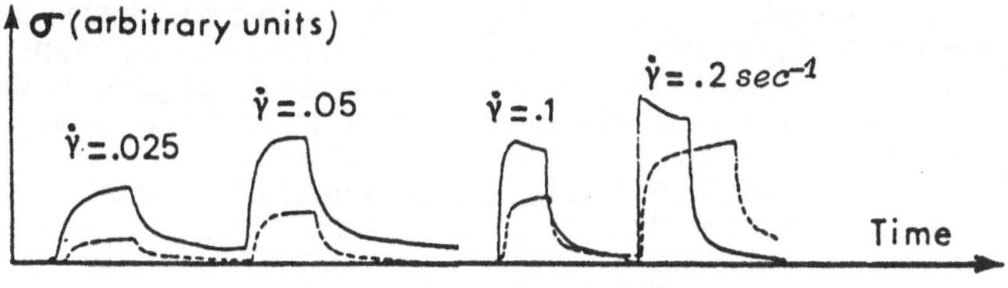

Fig.1.7- Time curves $\sigma(t)$ [—— normal blood,--- id. + salicylate].
(from Healy & Joly, 1975)

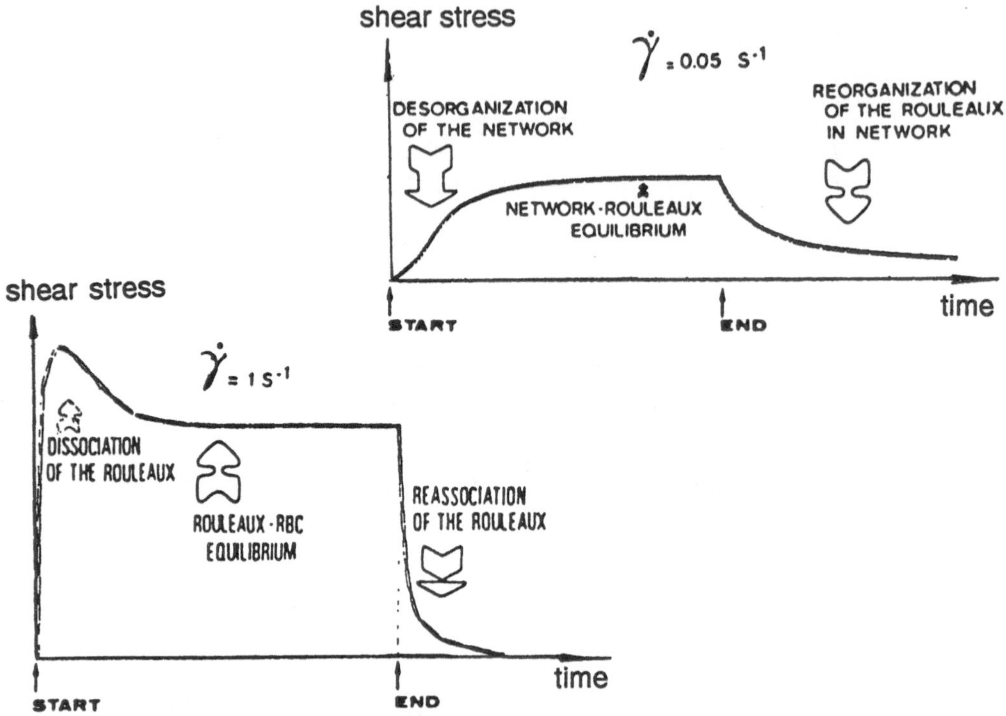

Fig.1.8- Illustration of structural changes in normal blood, during stress formation and stress relaxation experiments.(from Joly & al,1981)

More precisely, Fig.1.8 displays differences between two flow curves, one in the case of very small γ-amplitude (i.e. linear viscoelastic, without any "overshoot" in the response) and the other for higher amplitude (with an overshoot in the response). Expected structural changes "associated" to successive phases are illustrated on Fig.1.8, showing the major importance of RCA in the observed unsteady behaviour. Moreover, still related to RCA, time dependence of plasma layer thickness should be predicted. Such expectations seem in agreement with recent light transmission studies (Thurston,1990).

In summary, we have to keep in mind the main following properties of blood:

(i) blood essentially behaves as a highly concentrated suspension of RBC, in which reversible RCAggregation and RCDeformation play very important roles

(ii) kinetics of formation and rupture of cell aggregates governs the blood rheology at low shear rates . Kinetics of deformation-orientation of aggregates and of RBC plays an increasing role as shear rate is increased.

(iii) phase separation leads to a two-phase annular flow in small vessels

Such three "ingredients" should be of chief importance in blood modelling.

2. BLOOD AS A CONCENTRATED DISPERSION : Effect of particle concentration.

The particulate nature of blood suggests to begin with basic rheological models of concentrated suspensions and emulsions as a first approximation in blood rheological modelling. As up to now no exact theory of such concentrated media is available, we will discuss some conditions for the existence of two-phase flows and the consequences of applying a minimum principle to energy dissipation in such two-phase flows.

2.1 - Phase separation.

The presence of marginal layers in blood flows appears as a general feature, currently observed in number of systems. In non-homogeneous velocity fields, a flow-induced particle migration (of course, in absence of any direct particle-wall interaction), has been postulated as being responsible of this phase separation, although the origin of such a migration still remains conflicting in the absence of inertial effects. In the case of a polymer solution flowing through a pipe, one could expect that macromolecules migrate radially towards the tube axis in order to maximize the configurational entropy of the system, since the higher level of orientation is reached in the highest velocity gradient region, close to the tube wall. Such a speculation was quantified in the case of very dilute solutions (Tirrel & Malone,1977), but a more general calculation (Aubert & Tirrel, 1980) showed that, under the molecular assumptions used (i.e. a bead-spring model with zero bead volume), this kind of "thermo dynamical lift" does not occur in unidirectional non-homogeneous flows (channel and tube flows) . On the contrary, considering finite size dumbells led to predict cross-stream migration in such flows (Brunn & Chi, 1984).

For concentrated systems, however, the existence of such a migration is open to question. Some recent approaches on dispersions of deformable particles, based on Thermodynamics of Irreversible Processes, discovered extra-terms directly associated with particle deformation, which could be responsible of the expected migration (Lhuillier, 1990; Onuki,1990). In the case of rigid particles, aggregate deformability could play a role similar to the particle one.

Moreover, as the particle accumulation in the low shear region (close to the axis), and the corresponding particle depletion in the high shear region (close to the walls), should tend to increase the difference between the suspension viscosities in these two regions, we can expect to observe a flattening in the velocity profile. Such a flattening, in turn, will reinforce the viscosity difference (here, the suspension is assumed to be a shear-thinning medium), that ultimately amplifies the flattening and promotes a new increase in viscosity difference (through enhanced levels in aggregation and/or in aggregate deformation and orientation, in the low and high shear regions, respectively), and so on,... up to reaching a *dynamical equilibrium* between the different sub-units forming the internal structure of the suspension. We believe that, in concentrated systems, such effects could increase the difference in configurational entropy between wall and axial regions and allow us to postulate the existence of a cross-flow force as resulting from the non-homogeneity of the flow field. Moreover, in the case of deformable particles as droplets or cells, a radial drift in tube flows have been observed experimentally (Karnis & al, 1966)

If, in the presence of heterogeneous flows characterized by a spatial variation in shear rate $\dot{\gamma}$, a lift force of the form $(\beta/2)\nabla\dot{\gamma}^2$ (from symetry arguments) is postulated, we are able to analyse the flow stability of a homogeneous suspension (with a given volume fraction $\phi=nv_p$, n being the number density and v_p the particle volume). In the case of plane flows and if β is positive (e.g. if lift forces drag the particles towards the tube axis in a Poiseuille flow), density and shear rate fluctuations are found unstable (Nozières & Quemada, 1986) if the shear rate γ exceeds a critical value

$$\dot{\gamma}_{crit} = (\eta\mu'/\beta\eta')^{1/2} \qquad (2.1)$$

where η is the viscosity and μ , the chemical potential of the system, with $f'\equiv(df/dn)$ for any function of density n, $f=f(n)$.

Past this threshold, the original uniform state will break into domains: a first order transition appears, quite similar to the liquid-gas phase separation. In equilibrium state (that requires constant shear stress $\sigma=\sigma_o$ and chemical potential $\mu=\mu_o$), we will find, beyond γ_{crit} , two coexisting solutions $(n_1, \dot{\gamma}_1)$ and $(n_2, \dot{\gamma}_2)$ for each σ_o value.

Fig.2.1 & 2.2 illustrate the main results, with a spinodal line (where the μ-extrema A* and B* are located), a critical point C and a "Maxwell plateau" AB. We shall return later to these results, taking explicit n-dependences of μ and η.

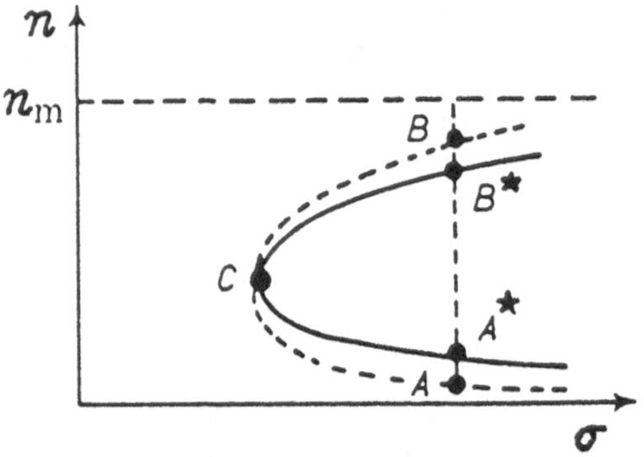

Fig.2.1- Phase diagram $\sigma(n)$
___ Spinodal line, with points A* and B*
- - - - Locus of stationary domains.

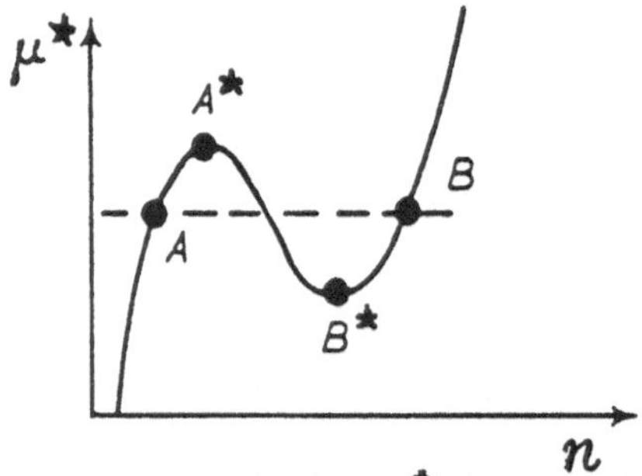

Fig.2.2- Sketch of $\mu^*(n)$ at fixed σ_o
The Maxwell plateau corresponds to a stationary wall

A recent Non-Equilibrium Molecular Dynamics (Loose & Hess,1989) led to a shear-induced ordering transition at a critical shear rate. These transition has been confirmed by linear stability analysis: past this threshold, coupled density and velocity fluctuations exhibit a velocity profile with shear minima corresponding to density maxima, i.e. plug formation. This more general approach could support the existence of the phase separation we found here as a consequence of assuming the presence of a lift force, although no prefered modes were observed in the NEMD data (see the Hess' lecture in this issue)

2.2 - Minimun energy dissipation.

For non-inertial flows of incompressible newtonian fluids, it is well-established that, under given conditions, the solution of the Stokes' equation is unique and satifies a minimum energy principle for viscous dissipation (see for instance, Batchelor,1970, p.227).

Such a principle was assumed to hold in viscometric flows of dispersions, taking into account possible cross-flow variations in volume fraction (Lighthill,1969). As hydrodynamic interactions (and lift forces, eventually) couple flow and particle motions, a self-consistent solution – that is, both velocity and concentration profiles **v**(**r**) and ϕ(**r**) – is expected to exist for fixed properties of the whole system (concentrations, temperature, suspending fluid characteristics,...) and well-defined flow conditions (flow rates or pressure gradient,...). Moreover, we assume that flow-induced structural changes can be described by a (structural) viscosity $\eta=\eta(\phi,\dot{\gamma})$, as a functional of ϕ and $\dot{\gamma}$.

Considering for instance a pipe flow (0< r< R), minimization of the rate of energy dissipation per unit tube length is achieved from applying the variational method to a functional with Lagrange multipliers corresponding to the constraints associated to given flow conditions . The resulting two Euler-Lagrange equations governs the coupled fields **v**(r) and ϕ(r) [Quemada, 1977].

If we assume the presence of the above-discussed phase separation (i.e. a diphasic annular flow, in the present case), decoupling of E-L equations is obtained with constant volume fractions in wall and core regions, respectively. For the corresponding (newtonian) high shear limit (hence $\eta=\eta(\phi)$ only), these two equations then reduce to

(i) Navier-Stokes equations for the classical two-fluid velocity profile (composed of two arches of parabola) ,

(ii) an equation for the relative fluidity, $F = \eta_F/\eta$ (the reciprocal of the relative viscosity), as a function of ϕ.

The latter relates values of F and its derivative $F' = dF/d\phi$ on each side of the domain wall (located at $r=\xi R$), i.e. in axial core ($0< r< \xi R$, s subscript) and marginal layer ($\xi R< r< R$, w subscript)

$$(F'_s + F'_w)(\phi_s-\phi_m) - 2(F_s - F_w) = 0 \tag{2.2}$$

Eq.(2.2) can be solved in two limiting cases which correspond to the following one-fluid limits (cf. Fig.2.3):

(i) an (infinitely thin) axial core,($\xi \to 0$), with $\phi_s=\phi_m$, a maximum volume fraction, that is a "packed"core (with zero fluidity)

(ii) a (vanishing) particle free marginal layer, ($\xi \to 1$), with $\phi_w=0$.

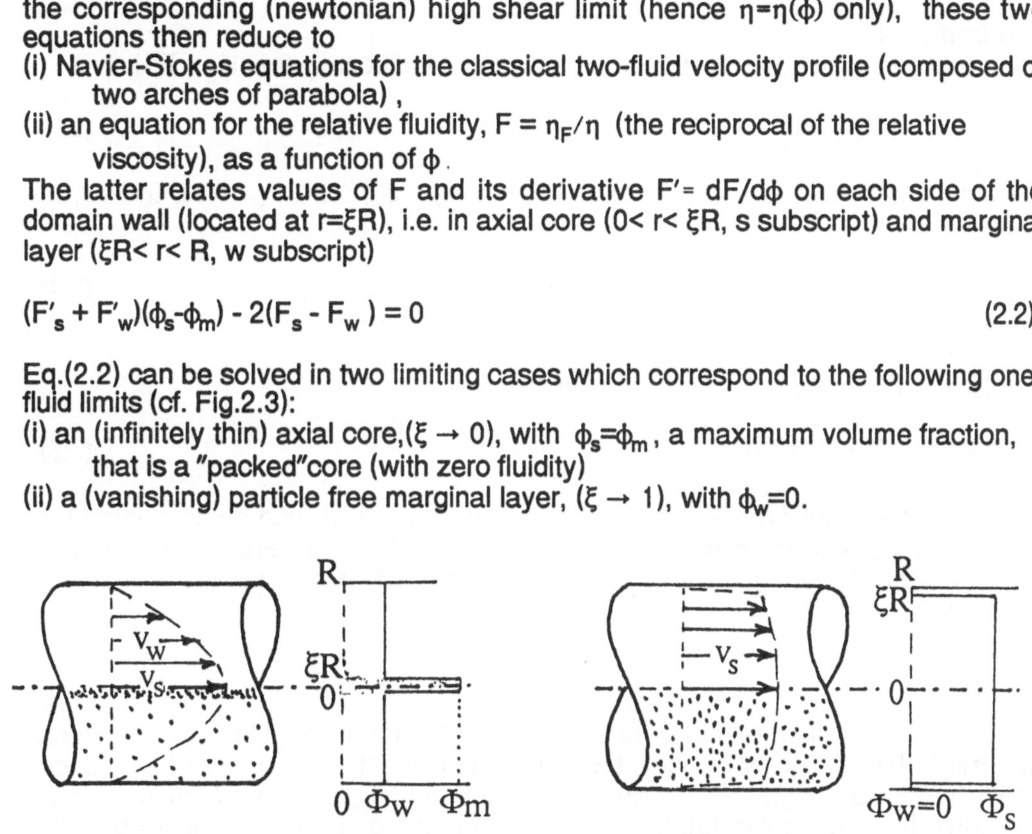

a) Packed axial core ($\xi \to 0$) b) Particle free marginal layer ($\xi \to 1$)

Fig.2.3 - One-fluid flow limits.

Let us add some comments on these limits.

As a consequence of the (postulated) radial migration in pipe flow, we have to assume that particle (at low concentration) should begin to accumulate on the pipe axis and that superimposition of diffusive and lift forces should give, at higher concentrations, a ϕ-profile having a maximum on the axis. Therefore, increasing the (total) feed concentration ϕ, we can expect to reach some critical value $\phi=\phi^*$ which corresponds to a packed core as a very thin file of particles on the axis, hence with $\phi_s=\phi_m$, the maximum volume fraction at rest ($\dot{\gamma}_s=0$ since $\eta_s=\infty$). As ϕ grows beyond ϕ^*, the core thickens at the expense of the marginal layer. This percolation-like transition at $\phi=\phi^*$ appears of the same type than the one already postulated in channel flow of non interacting rigid spheres (de Gennes, 1979), althouth we are only concerned here by percolation in the flow direction, not in the transverse one. Furthermore, as evidence, attractive particle interactions, if present, should greatly promote this percolation transition (in such a case, beyond ϕ^*, the domain wall location, $r=\xi R$, with ξ finite, will result from the balance between cohesive forces on particles forming an aggregate located at ξR and the shear stress the fluid exerts on it).

On the other hand, the second limit could be associated to the structure invoked to explain wall slip phenomena, as discussed yet, with a very thin lubricant wall layer and a sheared axial core filling almost the whole tube section (Quemada, 1982).

Packed core :
From eq.(2.2), with $F_s = F(\phi_m) = 0$ and assuming conditionally that $F'_s = F'(\phi_m) = 0$,

$$F'_w\,(\phi_m - \phi_w) + 2F_w = 0 \qquad (2.3)$$

the solution of which is (with C=const.) $F_w = C\,(\phi_m - \phi_w)^2$. As $F_w \rightarrow 1$ if $\phi_w \rightarrow 0$, we obtain

$$F'_w = (1 - \phi_w/\phi_m)^2 \qquad (2.4)$$

Particle free marginal layer.
For $\phi_w = 0$, hence $F_w = 1$, eq.(2.2) becomes

$$(F'_s - k_1)\,\phi_s + 2(1 - F_s) = 0 \qquad (2.5)$$

where $k_1 = \lim_{\phi \to 0}(dF/d\phi)_{\phi=\phi_w}$ is the intrinsic viscosity which can be calculated from highly dilute suspension theory (e.g. Batchelor, 1970), for instance for neutrally buoyant rigid spheres, the Einstein's result $k_1 = k_E = 2.5$. Integration (with K=Const.) leads to

$$F_s = 1 - k_1\phi_s + K\phi_s^2$$

For very diluted suspensions, the Einstein's limit (first order in ϕ) is recovered, $\eta_r = 1 + k_1\phi$. To second order in ϕ, the integration constant K cannot be deduced from theory (Batchelor & Green, 1972) which only gives the *high frequency* viscosity (without accounting for shear induced changes in the equilibrium distribution of particles, a problem still unsolvable) whereas we are here concerned with the high shear steady viscosity, except close to the tube axis.

Nevertheless, it is worth noting that, in principle, this second limit ($\xi \rightarrow 1$) remains valid as ϕ_s is increased up to $\phi_s = \phi_m$ (then giving the above-described plug flow, with a solid core sliding through the tube as a lubricated piston). If we assume that the function $F(\phi)$ is unique and reaches the same limits $F(\phi_m) = 0$ and $F'(\phi_m) = 0$ we took yet for F_w in the packed core limit, we have to put in eq.(2.6) $K = \phi_m^{-2}$ and $k_1 = 2\phi_m^{-1}$. Then, exactly the same ϕ-dependence in $F_s(\phi_s)$ than in $F_w(\phi_w)$ is found

$$F_s = (1 - \phi_s/\phi_m)^2 \qquad (2.6)$$

that could be considered as supporting a possible general character of an equation of the form

$$F = (1 - \phi/\phi_m)^2 \qquad (2.7a)$$

or, in terms of the relative viscosity η_r , a relation of the type

$$\eta_r= (1-\phi/\phi_m)^{-q} \qquad \text{with } q=2 \qquad (2.7)$$

Such a general character already appears in the fact that eq.(2.2) for the general case of diphasic flows (with $0<\xi<1$) is satisfied if we take of the form of eq.(2.7a).

Furthermore, eq.(2.7) is reminiscent of expressions of transport coefficients found in critical phenomena theory, with q as a critical exponent. The "minimun dissipation" approach appears as a kind of mean field approximation through the integer value q=2 found here.

Eq.(2.7) has been proved to be in fair agreement with number of viscosity data for rigid particle dispersions, especially in the case of (very well-characterized) suspensions of sterically stabilized silicate particles which were found to behave as hard spheres (de Kruif & al,1985). This viscosity equation appears identical to the Krieger's phenomenological equation (Krieger,1972), $\eta_r=(1-\lambda\phi)^{-k_1/\lambda}$ where λ is the Mooney's crowding factor (which was introduced to account for excluded volume type effects). The exponent value k_1/λ is found from the very dilute (Einstein) limit. Putting $\lambda=\phi_m^{-1}$, this equation reproduces eq.(2.7) if one takes $q=k_1\phi_m$. This relation also results from the condition $F'_s(\phi_m)=0$ applied to eq(2.6) once K was fixed by the condition $F_s(\phi_m)=0$.

Using eq.(2.7) and classical expression for chemical potential, $\mu=K_BT\text{Log}(n)$, in eq.(2.1) leads to $\gamma_{crit} = [K_BT(\phi_m-\phi)/\beta q\phi]^{1/2}$. The critical point (see Fig.2.2, point C) is defined by $\phi_C = \phi_m/2q$, $\gamma_C = (2-q^{-1})\, K_BT/\beta$. As the point at which the phase separation disappears, point C must represent the above-discussed percolation threshold ϕ^* (de Gennes,1979). Taking q=2 and ϕ_m=.74 (for hard spheres) leads to ϕ_C=.185, close to the critical volume fraction ϕ^*=.18, corresponding to the percolation threshold in 3D structures built by randomly packed spheres (Powell, 1979).

2.3 -Application to Blood and Red Cell suspensions.

As $.40 < \phi < .45$ and $.25< \phi <.70$ for normal and pathological blood respectively, using of viscosity equations for concentrated suspensions is uncontournable.

The most applied one has been the semi-empirical Arrhenius equation which gives a linear dependence of $\text{Log}(\eta_r)$ vs ϕ, although this equation fails at ϕ >.30 and low shear rates.

Application of eq.(2.7) to RBC suspensions is shown on Fig.2.3 for three different data, hardened RBC suspended in saline (Chien & al,1967), in Ringer (Brooks & al,1970) and in water (Chien & al,1971a), leading to slightly different values in ϕ_m (i.e. ϕ_m=.549 in water and ϕ_m=.583 in Ringer) which can be understood taking into account RBC swelling in each suspending medium. Indeed, the corresponding packing values are close to (i) ϕ_m=.58 calculated for geometrical packing of hexagonal discoid cells enclosing undeformed RBC (Burton,1960) and (ii) to ϕ_m=.60 found from sedimentation of hardened cells (Chien & al,1971b). Comparison with results of data fitting of the Brinkman-Roscoe equation (identical to eq.(2.7) but with q=2.5) is given for ϕ_m=.635 (Cokelet,1972).

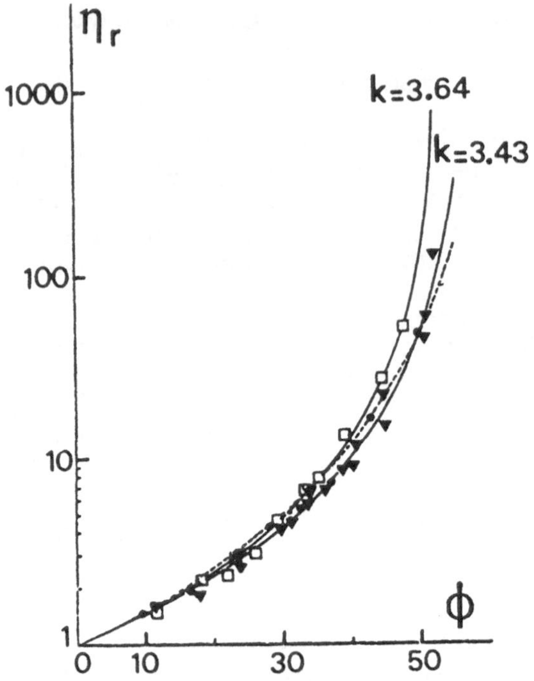

Fig.2.3- Relative viscosity vs particle volume fraction of hardened cells suspensions - Data : hardened RBC (•) in saline ,(▼)in Ringer and (□)in water.
Curves : eq.(2.7) —— , Brinkman & Roscoe eq. -----

2.4 - Blood as a concentrated emulsion.

Due to the very high deformability of normal RBC, modelling of blood as a concentrated emulsion has been proposed for a long time. As for suspensions of rigid particles, the very dilute limit is known (Taylor,1932), $\eta_r=1+k_1\phi$, with $k_1=Tk_E$ where $T=(k_E\lambda+1)/(\lambda+1)$ is the Taylor factor , with $\lambda = \eta_I/\eta_F$, the ratio of viscosities η_I and η_F of dispersed and continuous phase, respectively. However, such a result only holds in the limiting case if (viscous forces)/(surface tension) ~ $\eta_F\dot{\gamma}\, a/\Gamma << 1$ (with Γ=surface tension), i.e. is only valid for small deformations, thus it cannot be applied to blood.

On the contrary,recalling the observed "tank-treading" motion of sheared RBC, the membrane of which exhibiting a continuous rotation around the cell interior as a rigid body, we could regard each individual cell as composed of a hard core surrounded by a (more or less freely) rotating fluid shell.

As a first approximation, assuming this shell forms a part of the suspending fluid, we could tentatively consider a suspension of dispersed RBC (i.e. at high enough shear rate) as a suspension of hard cores, hence with a reduced volume fraction, $\phi_{eff} < \phi$.

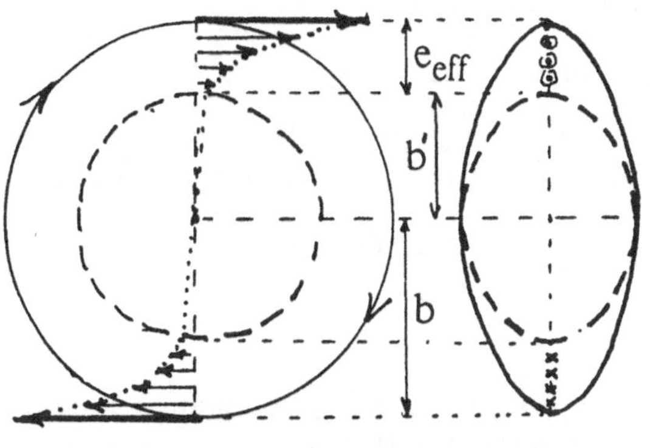

Fig.2.4- Expected velocity field associated with the tank-treading motion of a RBC.

Following the model of tank-treading motion (Keller & Skalak,1982; Sutera & al,1989), and retaining an ellipsoidal rotating RBC shape (volume $V=4\pi ab^2/3$), however assuming a finite extension of internal fluid rotation, we can estimate the reduced (no rotating) cell volume as $V'=4\pi ab'^2/3$, leading to an effective volume fraction $\phi_{eff}=\phi(b'/b)^2 = \phi(1-e_{eff}/a)^2$, where $e_{eff}= 1-b'/b$ is the maximum thickness of the rotating layer.

We shall return (§3.2) on this approximation.

3. BLOOD AS A SHEAR-THINNING FLUID : Effect of shear induced stuctural changes.

In the framework of two-phase flows, the observed non-newtonian blood viscosity could be simply related to shear dependent size of the domains the phase separation has produced, keeping the two phases as newtonian fluids (cf. §5). However, in the axial core of pipe flow for instance, shear induced changes in the "internal structure" likely occur, as resulting from either breaking (and forming) aggregates or modifying their microstucture. This picture has been confirmed in finding non-parabolic velocity profiles inside the axial core (see §5).Therefore, blood flow modelling must involve non-newtonian rheology for the dense phase.

3.1 - Some general aspects in rheological modelling of sheared concentrated suspensions.

Theoretical calculations (Russel & Gast,1986) of the equilibrium distribution function in sheared suspensions, including brownian effects, remain limited to their linear viscoelastic properties and, for steady state, to the low-shear limit in semi-diluted media since only pairwise hydrodynamic interactions were taken into account . Hence, we need approximate approaches which incorporate structural effects in a general manner and with, as far as possible, basic concepts .

We may distinguish four steps in improvement of any structural model. These steps correspond to either physical identification or choice of

(i) a set of "structural variables" S_i, i=1,2,3,... which as evidence should depend on both time and mechanisms (especially shear induced ones) leading to building up or breaking down the structure

(ii) kinetic equations which, through these mechanisms, govern the variables (S_i) and should involve shear dependent rate constants (or characteristic times)

(iii) shear rate (or shear stress) dependence of these rate constants which should also depend on other structural variables (as concentrations,...)

(iv) explicite S_i-dependences of rheological functions.

Step1 - We assume that the system is composed of $N_1+N_2+N_3...= N$ particles per unit volume. N_m means N_m aggregates of m particles ($1< m \leq N$), homogeneously distributed in space (in each phase, if phase separation occurs). As high levels in volume fraction result in formation of clusters, the size of which should depend on flow conditons, the particle distribution in size could be expected to contain the relevant information to describe the structure. In fact, such a structure characterization would introduce too many "micro-variables": only few of them - or, better, some averages of them - will have significant influence on macroscopic properties of the system, leading to a reduced set of variables , as either some moments of size distribution or/and some cluster characteristics (thus defined at the mesoscale). This reduced set is believed to constitute a (pertinent) set of structural variables.

As such variables, we may retain, for instance, the *average number of particles per aggregate*

$$p(t) = \sum_{m=1,N} mN_m / \sum_{m=1,N} N_m = N / \sum_{m=1,N} N_m \tag{3.1}$$

or the *aggregated fraction*, which can be expressed in terms of volume fractions ϕ_A and ϕ, ϕ_A being the volume fraction of aggregated particles

$$A(t) = \sum_{m=2,N} mN_m / N = \phi_A/\phi \tag{3.2}$$

However, as a primary consequence of clustering is to change ϕ_A into the corresponding effective volume fraction, ϕ_{Aeff}, which includes the amount of suspending fluid immobilized on the inside of aggregates, we have to take also the (shear dependent) ratio $\alpha(t) = \phi_{Aeff}/\phi_A$ as a structural variable. If the mean compactness of m-aggregates is ϖ_m (that is the averaged solid fraction in any aggregate of m particles), we get

$$\alpha(t) = \sum_{m=2,N} (mN_m / \varpi_m) / \sum_{m=2,N} mN_m = \phi_{Aeff}/\phi_A$$

Therefore, we can define an effective volume fraction for the whole suspension by $\phi_{eff} = \phi_1 + \phi_{Aeff}$, $\phi_1 = \phi - \phi_A$ being the volume fraction of individual particles, i.e.

$$\phi_{eff} = \phi + (\alpha - 1)\phi_A \tag{3.3}$$

where

$$\alpha = \alpha(t) = \phi_{Aeff} / A(t)\phi \tag{3.4}$$

Generally, since α mirrors the ϕ-dependence of aggregate distribution (in size, shape, internal structure...), it should depend on ϕ . Nevertheless, we may underline three special cases:

(1) if ϖ_m has the same value, say ϖ_o , for all aggregates (as in very concentrated systems with ϖ_o likely close to the maximum packing compatible with particle interactions), then ϕ_{Aeff} would reduce to ϕ_A/ϖ_o , i.e. $\alpha = \varpi_m^{-1}$

(2) if the system is composed of very polydisperse individual particles which hence can form very compact aggregates (where voids between biggest particles are filled with smaller ones, and so on...), ϖ_m should be close to one, for any m, leading to $\alpha \cong 1$, hence $\phi_{eff} \cong \phi$.

(3) if the m-aggregate has a fractal structure, $\varpi_m = m^{(D-3)/D}$, D being the fractal dimension.

Step 2 - The more general expression for reversible kinetics of the above-defined system of N particles is given by the system of equations

$$dn_m/dt = (1/2)\sum_{i+j=m} \kappa_A^{i,j} n_i n_j - \sum_{i=1,N} \kappa_A^{i,m} n_i n_m - \sum_{i=1,m-1} \kappa_D^{m,i,m-i} n_m + 2\sum_{i=m+1,N} \kappa_D^{i,i-m,m} n_i \qquad (3.5)$$

where n_m is the number fraction N_m/N of m-aggregates, $\kappa_A^{i,j}$ the kinetic constants for addition reaction (i)+(j)→(i+j) and $\kappa_D^{s,i,j}$, the kinetic constant for disruption reaction (s)→(i)+(j). In eq.(3.5), first and second terms correspond to the formation m-aggregate by addition of smaller ones and its disappearance by addition to any other, respectively. The last two terms represent the two kinds of formation and disappearance by splitting.
As kinetic constants $\kappa_A^{i,j}$ and $\kappa_D^{s,i,j}$ should depend on sizes of i-, j-, s-aggregates entering the reactions, solving these equations is a tremendous task : solutions are known only in the irreversible case [then eq.(3.5) reduces to the well-known Schmoluchovski equation (Schmoluchovski,1916)] and for constant values or very special dependences of kinetic constants (e.g. Friedlander,1964; Viksek & Family,1967). Such predictions have been recently verified experimentally on 2D suspensions of macroscopic spheres (Roussel & al,1989)

Introduction of reversibility, through the two last terms in eq.(3.5), led to quite different solutions (van Dongen & Ernst,1984). Numerical results in 2D (Botet & Jullien,1985) established that, after a long enough time, the system reaches a steady state in which the aggregate, although fluctuating in shape, has a mean radius of giration independent of its initial configuration. Extension to reversible cluster-cluster aggregation were also studied and showed similar features (Meakin & Deutch,1985).

A recent application of eq.(3.5) -after slight modifications- was devoted to RCAggregation in blood (Murata & Secomb,1988). A solution was obtained thanks to the following restritive assumptions on kinetic constants:

(1) all $\kappa_A^{i,j}$ and $\kappa_D^{s,i,j}$ are independent of i, j, s, thus reduced to common values κ_A and κ_D.

(2) $\kappa_A=K(\dot\gamma)S(\dot\gamma)$, where $K(\dot\gamma)=k_1\dot\gamma + k_2$ is a (shear dependent) collision rate and $S(\dot\gamma)$ a sticking probability such as $S(\dot\gamma)=1$ or $S(\dot\gamma)=(t_S\dot\gamma)^{-1}$ according to $t_S\dot\gamma \le 1$ or $t_S\dot\gamma \ge 1$, t_S being related to some minimum value of the post-collision time required for aggregate formation

(3) $\kappa_D \sim k_3\,\dot\gamma$

The average aggregate size <p> was found increasing monotonically with time up to a limiting value $\langle p\rangle_e$ for dynamical equilibrium. As a function of the shear rate, the latter can be written $\langle p\rangle_e = [1+(1+2\lambda)^{1/2}]$, where $\lambda = \kappa_A/\kappa_D = (k_1\dot\gamma + k_2)S(\dot\gamma)/k_3\,\dot\gamma$. Such a dependence was observed in fair agreement with experiments (see §3.2).

Alternatively, mechanics of flocs under shear, especially equilibrium floc size as a result of cohesive and shear forces (Adler & Mills,1979), including fractal structure of flocs as porous clusters (Sonntag & Russel,1986,1987) led to a (reduced) shear rate dependent floc radius $(R_G/a - 1) \sim \dot\gamma_R^{-s}$ with data fitted value s=.35 (close to s=1/3) and fractal dimension D=2.48 (close to the value D=2.5 for the Diffusion Limited Aggregation). The theory (Sonntag & Russel,1987) was shown in satisfactory agreement with data. Still considering fractal flocs, the rupture of which occuring if the shear force overpass some adhesive force (due to particle links in the breakage area) led to a similar expression (Mills,1985) however shear stress dependent, $(R_G/a - 1) = \sigma_R^{-s}$ with s = 1/(4-D). That leads to s=1/2 for D=2, which corresponds to the (chemical) Cluster-Cluster Aggregation, has been found from theoretical grounds, see later, eq(3.9).

In order to get very much simpler models, many authors took only one structural variable, either the number fraction of particles forming an aggregate (e.g. Denny & Brodkey,1962) or the number of links entering into the aggregated structure, as in a chain of particles (e.g. Cross,1965). For same aim of simple modelling, we may assume that processes involved in the reaction of building up and breaking down the aggregates are of the relaxation type, the number fraction (per unit volume) of aggregated particles, $n_A = N_A/N = \phi_A/\phi = A(t)$ should obeys the equation

$$dn_A/dt = (1-n_A)/\tau_A - n_A/\tau_D \tag{3.6}$$

where τ_A and τ_D are the (mean) relaxation times characterizing the formation and rupture of aggregates, respectively. (More generally, these times are associated to either organization or disorganization of the "structure", in a large sense, e.g. as orientation-disorientation processes). As evidence, assuming each process can be characterized by only one relaxation time could be though as a too crude approximation which should be replaced by a group of discrete times at least, if not a relaxation spectrum, as in polymers. Nevertheless, in presence of a steady shear stress which exerts given hydrodynamic forces on each aggregate, we must expect that aggregate sizes lie in a more or less narrow range of values, giving a (shear dependent) mean size, e.g. an "equivalent radius" R (as seen above from mechanical models of flocs), thus leading to a very reduced number of relaxation times.

For the sake of simplicity, we will continue hereafter with the only two times τ_A and τ_D , as it occurs in the case of suspensions of non-interacting spheres. In steady state conditions ($dn_A/dt=0$), the aggregated fraction number $n_A = A_{eq}$ will be, from eq(3.6)

$$A_{eq} = (1+\theta)^{-1} \qquad \text{where } \theta = \tau_A/\tau_D = \theta(\gamma). \tag{3.7}$$

and a shear dependent effective volume fraction, from eq(3.3),

$$\phi_{eff} = \phi\,[1+(\alpha-1)(1+\theta)^{-1}] \tag{3.8}$$

Step3 - In dilute suspensions of colloidal spheres of radius a, τ_A is the (translational) brownian diffusion time(Smoluchowski,1916), and τ_D can be taken as proportional to $\dot{\gamma}^{-1}$ (Goldsmith & Mason,1967) leading to write $\theta = \tau_{BR}/\tau_D = 6\pi\eta_F a^3\dot{\gamma}/K_B T = Pe$, the well-known Péclet number, with K_B=Boltzman constant and T= absolute temperature.

In concentrated systems, we may extend such a relation, in order to account for particles interactions, considering the suspension as an effective medium, i.e. changing the suspending fluid viscosity η_F into the suspension one, η. Now, the product $\eta\dot{\gamma}=\sigma$ appears in θ, that justifies the shear stress dependence, not the shear rate one, many authors assumed (e.g. Krieger & Dougherty,1959), leading to write, as a dimensionless variable, $\theta = \sigma/\sigma_c$, σ_c being a characteristic shear stress, proportional to $K_B T/a^3$. On the other hand, this may shed some light on non-analytical $\dot{\gamma}$-dependence, found theoretically (e.g. Dhont & al, 1989; Perez-Madrid & Rubi, 1990, with $\theta \sim \dot{\gamma}^{1/2}$), at least on a limited $\dot{\gamma}$-range. Indeed, if a power law $\sigma = C\dot{\gamma}^p$ is taken as an approximation of the bulk shear stress in this range, a shear rate dependence of θ is obtained

$$\theta = (\tau_c\dot{\gamma})^p \qquad , \qquad p=1/2 \tag{3.9}$$

τ_c being a characteristic time, by definition closely retated to relaxation times τ_A and τ_D. That leads to $\theta=(Pe)^{1/2}$ for hard spheres suspensions, in agreement with experiments (e.g. van der Werff & al,1989). It is worth noting that the Péclet number can be interpreted as the ratio of the work of viscous forces acting on a particle, W_H, to thermal energy, K_BT. Therefore, more generally, in the presence of a particular process (involving an interaction energy W_i), the dimensionless group $\tau_c\dot{\gamma}$ associated to this particular process can be taken as the ratio W_H/W_I, thus giving us the characteristic time τ_c, especially its explicit dependences in state variables, as temperature for instance. [As evidence, if the structure evolution involves several processes, the corresponding characteristic times, if their magnitudes lie on the inside of the $\dot{\gamma}$-range of interest , will be taken together].

It is worthy of note that considering the system as a suspension of $p=N/n$ fractal clusters (of mean radius R and fractal dimension D, with n single particles each) should lead to $(\phi_{eff}/\phi) = pR^3/Na^3= (R/a)^q$, where $q=3-D$. Taking any R vs. $\dot{\gamma}_R$ relationship (either from structure kinetics or from floc mechanics) will give (Quemada,1989) explicit dependence $(\phi_{eff}/\phi)=f(\dot{\gamma}_R)$, as in eq.(3.8). For instance, an expression of the form $(R/a)=(1+\dot{\gamma}_R^{-s})$ deduced from floc mechanics (see above, step 2) leads to $(\phi_{eff}/\phi)=(1+\dot{\gamma}_R^{-s})^q$, with $s=4-D$. For $D=2$, i.e. $q=1$, $s=1/2$, we recover the shear rate dependence close to that obtained putting θ from eq.(3.9) into eq.(3.8).

Step 4 - Finding the relation which gives the viscosity η as a function of structural variables remains the main difficulty in most of phenomenological models of non-newtonian behaviour. In the literature, one of the simplest forms has resulted from (arbitrary) assumption that η is a linear combination of the two fractions, stuctured and unstructured, present in equilibrium state. For instance, in a relaxation-type model, such an assumption would lead to $\eta = A\,n_A+B(1-n_A)$, the constants A and B being related to limiting η-values, at very low shear and very high shear, giving definite n_A-values (e.g. $n_A=1$ and $n_A=0$, respectively).

In order to avoid this arbitrariness, we can enter the effective volume fraction defined in eq.(3.3) into the viscosity equation drawn from minimization of viscous dissipation, eq.(2.7) thus leading to

$$\eta_r = (1-\phi_{eff}/\phi_m)^{-2} \qquad (3.10)$$

where the ratio ϕ_{eff}/ϕ_m , with eq(3.8), can be written $K\phi$, where

$$K = [1+(\alpha-1)/(1+\theta)]/\phi_m \qquad (3.11)$$

appears as a structural variable whose reciprocal value can be considered as an effective (shear dependent) packing fraction. In the simpler case α=const., this packing fraction takes the limiting values $\phi_\infty=K_\infty^{-1}=\phi_m$ as $\dot{\gamma}\rightarrow\infty$ (i.e. $\theta\rightarrow\infty$) and $\phi_o=K_o^{-2}=\phi_m/\alpha$, as $\dot{\gamma}\rightarrow 0$ (i.e. $\theta\rightarrow 0$), respectively. Written in terms of these limiting values, eqs.(3.10) and (3.11) become

$$\eta_r= (1-K\phi)^{-2} \qquad (3.12)$$

where $\quad K=K_\infty + (K_o-K_\infty)/(1+\theta) \quad$ and $\quad \theta=(\tau_c\dot{\gamma})^{1/2} \qquad (3.13)$

the corresponding limiting viscosities being

$$\eta_{ro}= (1-\phi/\phi_o)^{-2} = (1-K_o\phi)^{-2} \qquad \text{and} \quad \eta_{r\infty}= (1-\phi/\phi_\infty)^{-2} = (1-K_\infty\phi)^{-2} \qquad (3.14)$$

Hereafter, we shall refer to eqs.(3.10-11&12) as the "structural viscosity model" (SV-Model).

Alternatively, in unsteady conditions, with $K\phi = \phi_{eff}/\phi_m = [1+(\alpha-1)n_A(t)]\phi$,we can use the reduced (structural) variable

$$S(t) = (K-K_\infty)/(K_o-K_\infty) \equiv n_A(t) \qquad (3.15)$$

then the structure-dependent viscosity can be expressed as a time-dependent viscosity

$$\eta_r=\eta_{r\infty}[1-(1-\chi)S(t)]^{-2} \qquad (3.16)$$

where $\chi = (1-\phi/\phi_o)/(1-\phi/\phi_\infty) = \pm(\eta_{r\infty}/\eta_{ro})^{1/2}$ (3.17)

3.2 - Application to Blood and RBC Suspensions.

A very large number of empirical models of non-newtonian blood viscosity are still used. However in model evaluation, some confusion results between more or less unsatisfactory modelling, on one hand, and dispersion in data due to physio-pathological variations from one to another patient, on the other hand. Some tentatives (sometimes not well-justified) to apply structural models established for other materials have been carried out, as for instance the Casson model (Casson,1959) already developped to model the plastic behaviour of suspensions of paint pigments and previously applied to milk rheology. Two rheological variables, as "material constants", enter the Casson equation

$$\sigma^{1/2} = \sigma_y^{1/2} + (\eta_{cas}\dot{\gamma})^{1/2} \qquad (3.18)$$

the Casson viscosity η_{cas} and the yield stress σ_y , both concentration dependent, Fair data fittings have been obtained (cf. Fig.3.1). Unfortunately, for RBC suspensions with different concentrations in fibrinogen (thus changing the cohesive force between RBC in rouleaux), σ_y-values deduced from rheometry were found too large (at least one order in magnitude) if compared to expected ones from direct estimation of interaction forces between two RBC (Merrill & al, 1965).

Rouleau formation and equilibrium size under steady shear has been analysed with the kinetic equation, eq.(3.5) (Murata & Secomb,1988). Fair agreement with experiments (Shiga & al,1983) was observed for the initial rate of rouleau formation. Unfortunately, these results --especially the shear dependence of $\langle p \rangle_e$ -- were not used for blood viscosity modelling, especially in a further study of flow of aggregating blood (Murata & Secomb,1989), that we shall discuss later (see §5)

Applying the SV-Model to normal or modified blood (that can be easely obtained by temperature increase, or adding of chemical agents,...) or to pathological blood samples (corresponding to a specific pathology) could be performed by data fitting. As remarkable constancy and reproductibility are found in rheological parameter-values, ϕ_o ,ϕ_∞ and τ_c , for normal blood (especially for a given individual, even on very long periods of time), comparison of parameter-values for pathological blood to the ones deduced from normal blood, as reference values, can be carried out in order to obtain blood characterization.

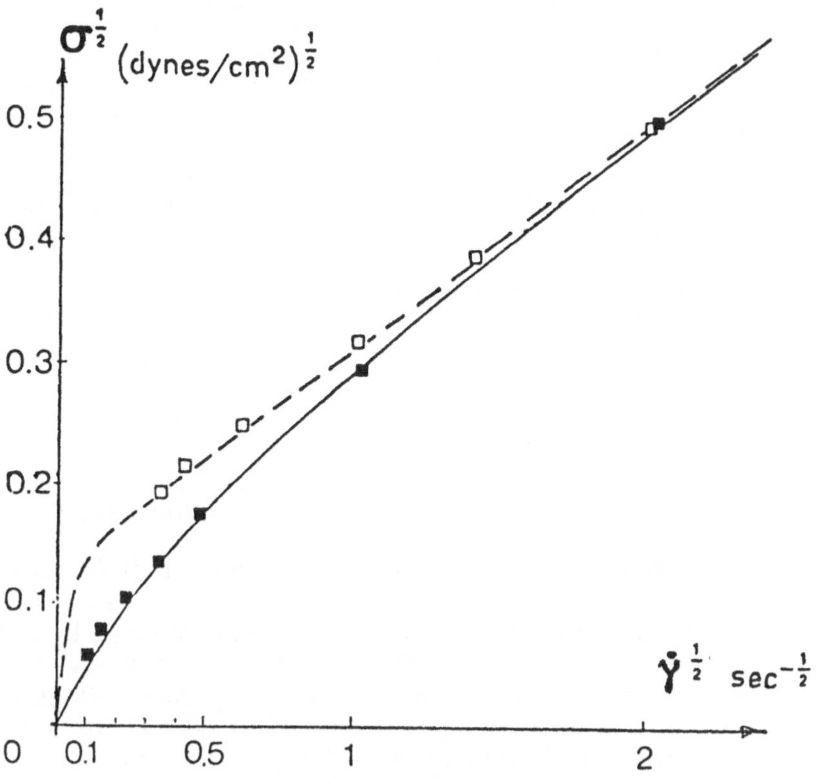

Fig.3.1- Casson plot $\sigma^{1/2}$ vs $\dot{\gamma}^{1/2}$ for whole blood (ϕ=.40) and RBC suspended in defibrinated plasma (ϕ=.39). (Data from Merrill & al, 1965)
Curves from SV-Model with rheological parameter-values

Whole Blood	ϕ_o=.43	ϕ_∞=1.11	τ_c=1.45 s
RBC in defibr. plasma	ϕ_o=.51	ϕ_∞=1.11	τ_c=0.01 s

Compared to the Casson Model, which exhibits an infinite zero shear viscosity η_o, the SV-Model involves finite η_o-values (see Fig.3.1). However, the Casson law is exactly recovered if $\eta_o \rightarrow \infty$, i.e. $\chi \rightarrow 0$ in eq.(3.17), with $S=S_{eq}=(1+\theta)^{-1}$.

From S.Chien's data (Cf. Fig.1.3), Table 3.1 gives the values of ϕ_o ,ϕ_∞ and τ_c after fitting eqs.(3.12-13) on data of NP, NA and HA suspensions. These values suggest the following interpretation (as a renewal of previous ones, Quemada, 1978) involving RC-Aggregation, RC-Deformation and RC-Orientation:

a) In (NA), zero shear packing $\phi_{o(NA)}$=.608 corresponds to suspension at rest, without neither aggregation nor deformation effects. Hence packing of hardened RBC have to take the same value. Indeed, that occurs by centrifuge of chemically hardened cells, that leads to packing ϕ_o=.61±.01(Chien & al,1971b). This fair agreement seems significant, since deformation of normal cells by crowding (in (NA) at rest) can be expected small enough to be neglected, with same order of approximation in neglecting hardened cell deformation under high speed centrifuge.

b) In (NP), zero shear packing $\phi_o(NP)$=.476, lower than $\phi_o(NA)$ results from RCA. At rest, branching of RBC rouleaux likely promotes the formation of a gel-like structure, although very labile, due to the very low interaction energy between RBC (that corresponds to a vanishing small yield stress, see Fig.3.1). Here we assume such a structure is a 3D-network, with a mesh (a rouleau composed in average of n cells). The structural unit is a "3-rouleaux" (see Fig.3.2), the compactness of which would be $\varpi=3\pi/(2r_e n)^2$ where r_e is the equivalent axis ratio of RBC as a disc, r_e=.26 (Goldsmith & Mason,1967). As the packing fraction of 3-Rouleaux in the "gel" is ϕ_m=1 , we can take $\varpi=\phi_o$= .476, that gives n close to n=10, in satisfactory agreement with the mean value from microscopic observation (Paulus & al,1986; see also Fig.1.2).

TABLE 3.1- Rheological Parameters of RBC Suspensions
ϕ=.45 η_F=1.2 cP

	ϕ_o	ϕ_∞	τ_c (sec)
NA	.608	1.124	.04
NP	.476	1.124	.2
HA	.552	—	---

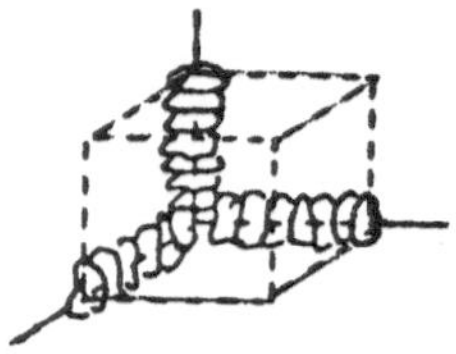
Fig.3.2-The "3-Rouleaux structural unit

c) Identical high shear packings are found in (NA) and (NP), $\phi_\infty(NA)=\phi_\infty(NP)$=1.124, as expected since both suspensions are composed of wholly dispersed normal RBC, with identical deformation induced by suspending fluids having the same viscosity η_F=1.2 cP. Such a packing-value, $\phi_\infty>1$, cannot be understood as a actual packing but as reflecting a particle effective volume smaller than the true one, as we suggested here-above (§2.4) due to RBC tank-treading motion. Indeed, if ϕ' is the effective volume fraction of rigid cores and ϕ'_m their packing , we may write $\phi_\infty=\phi'_m(1-e_{eff}/a)^{-2}$. Taking ϕ'_m=.58 and ϕ_∞=1.124 leads to a maximum thickness of the rotating layer thickness e_{eff} less than 30% of the cell radius, hence a about six times the RBC membrane one. This is a plausible value as a penetration depth of the layer rotatory motion (including the RBC membrane thickness.

d) High shear packing of RBC in (HA), $\phi_\infty(HA)$=.552, is lower than the closest packing of undeformable and non-aggregating RBC (i.e. $\phi_o(NA)$=.608). This lowering can be considered as a consequence of shear-induced rotation of hardened RBC (which do not undergo tank-treading motion). These rotating cells then behave as equivalent spheres with an effective volume $v_{peff}=\zeta v_p$, where ζ=2.52 for particles of axis ratio r_e=.26 (Goldsmith & Mason,1967, p.206) However, theory and experiments demonstrated that during its rotation, a disc spent more time with its faces aligned with the flow than normal to it. Considered as composed of n_o oriented cells (of volume v_p) and n_{rot}=(N-n_o) rotating cells (of volume v_{peff}), the suspension has an effective volume fraction $\phi_{eff}=[1+(\zeta-1)\alpha_{rot}]\phi$. Taking ζ=2.52 and α=.5 (Goldsmith,1971) leads to $\phi_m=\phi/\phi_{eff}$=.568, close to $\phi_\infty(HA)$. On the contrary, at low γ, almost all cells ($n_o \cong N$) exhibit impeded rotation (due to crowding at $\phi>\phi_D=\zeta^{-1}$=.40, Goldsmith & Mason,1967, p.206), thus giving $\phi_o(HA)$ close to $\phi_o(NA)$=.608. Now, n_o decrease as γ grows, leading to increasing HA packing. Such an increase in packing can be interpreted as a dilitant behaviour,

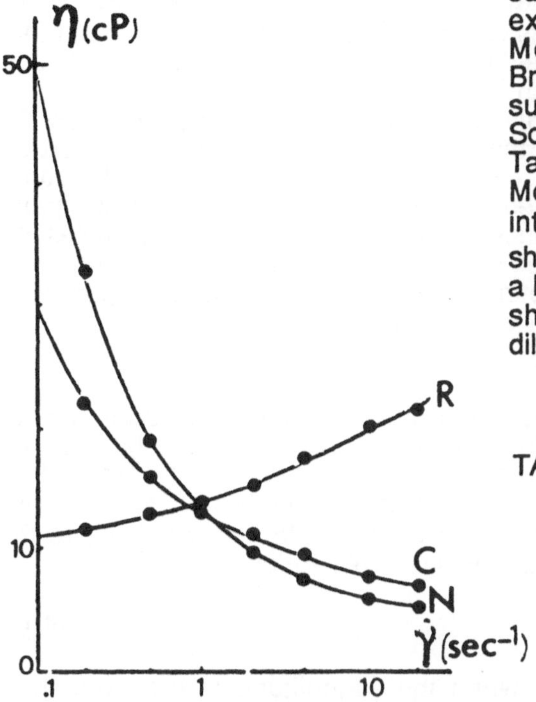

which has been found in concentrated suspensions of rigid spheres, either experimentally (Laun,1988) or from Molecular Dynamics (Bossis & Brady,1990) and already observed in suspensions of rigid RBC (Schmid-Schonbein,1975). (cf. Fig.3.3, with (cf. Table 3.2) data fitting values of the SV-Model in agreement with the given interpretation, especially $\phi_\infty < \phi_o$ i.e. a shear induced lowering in packing, with a high shear packing looser than the low shear one, in agreement with the dilatancy concept).

TABLE 3.2 Rheological Parameters of RBC Suspensions
Normal (N); Crenated (C); Rigid (R)

	ϕ	ϕ_o	ϕ_∞	τ_c (sec)
(N)	.40	.41	1.01	1.08
(C)	.40	.44	.76	.78
(R)	.39	.62	.50	1.34

Fig.3.3-Effects of RCDeformability on apparent viscosity vs. shear rate (RBC suspended in plasma)

e) As in the present model, we have assumed possible to characterize each reversible process responsible of changes in structure by only one characteristic time, values $\tau_{c(NA)}$ and $\tau_{c(NP)}$ are more difficult to be interpreted. As RBC (and a fortiori, rouleaux) are too large particles to exhibit significant brownian motion, these times are likely close to Maxwell relaxation times for non-rigid particles, here directly related to cell membrane viscoelasticity (and internal fluid properties). Indeed, τ_c-values about .2-.3 were found for $\tau_M = \eta_i/G_i$, taking viscosity η_i and elasticity modulus G_i from data of shear and swelling deformation (Skalak,1976)

4. BLOOD AS A THIXOTROPIC AND VISCOELASTIC FLUID : Effect of the time-dependent Structure

4.1 - Modelling of thixotropy and non-linear viscoelasticity

As dynamical equilibrium between structured and unstructured states was postulated to explain shear thinning behavior, thixotropy directly results from structure kinetics: if, for instance, a step function in shear rate is applied to the system in given steady conditions, the passage from the initial steady state to the final one should require a structural change, the time dependence of which is governed by the relaxation processes involved in the kinetic equation(s). As a consequence, shear viscosity becomes time-dependent, that is called thixotropy, which is by nature, closely related to the shear thinning behaviour observed in steady state. We do not therefore necessitate new ingredients in the previous models although we need now at least two relaxation times, as τ_A and τ_D instead of, (in the steady case), only one characteristic time τ_C related to the ratio of the formers. Nevertheless, additional effects could result from time dependence of the "macrostructure", for instance the time-dependent plasma layer thickness in two-phase flows (see e.g. Thurston,1990) and should require specific modelling.

However, intantaneous (reversible) deformation can be supported by the structure before to be broken down, that involves its pure elastic properties. Therefore, under unsteady conditions, the material exhibits the general features of viscoelastic materials with both stored energy (by deformation) and dissipated energy (by creeping flow). For dispersions of weakly interacting particles the domain of linear viscoelasticity often appears vanishingly small and both (retarded) elastic deformation and rupture or formation of the structure may be superimposed (i.e. leading to non-linear viscoelasticity).

Linear viscoelasticity

Linear viscoelasticity modelling is performed using mechanical models, as more or less complex networks of springs and dashpots. For fluid behaviour, the (simplest) one is represented by a Maxwell element (in series, a spring -with an elastic constant G, as shear modulus- and a dashpot -with friction η, as viscosity-) It is governed by the "Maxwell equation" $G^{-1}d\sigma/dt + \eta^{-1}\sigma = \dot{\gamma}$. Applying to the medium at rest a step in shear rate gives the stress relaxation

$$\sigma(t) = \eta\,\dot{\gamma}\,[1-\exp(-t/\tau_M)] \qquad (4.1)$$

in good agreement with experiments when the γ-amplitude is low enough to preserve linearity. In eq.(4.1), $\tau_M=\eta/G$ is the Maxwell relaxation time. More generally, a "n-component Maxwell fluid" (i.e. n Maxwell elements in parallel) can be taken, in order to account for the existence of several relaxation times (sometimes a large number, i.e. a relaxation spectrum, as in polymers). Stress relaxation hence exhibits a non-exponential behaviour as resulting from superimposition of single element responses of the type (4.1).

Non-linear viscoelasticity

Alternatively, in order to describe non-linear properties, as thixotropy and non-linear elasticity, we can generalize the Maxwell equation --after adding a term $\tau_O(d\dot{\gamma}/dt)$ to account for retardation effects (i.e. for shear rate relaxation in creep experiments)--, considering the coefficients G^{-1}, η^{-1} (and also the retardation time τ_O) are structural-dependent variables. In order to account for possible plastic behaviour (with a yield stress σ_y) we assume that the Maxwell equation holds for the shear stess difference $\sigma'= \sigma - \sigma_y$, hence leading to

$$G^{-1}([S])\,d\sigma'/dt + \eta^{-1}([S])\,\sigma' = \dot{\gamma} + \tau_O([S])d\dot{\gamma}/dt \qquad (4.2)$$

where [S] means a set of structural variables, which can be reduced, for the sake of simplicity (see §3.1) to only one scalar variable $S=S(\phi,\dot{\gamma},t)$. The general solution of eq.(4.2) may be written in the form (with $\ddot{\gamma} = d\dot{\gamma}/dt$)

$$\sigma(t) = \sigma_y + [\sigma(0)-\sigma_y]\exp[-F(t,0)] + \int_0^t G(t')\exp[-F(t,t')][\dot{\gamma}(t')+ \tau_O(t')\ddot{\gamma}(t')]dt' \qquad (4.3)$$

where $F(t,t') = \int_{t'}^{t} [G(t'')/\eta(t'')] \, dt''$

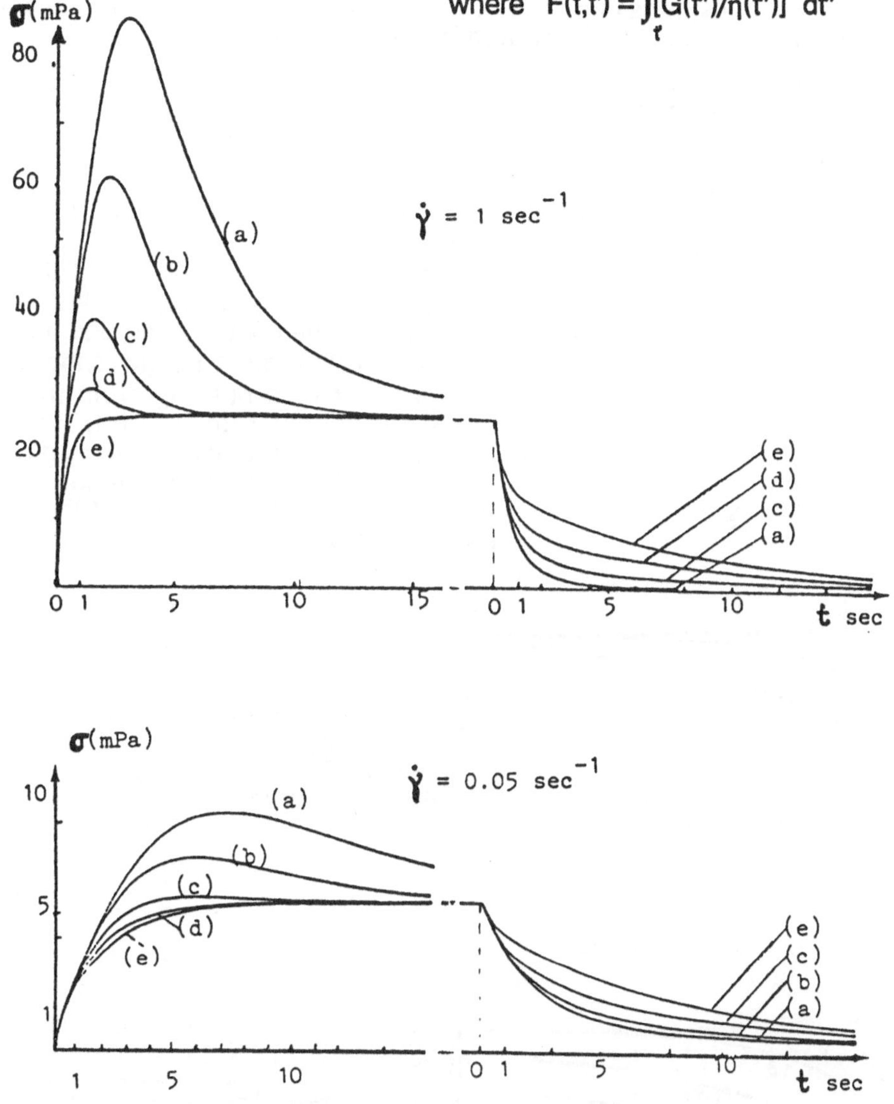

Fig.4.1- Theoretical curves for stress formation and stress relaxation: effects of τ_A-variations (τ_M=.1s; m=1)

τ_A-values: (a)=10s; (b)= 5s; (c)= 2s; (d)= 1s; (e)= .5s

(SV-Model steady parameters: ϕ=.45; η_P=1mPa.s; η_∞=5mPa.s; η_o=.5Pa.s; τ_c=.5s; σ_Y=1mPa)- (from Quemada & Droz, 1983)

Comparing this solution to constitutive equations $\sigma = \sigma(t)$ of the integral type in theories of non-linear viscoelasticity, it is worthy of note that the function $G(t')\exp[-F(t,t')]$ works like a memory function. Indeed, the linear viscoelastic solution (cf eq.(4.1)) is recovered since $\exp[-F(t,t')]$ reduces to $\exp[-(t-t')/\tau_M]$ if the ratio $\eta(t)/G(t)$ is constant, $(\eta/G)=\tau_M$. Moreover such a form $G(t')\exp[-F(t,t')]$ is reminiscent of many expressions several authors (e.g. Meister,1971) proposed for the memory function,which can be written in the general form

$$m[t,t',II(t')] = \sum_{p=1,N} G_p \exp\left(-\int_{t'}^{t} M[II(t''), \tau_p]\, dt''\right) \tag{4.4}$$

where $II(t)$ is the second invariant of the deformation rate tensor and τ_p a spectrum of relaxation times.

A basic difference between the memory function in eq.(4.3) and the one defined in eq.(4.4) lies in their time dependences. Through M, the latter only depends on $II(t)$, i.e. the given shear rate history. On the contrary, through $\eta(S)$, $G(S)$ and $\tau_O(S)$, the time dependence of the former results not only from the time dependent shear rate, i.e. $II(t)$, but also from the structure kinetics which governs the material history in response to the shear rate one. We can say that the model based on the former introduces only one time dependent Maxwell time, $\tau_M(t)$, as the ratio $\eta(S)/G(S)$, instead of a permanent spectrum of relaxation times, τ_p involved in the latter.

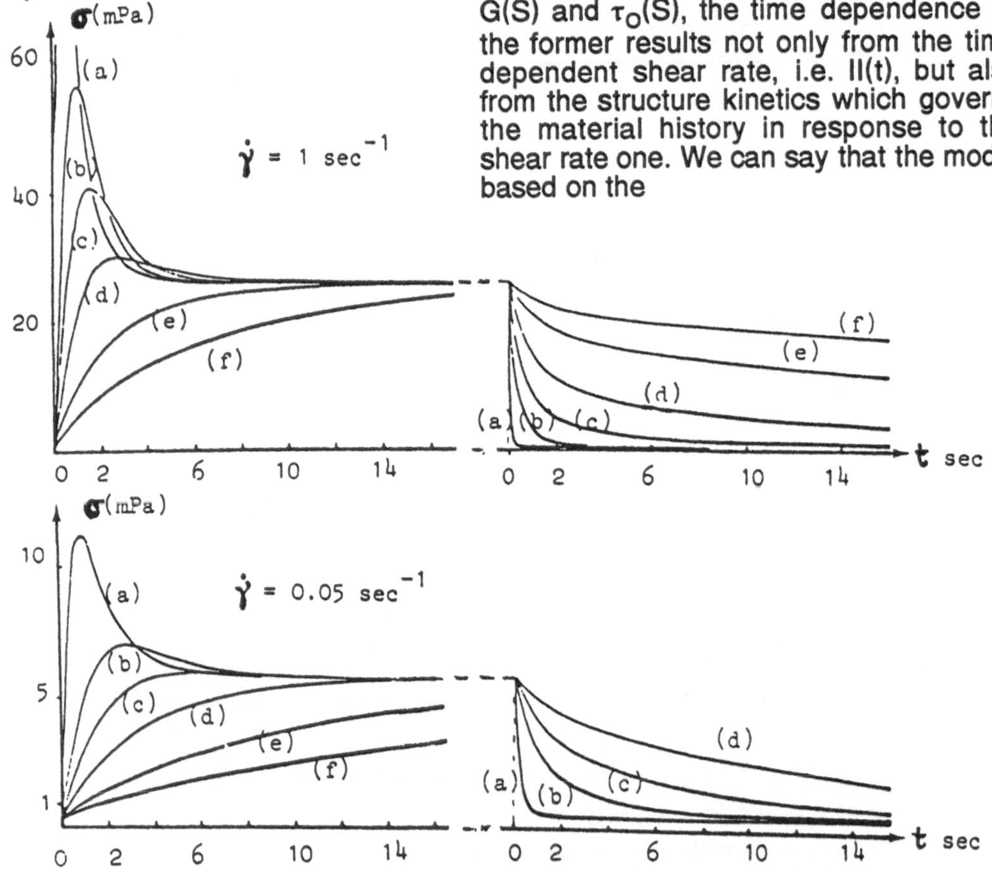

Fig.4.2-Theoretical curves for stress formation and stress relaxation: effects of τ_M-variations (τ_A=2s; m=1)

τ_M-values: (a)=.01s; (b)=.05s; (c)=.1s; (d)=.2s; (e)=.5s; (f)=1s

(SV-Model steady parameters values as in Fig.4.1) (from Quemada & Droz,1983)

Fig.4.1& 4.2 displays the variations $\sigma(t)$ given by eq.(4.3), when we take $\eta(S)$ from the SV-Model, eq.(3.15), and taking (in the absence of elastic model) $G(S)$ from an empirical expression of the form $G_o S^m$, assuming all structural elements responsible of the elastic behaviour are broken down at very high shear rate ($G \rightarrow 0$ as $\dot{\gamma} \rightarrow \infty$). We can see how much the shape of these theoretical curves strongly depends on parameter values, although the unsteady model depends on a very small number of parameters.

4.2 - Application to Blood and RBC suspensions

Modelling oscillatory blood flow through small tubes has been proposed (Thurston,1979) from a "(n+1)-component viscoelastic fluid" model (composed of n Maxwell elements and a dashpot, in parallel). Under sinusoidal excitation, $\dot{\gamma}\sim \exp(i\omega t)$, a dynamic viscosityis found

$$\eta^* = \eta_\infty + \sum_{k=1,n} \eta_k / (1+i\omega\tau_k) \qquad \tau_k = \eta_k / G_k \qquad (4.5)$$

leading to the steady (shear thinning) viscosity as ω->0 , $\eta_S = \eta_\infty + \sum_{k=1,n} \eta_k$.

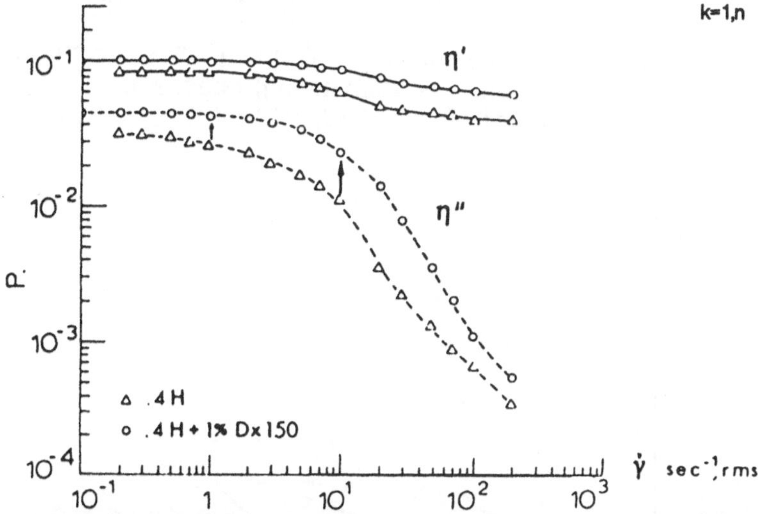

Fig.4.3- Thurston's Data : Variations of η' and η'' vs. $\dot{\gamma}$ and frequency
Effects of increased RBC aggregation by adding of 1% of Dextran 150

In order to describe "the manner in which the model elements change with state of dynamic equilibrium", η_k and G_k are assumed to depend on $\dot{\gamma}$, according to $\eta_k = \eta_{ok} H(\dot{\gamma}\tau_k)$ and $G_k = G_{ok} H(\dot{\gamma}\tau_k)$, with $H(\dot{\gamma}\tau_k) = [1+(\dot{\gamma}\tau_k)^2]^A$. Fig.4.4 illustrates a satisfactory data fitting obtained with 5 relaxing elements. However, taking the same degradation function $H(\dot{\gamma}\tau_k)$, hence constant Maxwell times, discards any proper dependence in time of the structure as in eq.(4.4).

As underlined above, effects of time dependent structure are included in eq.(4.2) through the structure dependent quantities $\eta([S])$, $G([S])$ and $\tau_o([S])$. Therefore improvement of this non-linear viscoelastic behaviour requires prior modelling of viscosity, shear modulus and, eventually, retardation time, i.e.definition of a structural variable S, knowledge of its kinetics as $S= S[\dot{\gamma}(t),t]$ and

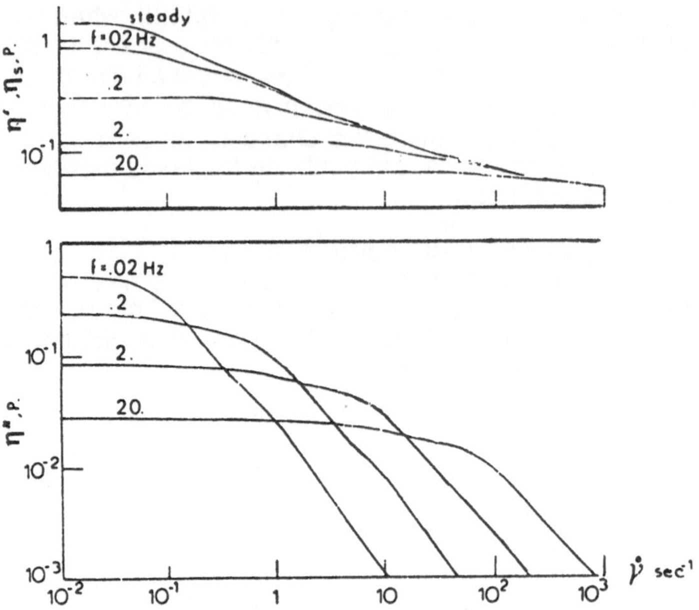

Fig.4.4- Thurston's Model: Variations of η_s, η' and η'' vs. $\dot{\gamma}$ and frequency
Model parameters (five relaxation processes):
τ_k=10, 1, .1, .01, .001 sec; η_{ok}=1, .32, .1, .032, .01 Poise; η_∞=.04 Poise, A=1

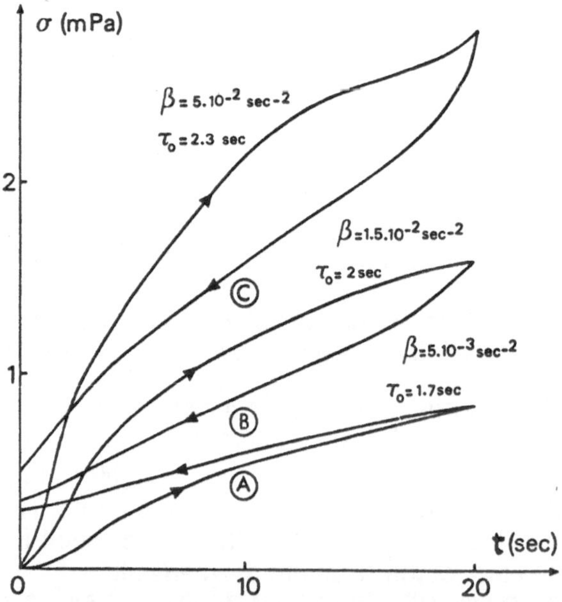

Fig.4.5- Theoretical hysteresis cycles $\sigma(t)$
(in response to triangular steps in shear rate) -(Quemada,1984)
Parameter values: τ_A=2 sec; τ_M=.1 sec; m=2; σ_y=0

explicit forms of $\eta(S)$, G(S) and $\tau_o(S)$. Satisfactory data fittings were obtained (Quemada,1984) for stress formation and stress relaxation experiments on normal blood, using the SVModel, eq.(3.15), the relationship $G(S)=G_oS^m$ and $\tau_o(S)=\tau_o$=Const. We show here an example of hysteresis curves as time curves $\sigma(t)$, (Fig.4.5), in response to linear $\dot{\gamma}(t)$-growth immediatly followed by a symetric decrease, applied to normal blood. Values of all rheological parameters but τ_o were taken from steady and relaxation data fittings independently performed (Quemada & Droz,1983). In comparing these model curves to experiments, an early data resulting from same conditions was taken(see Fig.4.6). Using in the model same $\ddot{\gamma}$-values than in experiment, only slight adjustment in τ_o were necessary to get maximum values of model curves close to experimental measurements. The agreement is satisfactory, taking into account the crudeness of the assumed elastic model.

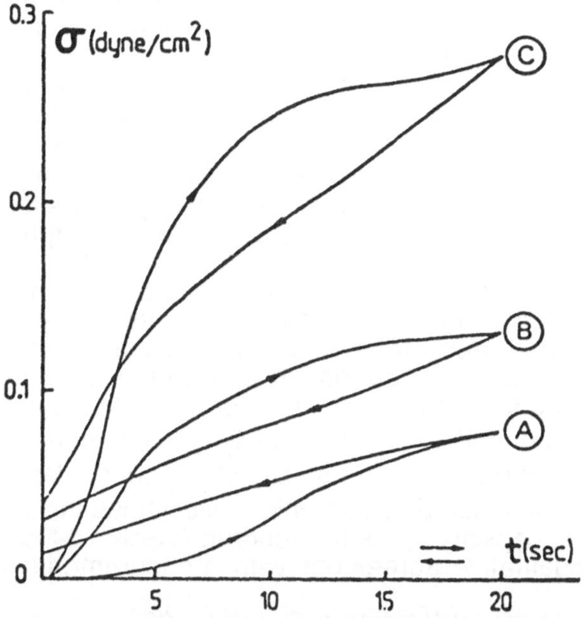

Fig.4.6- Time curves $\sigma(t)$ of hysteresis cycles (triangular steps in shear rate) Normal blood Data (Bureau & al,1978).
$\ddot{\gamma}$-values(sec^{-2}): (A): 5.10^{-3}; (B): 1.45.10^{-2}; (C): 4.2.10^{-2}

5. MODELLING OF BLOOD MICROCIRCULATION.

In principle, flow modelling of any fluid through a given tube can be predicted once the rheological equation of this fluid, as $\sigma=\sigma(\phi,\dot{\gamma},t)$ is known. Number of studies has been devoted to blood flow modelling in cylindrical tubes, some of them taking into account the annular diphasic structure.

In early works on blood flow through narrow vessels (e.g. Bayliiss,.1952; Haynes,1962; Thomas,1962), both phases were considered as newtonian, the core ($0<r<\beta R$, with viscosity$=\eta_s$) and the plasma layer ($\beta R<r<R$, with thickness

$\delta=(1-\beta)R$ and viscosity=η_P). The apparent viscosity is defined from the Poiseuille law applied to a one-fluid flow, i.e. $\eta_{app}=\pi R^4 \Delta p/8QL$. In terms of the relative apparent fluidity $F_a=\eta_P/\eta_{app}$ and relative core fluidity $F_s=\eta_P/\eta_s$ one get

$$F_a=1-\beta^4(1-F_s)\,, \qquad (5.1)$$

In the limit $\delta<<R$, eq.(5.1) becomes $\eta_{app}=\eta_P[1-4(\delta/R)(\eta_P/\eta -1)]$, which described qualitatively the Fahraeus-Lindqvist effect since in this limit δ cannot depend on R (indeed, it should be close d-value in a flow along a plane). In normal blood, for "normal" conditions, δ-values were found close to about 0.7µm(see for instance Middelman,1972).

Later, more realistic models kept newtonian the plasma layer whereas the core was assumed to be non-newtonian, for instance a Casson fluid (Charm & Kurland,1962; Charm & al,1968). Although fitted δ-values obtained from numerous pressure-flow rate measurements were found rather scattered, the $\delta=\delta(\phi)$ variation exhibits the expected decreasing. Shear stress dependence $\delta=\delta(\sigma)$ was not clearly known.

However, it is worth noting that, in the absence of any theory of the wall layer in concentrated suspensions, such approaches remain limited in interest. Indeed, since the plasma layer thickness $\delta(\sigma)$ enters the model as an unknown variable, one will obtain from pressure-flow rate data the (ad-hoc) δ-value leading to the best data fitting, whatever the rheological model may be. As direct δ-measurement is often very difficult to carry out, this difficuty could be overpassed if simultaneous verifying of another prediction of the model was possible.

Such a tentative was carried out in blood flow through narrow slits, comparing the predictions of the SV-Model to both velocity profiles (from Doppler-laser velocimetry) and pressure-flow rate data (Dufaux & al,1980; Quemada & al,1980). Eq.(5.1) ,with equation of continuity of the disperse phase $\phi_s Q_s=\phi_a(Q_w+Q_s)$, led to a core volume fraction $\phi_s=\phi_s(\beta)$ if, for given slit thickness, pressure gradient and feed volume fraction ϕ_a , we equal the theoretical fluidity F_a to the experimental one F_a' (see Fig.5.1). A shear stress dependent plasma layer thickness $\delta(\sigma)$ was obtained (Fig.5.1) which ressembles the similar variation $\delta(\sigma)$ directly observed in capillary tubes (Devendran & Schmid-Schonbein,1975)hence leading to only qualitative agreement with measurements performed in very narrow slits.

A very recent model (Murata & Secomb,1989) was based on a 3-phasic annular flow, composed of a rigid (aggregated) core (region C), surrounded by a suspension of single RBC(region S, with viscosity η_S), both regions being surrounded by a plasma layer (region P, with viscosity η_P). The model is oversimplified assuming the suspension is newtonian, with a viscosity given by the Arrhenius law, $\eta_S=\eta_P\exp(a\phi_S)$. The novel aspect in this model comes from a kinetic equation introduced for aggregated core formation. A very simple form was taken for this equation as governing the number of cells in the core, n_C

$$dn_C/dt = K_g - K_d$$

where K_g and K_d are growth and degradation rates of the core, respectively.The former was assumed proportional to the number of cells contained within a thin layer (of thickness of the order of the cell size) close to the core surface (at $r=r_C$)

and to the shear rate at $r=r_C$. The latter is proportional to (i) the core surface and (ii) the stress difference $\tau_C-\tau_0$ between the shear stress value τ_C at $r=r_C$ and some critical stress τ_0 beyond which RBC aggregates are broken, if $\tau_C \geq \tau_0$, whereas $K_d = 0$ if $\tau_C \leq \tau_0$.

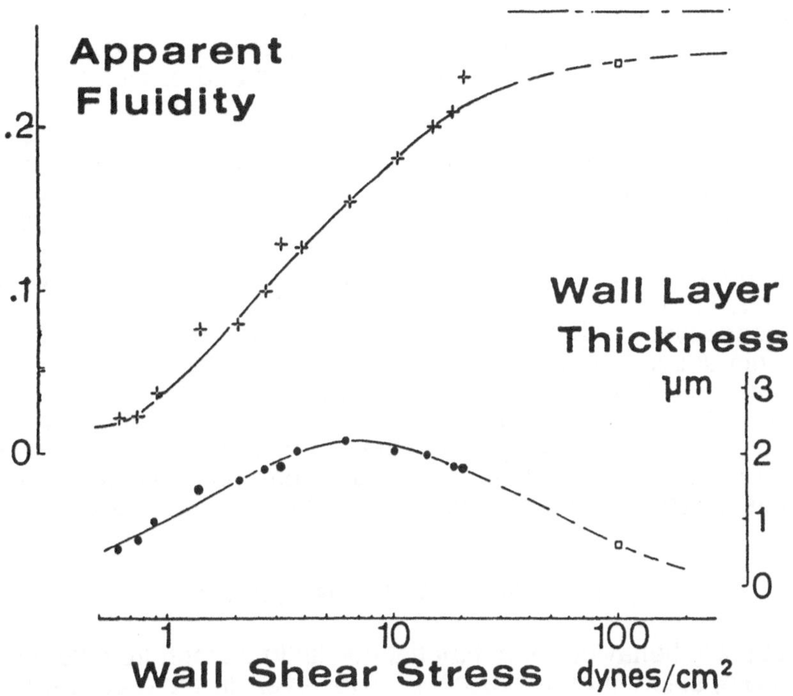

Fig.5.1- Apparent fluidity F_a' and Wall layer thickness δ vs. shear stress σ. Normal human blood flow (ϕ_a=.57) through a slit (350µm x 1.1cm x 10cm). T=23°C. Rheological variables (from Couette Viscometry): K_o=1.62 ; K_∞=.84 ; τ_c=.135sec (●-Points from F_a-curve fitted on F_a' -data ; □-Theoretical point from F_a-curve)-
(from Quemada & al,1980)

A stationnary solution was discussed and compared to experimental data (Reinke & al,1987). Despite the fact that model predictions for relative viscosity are consistent with experimental results, the many simplifying assumptions, especially a cell-free layer thickness independent of flow rate, made questionable the author's conclusions. Furthermore, it is a pity that instead of building this 3-phase model, these authors did not try to incorporate the aggregate model they developed (see §3.1, step2) into some structural model. In fact, the SV-Model (see above) with 2-phases, a (newtonian) plasma layer and a shear thinning core (with the possibility to have a rigid axial core, i.e. to change the shear thinning behaviour into a plastic one -the Casson limit-) does not appear very more complicated than the Murata-Secomb model and indeed, it gives a flow rate dependent plasma layer thickness, as schown on Fig.5.1.

As evidence, it could be possible to built more complex models. Nevertheless, practice requires the knowledge of blood microcirculation in bulk, not limited to flow through an only tube, that rends of limited interest too sophisticated descriptions of the latter. Even if strong simplifications are necessary, new classes of difficulties (here, we shall limit us to the main (physico-mechanical) ones) arise from:

(i) geometry of vessels, as *curvature, tapering* and more especially *branching* , which leads to an averaged tube "length" (between two successive branchings) almost always too short to justify discarding of any entrance effects (these effects contribute to blunt the velocity profile).

(ii) time dependence in "external" conditions, as *driving pressure* (a pulsatile one as resulting from superimposition of continuous and oscillating components; the larger the downstream distance, the smaller is the latter, that results into intermittency at the precapillary level) and *vessel diameter* (the vasomotricity, which is governed by number of control parameters which maintain constant vital functions, e.g. blood flow rate in the physiological range). It is worthnoting that viscoelasticity of vessel walls (essentially here small arteries and arterioles) are quite negligible.

(iii) *elongational effects* , especially due to branching and only in part to (generally small) to vessel tapering.

As a consequence, there is a great probability to overshadow a large part of the details introduced in modelling "one-tube" flow. Moreover, in measuring bulk properties of blood flow through the very complex microcirculatory network (necessarily including the (true) capillary network, for which we need another modelling, for single RBC rheology), On the contrary, due to the above-discussed variability, in both space and time, we expect some improvements could result from applying statistical methods (Guiffant & al,1988; Arhaliass & al,1989).

6. CONCLUDING REMARKS

Macroscopic behaviour of a material results in general from its properties at a smaller scale, sometimes microscopic. However, in most cases, large scale properties are not the simple addition of small scale ones: "collective" properties can be observed, which may be entirely new and necessarily involves a small number of (macroscopic) variables. In such a meaning, one may expect that some "universal" properties of disperse systems may exist and may be described by simple models. Therefore, Blood and RBC suspensions, as disperse media, should exhibit (at least in part) these "universal" properties, i.e. general models can be applied to them: in this lecture we moderately succeed in this goal, although some fair agreement in data fitting were observed and satisfatory physical interpretations were given. The more interesting aspect seems to lie in the phase separation (which lead to annular diphasic flow through a tube), not only because the lubricant property of the wall phase is of prior importance in practice, as reducing the resistance to flowing, but also as the existence of such a transition remains an open question. Furthermore, beyond the proof of a (particle depleted) wall layer, the problem of its time dependent properties should be solve in order to improve our understanding in measurements of non-linear viscoelastic properties of Blood and, more generally, of any disperse media.

REFERENCES

Adler & Mills (1979) J.Rheol. 23: 25-30
Alonso C, Pries AR & Gaehtgens P (1989) Biorheol. 26: 229-246
Arhaliass A, Guiffant G & Dufaux J (1989) Biorheol. 26: 552-566
Aubert JH & Tirrel M (1980) J.Chem.Phys.72: 2694-2701
Batchelor GK & Green JT(1972) J. Fluid Mech. 56: 401-427
Batchelor GK (1970) *An Introduction to Fluid Dynamics*, Cambridge Univ. Press
Bayliss L (1952) in *Deformation and Flow in Biological Systems* (Frey-Wissling ed, North Holland Publ., Amsterdam), 355-418
Bell G (1978) Science 200: 618-627
Bell G, Dembo M & Bongrand P (1984) Biophys.J. 45: 1051-1064
Bossis G & Brady JF (1990) in *Hydrodynamics of Dispersed Media* (Hulin JP, Cazabat AM, Guyon E & Carmona F, eds, North Holland Publ., Amsterdam) 119-137
Botet R & Jullien R (1985) Phys.Rev. Lett. 55: 1943-1952
Boynard M (1986) VIth Intern. Congr. Biorheology, Vancouver, in Biorheol. 23: 243 (abstract)
Boynard M & Hanss M(1981) in: *Hemorheology and Diseases* (Stoltz JF & Drouin P eds,Doin Paris), 185-188
Brooks DE, Goodwin JW & Seaman GVF (1970) J.Appl.Physiol. 28: 172-177
Brooks DE, Goodwin JW & Seaman GVF (1974) Biorheol. 11: 69-77
Brunn PO & Chi S (1984) Rheol.Acta 23: 163-171
Bucherer C, Lelièvre JC & Lacombe C (1988) Biorheol. 25: 639-649
Bureau M, Healy JC, Bourgoin D & Joly M (1978) Rheol. Acta 17: 612-625
Burton AC (1966) *Physiology and Biophysics of the Circulation*, Year Book Medic.Publ., Chicago
Casson N (1959) in: *Rheology of Disperse Systems* (CC Mill ed. Pergamon, London) 84-102
Charm SE & Kurland GS (1962) Trans. Soc. Rheol. 6: 25
Charm SE, Kurland GS & Brown SL (1968) Biorheology 5: 15-43
Chien S (1970) Science 168: 977-979
Chien S & Jan KM (1973) J. Supramol. Struct. 1: 385
Chien S, Usami S, Dellenbach RJ & Gregersen MI (1967) Science 157: 827-829
Chien S, Usami S, Dellenbach RJ & Bryant CA (1971a) Biorheology 8: 35-37
Chien S, Usami S, Dellenbach RJ, Bryant CA & Gregersen MI(1971b) *Theorical and Clinical Hemorheology* (Hartett HH & Copley AL, eds, Springer-Verlag, Berlin) 136-143
Chien S, Luse SA, Jan KM, Usami S, Miller LH & Fremount H (1971c) Proc. 6th Europ. Conf. on Microcirculation (Ditzel & Lewis, eds, Karger, Basel), 29-34
Cokelet GR (1972) in: *Biomechanics. Its Foundations and Perspectives*, (Fung YC & al, eds, Prentice-Hall,Englewood Cliffs, N.J.) 63-103
Cokelet GR (1976) in: *Microcirculation* (Graysson J & Zingg W,eds., Plenum Press, NY) 1: 9-32
Copley AL (1984) Biorheol. 21: 135-153
Cross MM (1965) J.Colloid Sci. 20: 417-437
De Gennes PG (1979) J. de Physique 40: 783-787
Dembo M, Torney DC, Saxman K & Hammer D (1988) Proc. R. Soc. Lond. B. 234: 55-83
Denny DA & Brodkey RS (1962) J. Appl. Phys., 33: 2269-2274
Devendran T & Schmid-Schönbein H (1975) Pflügers Arch. 355: R20 (abstract)
Dhont JKG, van der Werff & de Kruif (1989) Physica A 160: 195-204
Dufaux J, Quemada D & Mills P (1980) in *Rheology* (Astarita G, Marrucci G & Nicolais L, eds, Plenum Press, NY) Applications(3): 561-566
Evans EA & La Celle PL (1975) Blood 45: 29-42
Evans EA (1985) Biophys. J. 48: (I) 175-183; (II) 185-192
Farhaeus R (1929) Physiol. Rev. 9: 241-274
Farhaeus R & Lindqvist T (1931) Amer. J. Physiol. 96: 562-568
Fischer T & Schmid-Schönbein H (1977) Blood Cells 3: 351-365
Fischer T, Stöhr-Liesen & Schmid-Schönbein H (1978) Science 202: 894-896
Friedlander S (1964) *Smoke, Dust and Haze* , J Wiley and sons, NY

Gaehtgens P (1979) in: *Hemorheology and Diseases* (Stoltz JF and Drouin P, eds, Doin Paris) 241-248
Goldsmith HL (1971) Biorheology 7: 235-242
Goldsmith HL & Mason SG (1967) in: *Rheology :Theory and Applications.* (Eirich FR, ed., Acad. Press, NY), 85-250
Guiffant G, Arhaliass A & Dufaux J (1988) Int. J. Heat Mass Transfert 13: 1952-1954
Hanss M (1983) Biorheol. 20: 199-211
Haynes RH (1962) Biophys. J. 2: 95
Healy JC & Joly M (1975) Biorheol. 12: 335-340
Joly M, Lacombe C & Quemada D (1981) Biorheol. 18: 445-452
Karnis A, Goldsmith HL & Mason SG (1966) J.Colloid.Int.Sci. 22: 531-553
Keller SR & Skalak R (1982) J.Fluid Mech. 120: 27-47
Kon K, Maeda N & Shiga T (1983) J.Physiol.(London) 339: 573-582
Krieger IM & Dougherty TJ (1959) Trans. of Soc. of Rheology (III), 137-152
Krieger IM (1972) Ad.Colloid Int.Sci. 3: 111-136
de Kruif CG, van Iersel EMF, Vrij A & Russel WB (1985) J.Chem.Phys.83: 4717-4725
Laun HM (1988) in: *Progress and trends in Rheology* (Giesekus H & al, eds., Steinkopff Verlag, Darmstat) II, pp 287-290
Lewis AE (1970) *Principles of Hematology* , Butterworths, London
Lhuillier D (1990) Physica A 165: 303-319
Lighthill MJ (1969) in: *Circulatory and respiratory Mass Transport* . Ciba Foundation Sympos. (Wolastenholme GEV & Knight J, eds., London) 85-98
Loose W & Hess S (1989) Rheol.Acta 28: 91-101
Meakin P & Deutch J (1985) J.Chem.Phys. 83: 4086-4094
Meister BJ (1971) Trans.Soc.Rheol. 15: 63-89
Merrill EW, Margetts WG, Cokelet GR, Britten A, Salzman EW, Pennel RB & Melin M (1965) in: *Symposium on Biorheology* (Copley AL, ed., Wiley & sons, NY) 601-611
Merrill EW, Margetts WG, Cokelet GR & Gilliland ER (1965) in: *Symposium on Biorheology* (Copley AL, ed., Wiley & sons, NY) 135-143
Middleman S (1972) *Transport Phenomena in the Cardiovascular System* . Wiley-Intersc. NY
Mills P, Quemada D & Dufaux J (1980) in *Rheology* (Astarita G, Marrucci G & Nicolais L, eds, Plenum Press, NY) Applications (3): 567-572
Mills P (1985) J.Physique Lett. 46: L301-L309
Murata T & Secomb TW (1988) Biorheol. 25: 113-122
Murata T & Secomb TW (1989) Biorheol. 26: 247-259
Nozières P & Quemada D (1986) Europhys.Lett. 2: 129-135
Onuki A (1990) J.Phys.Soc.Japan. (submitted)
Paulus F, Dixneuf P & Stoltz JF (1986) in *Séminaire INSERM: Techniques en Biorhéologie* (Stoltz JF, Donner M & Puchelle E, eds, Editions INSERM, Paris) 143: 245-250
Perez-Madrid A, Rubi JM & Bedeaux D (1990) Physica A 163: 778-790
Poiseuille M (1835) Compt.Rendus.Acad.Sciences (Paris), 554-560
Powell MJ (1979) Phys.Rev. B20: 4194-4198
Quemada D (1977) Rheol. Acta 16: 82-94
Quemada D (1978) Rheol. Acta 17: 632-642 & 17: 643-653
Quemada D, Dufaux J & Mills P (1980) in *Rheology* (Astarita G, Marrucci G & Nicolais L, eds, Plenum Press, NY) Fluids (2): 633-638
Quemada D (1982) Lect.Notes in Phys.(Casas-Vazquez J & Lebon J (eds),Springer, Berlin) 164: 210-247
Quemada D & Droz R (1983) Biorheol. 20: 635-651
Quemada D (1984a) in *Advances in Rheology* (Mena B & al ,eds.,UNAM Mexico-City) (II): 571-582
Quemada D (1984b) Biorheol. 21: 423-436
Quemada D (1989) Progr.Colloid Polym.Sci. 79: 112-119
Reinke W, Gaehtgens P & Johnson PC (1987) Am.J.Physiol. 253: H540-H547
Roussel JF, Blanc R & Camoin C (1989) J.Phys.France 50: 3269-3283
Russel WB & Gast AP (1986) J.Chem.Phys. 84: 1815-1826
Schmid-Schönbein H, Volger E & Klose HJ (1972) Pflügers Arch. 333: 140-155

Schmid-Schönbein H (1975) Blood Cells 2: 285-306
Schmoluchovski M (1916) Kolloid Z. 190-195
Shiga T, Imaizumi K, Maeda N & Kon K (1983) Am. J.Physiol. 245: H259-H264
Shiga T, Maeda N & Kon K (1990) Critic.Rev.Oncol.Hematol. 10: 9-48
Skalak R & Zhu C (1990) Biorheol. 27: 309-325
Skalak R (1973) Biorheol. 10: 229-238
Skalak R (1976) in *Microcirculation* (Grayson J & Zingg, eds., Plenum Press,NY) 457-499
Skalak R (1990) Biorheol. 27: 277-293
Snabre P, Bitbol M & Mills P (1987) Biophys.J. 51: 795-807
Sonntag RC & Russel WB (1986) J.Colloid Int.Sci. 113: 399-413
Sonntag RC & Russel WB (1987) J.Colloid Int.Sci. 115: 378-389
Sperry PR (1984) J.Colloid Int.Sci. 99: 97-108
Stadler AA, Zilow EP & Linderkamp O (1990) Biorheol. 27: 779-788
Sutera SP, Pierre PR & Zahalak GI (1989) Biorheol. 26: 177-197
Taylor GI (1932) Proc. Roy. Soc. London, 138A: 41-45
Thomas HW (1962) Biorheol. 1: 41-56
Thurston GB (1979a) in:*Hemorheology and Diseases* (Stoltz JF and Drouin P,eds,Doin Paris), 51-66
Thurston GB (1979b) J.Rheol. 23: 703-719
Thurston GB (1990) Biorheol. 27: 685-700
Tirrel M & Malone MF (1977) J.Polym.Sci. 15: 1569-1578
Van der Werff JC, de Kruif CG & Dhont JKG (1989) Physica A 160: 205-212
Van Dongen P & Ernst M (1984) J.Stat.Phys. 37: 301-311
Viksek T & Family F (1984) Phys.Rev.Lett. 52: 1669-1674

PHASE SEPARATION OF FLOWING POLYMER SOLUTIONS

B.A. Wolf
Institut für Physikalische Chemie, Universität Mainz
Jakob-Welder-Weg 13, D-65 Mainz, FRG

INTRODUCTION

To study the influence of shear or extensional flow on the miscibility of liquids of ***low molecular weight***, highly sophisticated apparatus is needed or no effects will be observed. With ***polymer***-containing systems, on the other hand, deviations from the equilibrium behavior can be seen even with the naked eye. To the knowledge of the author, the pioneering work in this field was done by A. Silberberg and W. Kuhn [1,2], who reported that the demixing, observed with solutions of the (incompatible) polymers polystyrene and ethylcellulose in benzene upon cooling, can be shifted to temperatures about 10 °C lower by shear rates that are only of the order of 200 s^{-1}.

As long as two high molecular weight components are present in the mixture, the effects are particularly large, but even with ordinary polymer solutions they cannot escape one's notice, for instance in the visual determination of their cloud points[3]: Depending on the particular system and conditions, the demixing temperatures measured at rest and while stirring differ typically by a few tenths of a degree. Some solutions become less turbid as they flow (i.e. they behave like the above solutions of polymer mixtures), but others become more cloudy. This means that shear-demixing is observed in addition to shear dissolution. Meanwhile many additional examples[4-17] for the effects of flow have been published, including reports on a flow-induced phase separation of polymer mixtures in the absence[15] and in the presence[17] of a solvent.

It should be noted in passing, that the diversity of phenomena outlined above is not only of theoretical interest, but also of great practical importance. This is particularly true for the mastering of the processing of polymer blends, since it is essential to know whether the system changes from the homogeneous to the two-phase state or *vice versa* during extrusion.

The present paper reports on typical experimental results and deals with the question how the great variety of findings can be well enough understood theoretically to perform predictive calculations. It is confined to shear flow and does not deal with extensional flow. It should, however, be noted that the effects of both types of stress and their qualitative theoretical explanation are very similar.

PROCEDURES AND OBSERVATIONS

Methods

In principle any measurable physico-chemical quantity that is sufficiently different for the phases formed upon the demixing of the solutions of present interest can be used to study the influence of shear. In practice, however, only the refractive indices and viscosity coefficients are used for this purpose.

Optical methods Most of the early experiments were carried out with cells in which the flowing solutions can be seen. Silberberg and Kuhn[1,2] have used a Couette-type apparatus to shear their systems. The change from the homogeneous state into the heterogeneous state and *vice versa* was monitored by means of a specially devised optical method that utilizes the statistical addition of the deflections introduced into parallel light by the presence of drops. In similar experiments with a Searl-type shearing cell, the phase behavior of flowing solutions was measured via turbidity[6,9,12]. In other cases the scattered light was used[16]. Sometimes the appearance[3,4,8] or disappearance[3] of cloudiness produced by shear was directly determined visually.

Viscometric methods For polymer/solvent systems[5,6,12,17] it is often far more convenient to monitor the entrance from the homogenous into the two-phase region by utilizing the discontinuities[18] of the apparent viscosity η at a given shear rate $\dot{\gamma}$ or shear stress $\dot{\gamma}\eta$ upon the variation of temperature or pressure. In fact it was with this method that shear-dissolution was (at elevated pressures) observed[19] for the first time. Although the changes in viscosity are normally less pronounced when polymer mixtures are involved, they can also be used for the studies of such systems[12-15].

As a consequence of the alterations in η described above, a discontinuity is seen as one plots the temperature inside the shearing cell *versus* that of the thermostatting fluid. This means that the entrance into the two phase region can also be monitored by this thermometric method[12].

Effects

In the following, typical effects observed with the sheared solutions of one or two polymers in a low molecular weight solvent, and with polymer mixtures are reported. In the case of polymer/solvent systems, the demixing conditions, as determined from turbidity and viscosity, respectively, differ in the region of higher shear rates[12]. It turns out that the segregation of a second phase is, under these circumstances, detected at a later stage in terms of increased cloudiness than in terms of reduced flow resistance.

Polymer/solvent systems Fig. 1 shows an example[12] for turbidimetric measurements. The fraction I/I_0 of the intensity I_0 of primary beam that passes the solution at different shear rates is plotted as a function of temperature. As can be seen from this diagram, the transition from slightly turbid to non-transparent takes place in a very narrow T-interval for the present concentration at all $\dot{\gamma}$ values. In the vicinity of the critical point, however, this range can become much larger[5]. For stagnant solutions the opalescence within the homogeneous region was used to determine the binodal temperatures by means of a generalized Debye plot ($\ln(I/I_0)$ *vers.* T). The critical composition can than be obtained from the minimum in $\ln(I/I_0)_{binodal}$ as a function of polymer concentration.

The first instance[19] for an influence of shear rate on viscometrically determined demixing conditions is given in Fig. 2. The entrance into the two-phase region upon application of pressure (noticeable from the reduction of η as a result of the transference of considerable amounts of polymer into the suspended phase) is shifted by more than 100 bar to higher values if $\dot{\gamma}$ is only moderately increased.

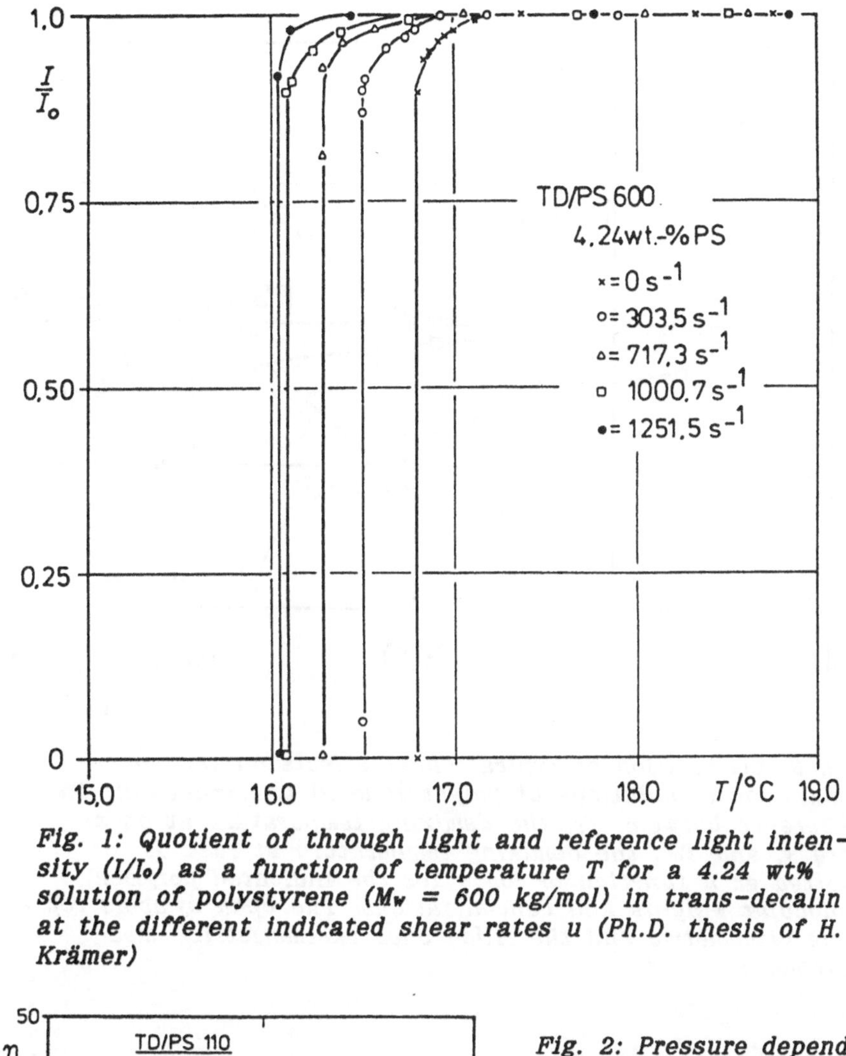

Fig. 1: Quotient of through light and reference light intensity (I/I_0) as a function of temperature T for a 4.24 wt% solution of polystyrene (M_w = 600 kg/mol) in trans-decalin at the different indicated shear rates u (Ph.D. thesis of H. Krämer)

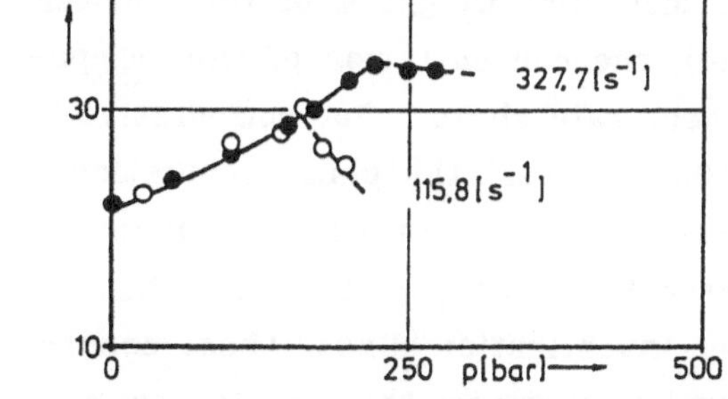

Fig. 2: Pressure dependence of the viscosity η of an 8 wt% solution of polystyrene (M_w = 110 kg/mol) in trans-decalin at the indicated shear rates[19]; a reduction of η upon an increase of p indicates demixing.

The essential viscometric information for the probably best studied system *trans*-decalin/polystyrene[9,12] (TD/PS) is collected in Fig. 3 in terms of ΔT_{visc}, the difference of $T_{\dot\gamma}$, the demixing temperature at shear rate $\dot\gamma$, and T_0, that under equilibrium conditions ($\dot\gamma$ = 0).

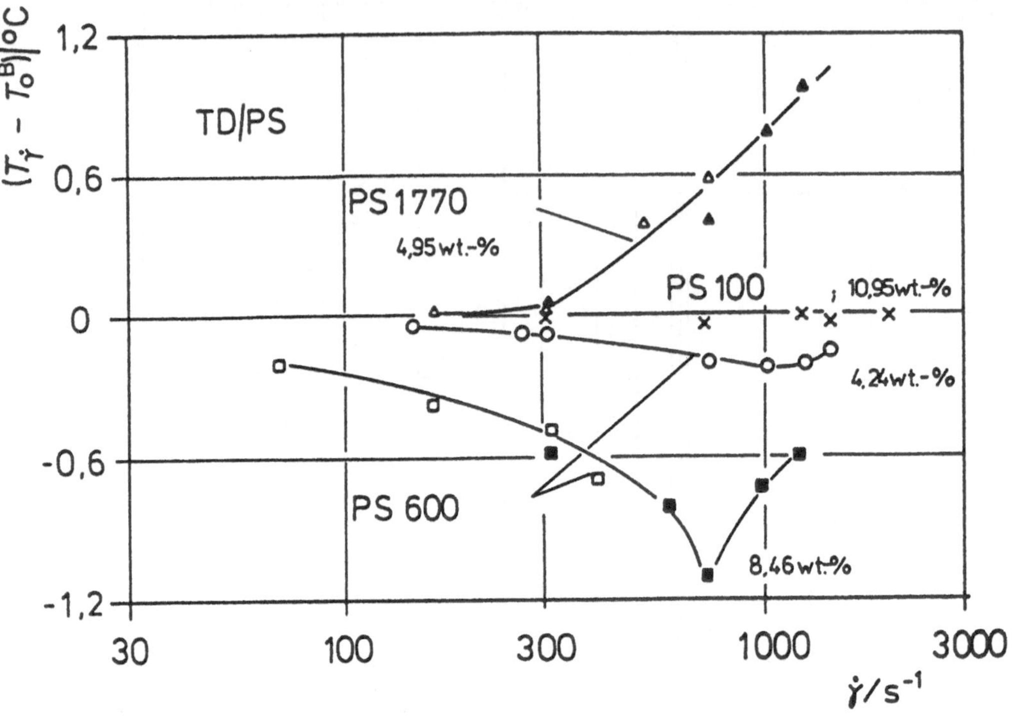

Fig. 3: Compilation of typical shear effects on the phase separation of solutions of polystyrene in trans-decalin. The difference between $T_{\dot\gamma}$, the demixing temperature at shear rate $\dot\gamma$, and T_o^B, the demixing temperature at rest, is plotted as a function of log $\dot\gamma$ for the indicated polymer molecular weights and concentrations. The open symbols represent viscometric and the filled ones thermometric[12] measurements.

This diagram demonstrates that all possible phenomena can be observed with that system, depending on the molecular weight M of the polymer (given in its abbreviation as kg/mol), the concentration of this component in the solution, and on the shear rate chosen. For comparatively low molecular weight polymer (PS 100 and less) the phase separation temperatures are within experimental error not changed by $\dot\gamma$, independent of concentration. For moderately large M values (e.g. PS 600) ΔT_{visc} is negative at all shear rates under investigation. Under these conditions shear-dissolution is the general phenomena; the effects become more pronounced at higher concentrations. The connection of the measuring points in Fig. 3 to yield two intersecting lines will be justified in the theoretical section. For very large M (≥ PS 1770), finally,

ΔT_{visc} is positive over the entire range of $\dot{\gamma}$, i.e. the solutions show shear-demixing behavior.

If the chain length is chosen appropriately (e.g. PS 1180), it is even possible to observe an inversion of shear effects[12] as shown in Fig. 4. In the region of low $\dot{\gamma}$ values the demixing temperature of a given solution remains practically unchanged. As $\dot{\gamma}$ is raised, shear dissolution sets in, reaches a maximum extent and decreases again. Eventually the effect changes sign at a characteristic shear rate above which shear-demixing is observed.

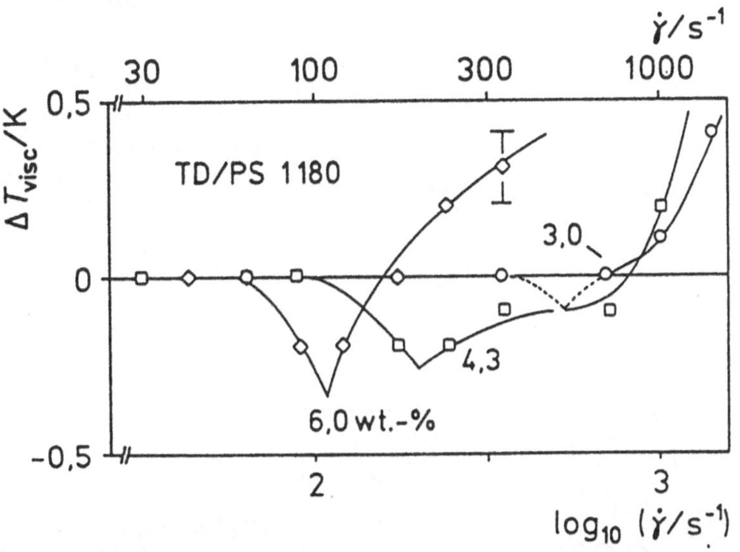

Fig. 4: Plot of $\Delta T_{visc} = T_{\dot{\gamma}}^{visc} - T_0^B$ vers. log $\dot{\gamma}$ (cf. Fig. 3) for three solutions of polystyrene (M_w = 1180 kg/mol) of the indicated concentrations in trans-decalin[12].

Polymer-1/polymer-2/solvent systems In order to keep two incompatible polymers in solution, large amounts of solvent are required and one ends up with systems which differ, roughly speaking, from those studied in the last section only by the altered chemical nature of part of the high molecular weight material. Indeed, all phenomena described for the binary systems can also be found with the present ternary ones; the effects, however, are normally larger by almost one order of magnitude. Turbidimetric results[17] for the system TD/PIB 80/PS 150 (PIB: polyisobutene) given in Fig. 5 illustrate shear-dissolution. The viscometric results[17] for the system D/PIB 80/PS 2000 (D: equilibrium mixture of the *cis* and *trans* isomers of decalin) shown in Fig. 6 document the inversion of shear effects.

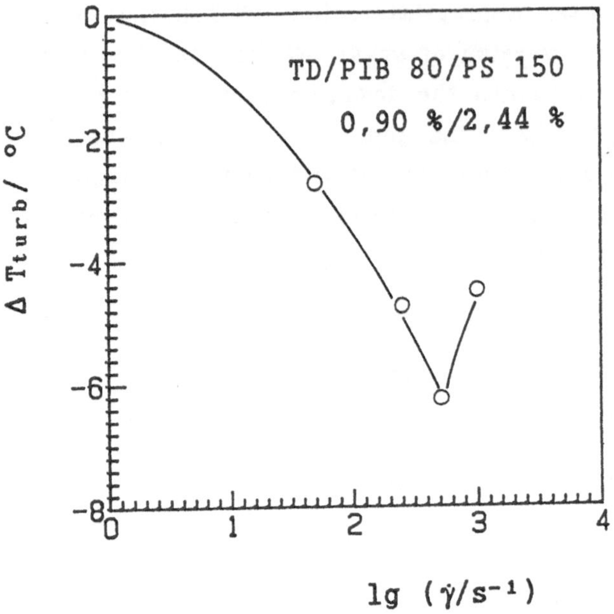

Fig. 5: Plot of $\Delta T_{turb} = T_{\dot{\gamma}}^{turb} - T_0^B$ vers. log $\dot{\gamma}$ for one composition of the ternary system trans-decalin/polyisobutene/polystyrene[17]; the molar masses (kg/mol) and polymer concentrations (wt%) are indicated.

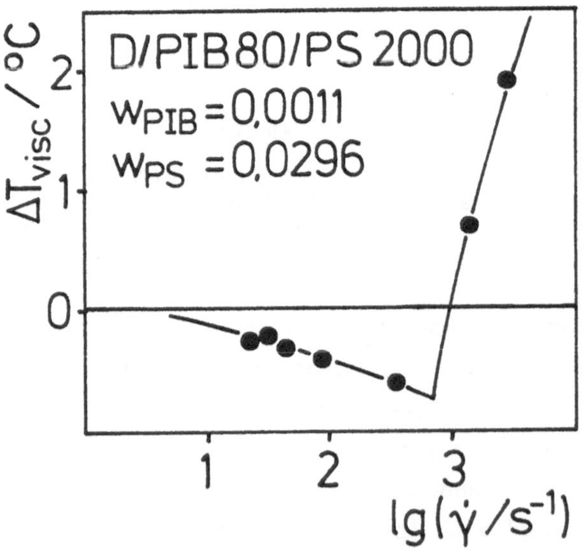

Fig. 6: Plot of $\Delta T_{visc} = T_{\dot{\gamma}}^{visc} - T_0^B$ vers. log $\dot{\gamma}$ for one composition of the ternary system decalin/polyiso-butene/polystyrene[17]; the molar masses (kg/mol) and polymer concentrations (wt-fractions) are indicated.

Polymer-1/polymer-2 systems The study of the phase separation of flowing liquids is naturally restricted to partially compatible pairs of macromolecules in their molten state. The present mixtures differ from the above cases by the fact that not only the solute but also the solvent is high molecular in its nature. The system PS/PVME (PVME: poly(vinyl methyl ether)) exhibiting lower critical solutions temperatures (LCSTs) is probably best studied. How the entrance into the two-phase region upon heating is monitored by the viscosity of the system in this case[14] can be seen from Fig. 7.

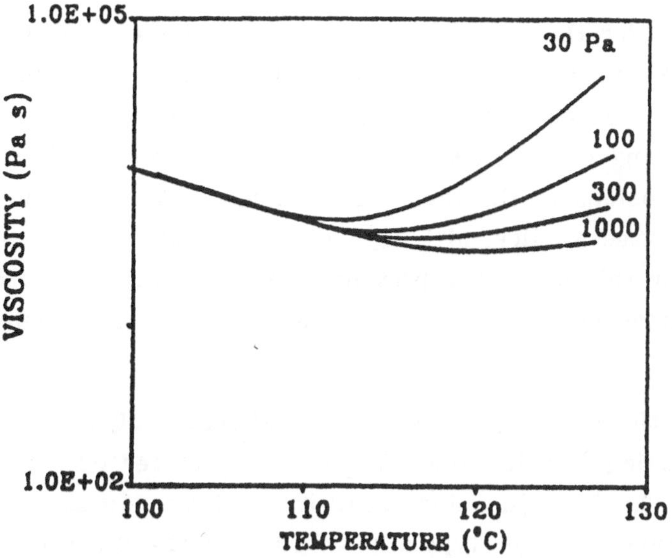

Fig. 7: Temperature dependence of the viscosity of a mixture of polystyrene (PS, 44 wt%)) and poly(vinyl methyl ether) (PVME) at the indicated shear stresses[14]; PS: M_w = 348 kg/mol and $U = (M_w/M_n)-1 = 0.9$; PVME: M_w = 48 kg/mol and $U = 0.9$.

Although the discontinuities are not as sharp as with the solutions of polymers in low molecular weight liquids, the minima in $\eta(T)$ indicate phase separation beyond doubt[14]. The above diagram corresponds to an extension of the homogeneous region by the application of shear. Examples have, however, also been found for the opposite effect[15]. So far no inversion of effects has been reported to the knowledge of the author. From the material published so far one can conclude that the magnitude of shear effects is comparable to that observed with the solutions of incompatible polymers in a common solvent.

CALCULATION OF PHASE DIAGRAMS

Different theoretical approaches have been taken to explain the influences of shear on demixing. In the thermodynamic treatment, the frictional forces can be considered in at least two ways: Either in terms of the stored energy which is added to the Gibbs energy of mixing[20,21], or in terms of chain stretching which makes a flexible polymer "semi"-flexible[21].

The conditions for the transition between the phase separated and the homogeneous state of flowing polymer/solvent mixtures can also be calculated on the basis of the stationary size of the droplets of the suspended coexisting phase[22]. As soon as these droplets are broken down to the dimensions of the individual polymer molecules, the system is homogeneous. Some studies, although not ending up with phase diagrams, are closely related to the present topic, particularly those dealing with scattering functions and the dynamics of phase separation under flow[23,24].

In this work, the demixing conditions for sheared polymer solutions are exclusively calculated in terms of the energy stored during stationary flow. The fact that this recoverable shear normally constitutes only a minute fraction (typically $\ll 10^{-3}$) of the Gibbs energy of mixing[20] justifies the application of equations which are strictly speaking only valid for equilibrium conditions.

Generalized Gibbs energy of mixing $\Delta G_{\dot\gamma}$

The starting point of the present calculations is Eq. (1)[20]

$$\Delta G_{\dot\gamma} = \Delta G_z + E_s \qquad (1)$$

in which ΔG_z is the molar Gibbs energy of mixing (zero shear, equilibrium conditions) and E_s the energy the solution can store per mol in the stationary state while it flows. In the two subsequent sections it is shown how the two terms of the sum of Eq. (1) are accessible.

Expressions for ΔG_z The thermodynamic properties of stagnant polymer solutions are normally described by

(2) $$\Delta G_z/(RT) = x_1 \ln\varphi_1 + x_2 \ln\varphi_2 + g x_1 \varphi_2$$

where x_i and φ_i are the mol fraction and the volume fraction of the components (index 1 standing for the solvent and 2 for the polymer). Using this relation, all characteristics of an actual system are incorporated into the (concentration and temperature dependent) interaction parameter g.

The measured thermodynamic mixing behavior is normally expressed in terms of either Eq. (3) or (4). The former relation[25] is theoretically justified and reads

(3) $$g = \alpha + \beta_o/(1-\gamma\varphi_2).$$

α and γ are constants for a given system and β_o varies in a characteristic manner with temperature; for UCSTs it is normally a linear function of T^{-1}.

Eq. (4) constitutes an empirical series expansion

(4) $$g = g_{00}/T + g_{01}T + (g_{10}/T + g_{11}T)\varphi_2 + (g_{20}/T)\varphi_2^2$$

for which the individual coefficients g_{ij} are again accessible from measurements of the chemical potential of the solvent as a function of composition and temperature. Depending on the precision wanted, four and more parameters are required for the description of $g(\varphi_2,T)$.

Expressions for E_s The stored energy is accessible from the solution elasticity by means of a theoretical relationship[26] which, however, becomes invalid for too pronounced non-linear viscoelastic behavior. According to experimental experience[27] E_s can be more conveniently also obtained from flow curves according to[12]

(5) $$E_s = (x_1V_1 + x_2V_2)\ \eta\tau_o\dot{\gamma}^2\ (\eta/\eta_o)\ |\ \eta\dot{\gamma}\ |^{-2d^*}.$$

η_o is the zero shear viscosity and τ_o the characteristic viscometric relaxation time, introduced by Graessley[28] for the description of flow curves by means of the following equations:

(6) $$\eta/\eta_o = g^{1.5}(\Theta)\ h(\Theta)$$

(7) $$\Theta = \eta/\eta_o\ \dot{\gamma}\tau_o/2$$

(8) $g(\Theta) = 2/\pi\ [\text{arccot}\ \Theta + \Theta/(1+\Theta^2)]$

(9) $h(\Theta) = 2/\pi\ [\text{arccot}\ \Theta + \Theta(1-\Theta^2)/(1+\Theta^2)^2]$.

From experience one knows[29] that τ_0 is normally on the order of the theoretically calculated Rouse relaxation time[30] τ_R

(10) $\tau_R = 6/\pi^2\ [(\eta_0-\eta_s)M/(c_2RT)]$;

η_s is the viscosity of the pure solvent and c_2 the concentration (mass per volume) of the polymer with the molar mass M.

The parameter d*, appearing in Eq. (5), constitutes a generalized power-law exponent defined as

(11) $d^* = -(\partial \ln \eta)/(\partial \ln \dot{\gamma})$

and accounts for the fact that the ability of a solution to store energy becomes less as the disentanglement processes proceed. For known values of η_0 and τ_0 it can be calculated by means of the Eqs. 6–10.

The temperature and concentration dependence of η_0 can frequently be described[29] by the Arrhenius Eq. (12), in which the parameters vary with φ_2 as formulated in Eqs. (13) and (14):

(12) $\ln(\eta_0/\text{Pas}) = A + E^{\neq}/RT$

(13) $A = A_0 + A_1\ \varphi_2$

(14) $E^{\neq} = E^{\neq}_0 + E^{\neq}_1\ \varphi_2$

For the mathematical description of E_s (at given M and setting $\tau_0 \approx \tau_R$) as a function of composition, temperature and shear rate, four parameters (A_0, A_1, $E^{\neq}_0$ and $E^{\neq}_1$) are required, as long as η_s can be neglected with respect to η_0 in Eq. (10).

Shear-dissolution and shear-demixing

On the basis of the generalized Gibbs energy of mixing (Eq. (1)) the occurrence of opposite shear effects, depending on the particular system and the conditions chosen, can be immediately rationalized. In plots of ΔG_z or $\Delta G_{\dot{\gamma}}$ as a function of the mol fraction of polymer, demixing is indicated by the possibility to construct double tangents. From the diagrams of Fig. 8 (not drawn to scale) it can be seen how $\Delta G_{\dot{\gamma}}(x_2)$ differs from $\Delta G_z(x_2)$ as a result of the addition of the stored energy E_s.

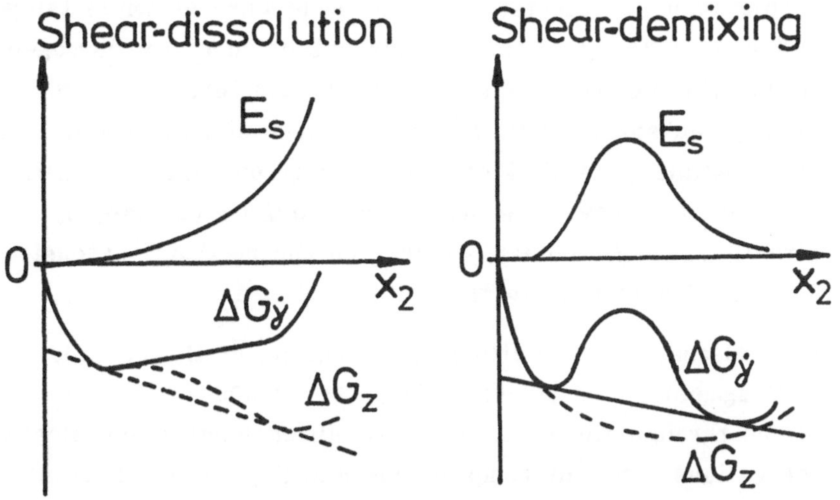

Fig. 8: Scheme[12] of the situations leading to shear-dissolution or to shear-demixing in terms of the generalized Gibbs energy of mixing, as defined in Eq. (1).

In the case of linear viscoelastic behavior ($d^* = 0$; l.h.s. of Fig. 8), E_s increases so drastically with rising polymer concentration that this contribution can undo the "hump" in $\Delta G_z(x_2)$ causing the demixing of the stagnant system. This means that shear-dissolution is typical for linear viscoelastic circumstances. Shear-demixing, on the other hand, is bound to non-linear viscoelasticity ($d^* > 0$; r.h.s. of Fig. 8). In that case $E_s(x_2)$ exhibits a maximum and can consequently cause a "hump" in $\Delta G_{\dot{\gamma}}(x_2)$: The flowing solution phase separates in contrast to the quiescent one.

Quantitative determinations of the composition of the phases which coexist at different shear rates were performed along the above lines either graphically[20] (using an appropriate auxiliary function) or on a computer[31]. The next two paragraphs deal with special phenomena observed in this context.

Eulytic points

One particularity resulting with appropriate systems (for linear viscoelastic behavior as well as for non-linear) is the coexistence of three liquid phases in the sheared state under special conditions. This situation, which can be detected from the occurrence of triple tangents to $\Delta G_{\dot\gamma}(x_2)$, indicates the greatest possible extension of the homogeneous region (i.e. the maximum extent of shear-dissolution that can be reached for a given shear rate). The values of T and x_2 at which this happens for constant $\dot\gamma$, or that of T and $\dot\gamma$ for constant x_2, constitute characteristic points. They have been termed *eulytic* points[20], by analogy with the *eutectic* points, which give the maximum extension of the liquid state with respect to the solid.

For linear viscoelastic conditions the demixing of the sheared system TD/PS (5000 segments) was calculated[20] up to 10 000 s^{-1}. Fig. 9 shows sections through the obtained three-dimensional phase diagram $T_{\dot\gamma}(x_2,\dot\gamma)$ at various constant compositions and Fig. 10 at different shear rates. The eulytic points are in both cases indicated by full circles.

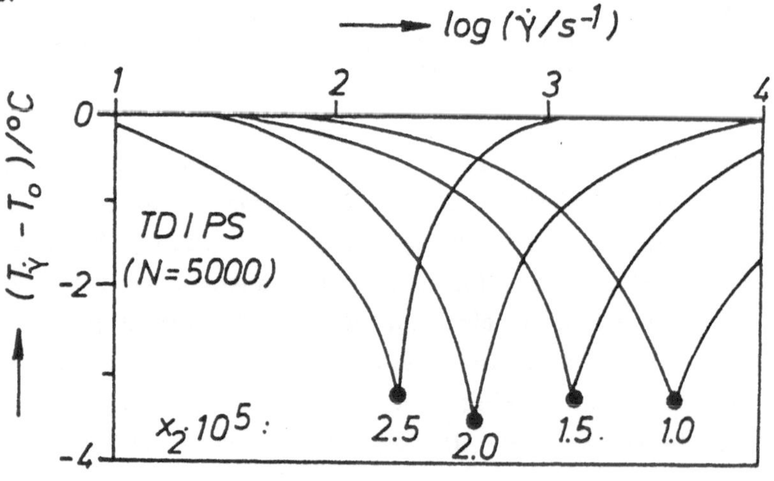

Fig. 9: Difference between $T_{\dot\gamma}$, the demixing temperature of a given solution at the shear rate $\dot\gamma$, and T_0, that of the solution at rest, as a function of log $\dot\gamma$, calculated for the system trans-decalin/polystyrene (5000 segments) and the indicated mol fractions of the polymer[20]; the filled circles represent the eulytic points.

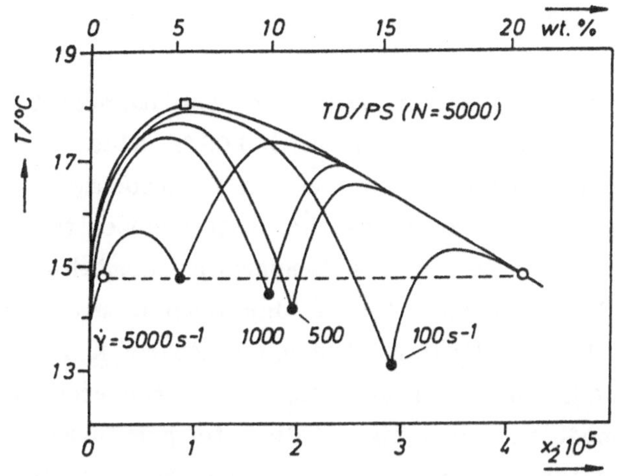

Fig. 10: Phase diagram of the system trans-decalin/polystyrene (5000 segments) calculated[20] for the indicated constant shear rates $\dot{\gamma}$; the solid circles give the eulytic points, the dashed line shows, as an example, the coexistence of three phases at $\dot{\gamma}$ = 5000 s^{-1}.

The non-linear case was studied[12] with another representative of the above system, where the number of segments of the polymer is approximately three times larger. The results of these calculations are shown in Fig. 11.

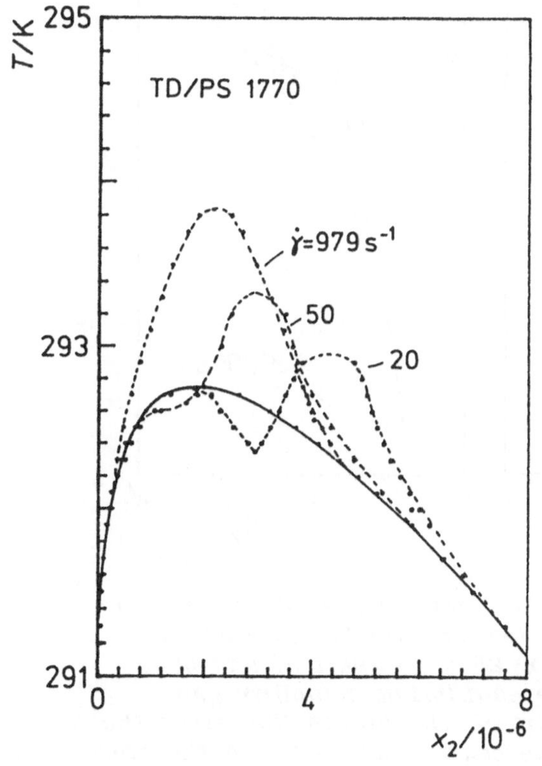

Fig. 11: Phase diagram of the system trans-decalin/polystyrene (M_w = 1770 kg/mol) calculated[12] for different shear rates $\dot{\gamma}$.

Flow-induced closed miscibility gaps

So far, all considerations concerned UCSTs; except for one important item the situation should be completely analogous with LCSTs. This essential difference lies in the fact that the increase in E_s associated with the reduction of T at constant $\dot{\gamma}$ brings the system closer to its equilibrium miscibility gap in the case of low temperature demixing, whereas the distance becomes larger in the case of high temperature demixing. As a consequence of this particular situation a totally new phenomenon turns up if the conditions are favorable, namely the creation of a closed solubility gap by moderate shearing within a temperature-composition range where the system at rest is homogeneous. Fig. 12 gives an example[31] for the results obtained from model calculations with a polymer consisting of 5000 segments and an exothermal theta temperature of 333.3 K.

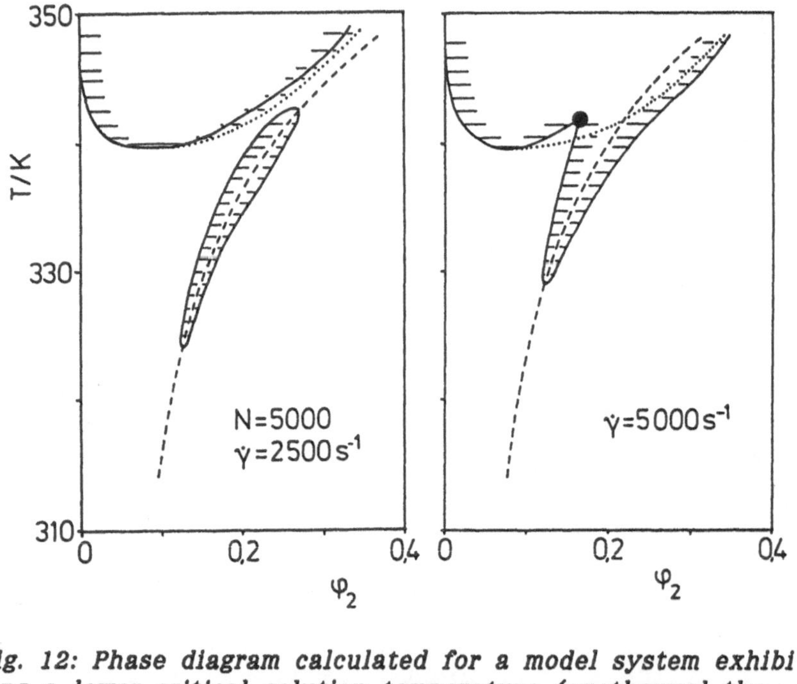

Fig. 12: Phase diagram calculated for a model system exhibiting a lower critical solution temperature (exothermal theta temperature at 333.3 K) for 2500 s^{-1} (part a) and 5000 s^{-1} (part b) respectively. The equilibrium solubility gap is indicated by the dotted curves. The broken line gives the temperature dependence of the composition at which the stored energy becomes maximum for the particular shear rate. The solid circle of part b represents an eulytic point.

For the occurrence of shear induced closed solubility gaps of the above type it is essential that the curve describing the temperature dependence of the composition at which the maximum amount of energy can be stored for a given shear rate does not intersect the equilibrium miscibility gap like shown in part a of Fig. 12. As with upper critical solution temperatures, shear-dissolution and eulytic points, as well as shear-demixing, can also be observed with LCSTs as demonstrated in part b of Fig. 12.

DISCUSSION

Detailed theoretical calculations ***and*** systematic experiments have so far only been performed for polymer/solvent systems. For this reason the comparison of predicted and actual behavior must in the main be confined to this case, for which all required data are available. Only some qualitative arguments can be given for the other systems.

With polymer mixtures (in the absence and in the presence of a low molecular weight liquid) one problem for the application of the present theoretical approach consists in the lack of experimental or theoretical information on the stored energy as a function of temperature, composition and shear rate. With solvent/polymer-1/polymer-2 systems an additional complication is encountered: Not even the equilibrium phase behavior can at present be described theoretically by means of physically significant interaction parameters.

In those cases, however, in which all required information concerning $\Delta G_z(x_1,T)$ and $E_s(x_1,T,\dot{\gamma})$ is available with sufficient accuracy, shear influences on phase separation can be predicted very accurately, as will be documented in the following sections for the most interesting representative of the system *trans*-decalin/polystyrene (M = 1180 kg/mol), for which the effects of shear change sign upon the variation of $\dot{\gamma}$. Fig. 13 shows the experimental results[12] in terms of the phase diagram at different shear rates and Fig. 14 the corresponding predictions[12].

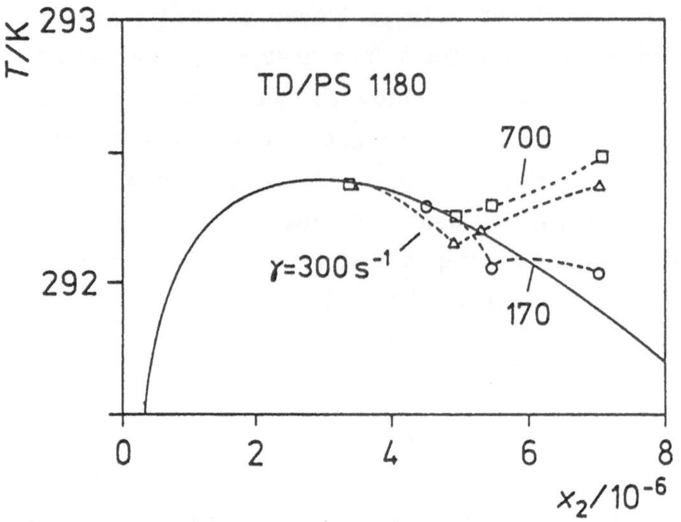

Fig. 13: Phase diagram of the system trans-decalin/polystyrene (M_w = 1180 kg/mol) determined viscometrically[12] at the indicated shear rates (dashed lines) and equilibrium solubility gap (solid line).

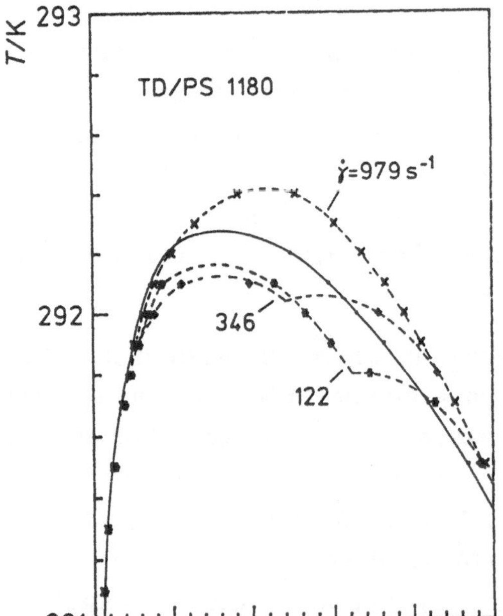

Fig. 14: Phase diagram of the system trans-decalin/polystyrene (M_w = 1180 kg/mol) calculated theoretically[12] for the indicated shear rates (dashed lines) and equilibrium solubility gap (solid line).

For technical reasons, no measurements can be made at very small and very large $\dot{\gamma}$ values, but where the experimental information is accessible, it agrees almost quantitatively with prediction. In particular it can be seen from the two diagrams that the eulytic point vanishes at shear rates between 346 and 979 s^{-1} and that shear-demixing is observed exclusively above 700 s^{-1}. Experiment and theory can be more directly compared by means of Fig. 15.

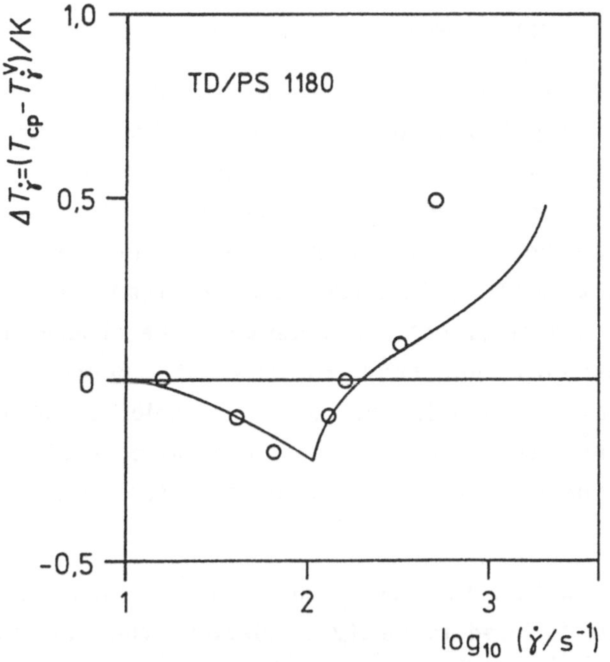

Fig. 15: Comparison[12] between the experimentally determined (symbols) and the calculated (solid line) effects of shear on the demixing temperature of a 6 wt.% solution of polystyrene (M_w = 1180 kg/mol) in trans-decalin in terms of $\Delta T_{\dot{\gamma}}$ as a function of log $\dot{\gamma}$.

In this graph[12] the effects are depicted in terms of the displacements of the demixing temperatures of the flowing solution with respect to the stagnant one as a function of shear rate. Fig. 15 demonstrates for one particular solution (6 wt.%) how shear-dissolution changes to shear demixing; for the present polymer concentration, the eulytic point is situated at approx. 100 s^{-1}. The measuring data agree well with the calculated curve, except for the highest shear rate, at which the theoretical equation used for the calculation of E_s has already become invalid.

The almost quantitative agreement between experiment and theoretical prediction is not surprising in view of the fact that phase separation under shear can be treated as a near equilibrium phenomenon. Despite the simplicity of the approach, however, additional insight can be gained. So far it primarily concerns the coexistence of three liquid phases at the eulytic points and the occurrence of shear induced closed miscibility gaps in the case of high temperature demixing.

All effects predicted or observed for polymer solutions under shear should also occur in elongational flow and with polymer mixtures, as indicated by published experimental results. It is, for example, easily possible to connect the phase separation temperatures of polymer mixtures during elongational flow as a function of compositions in a published diagram[15] such that eulytic points (intersection of two branches of the demixing curves) are obtained[15]. Furthermore, the observation[15] of a flow induced miscibility gap below the LCST of a polymer mixture, which is separated from the equilibrium gap by a region of homogeneity, can be interpreted along the same lines as the occurrence of a closed miscibility gap with sheared polymer solutions near their high temperature demixing.

In order to check whether the above conjectures are actually correct, additional experiments tailored to verify or disprove them are necessary. Another interesting item concerns the actual observation of the three phases which should coexist according to theory at the eulytic points.

Some generally valid statement can, however, already be made now, since they are only based on phenomenological thermodynamics: The occurrence of shear-dissolution or shear-demixing is for any system bound to the situations outlined in Fig. 8. If the flow conditions are such that the liquids behave linearly viscoelastic at all compositions, no measurable effect or an enlargement of the homogeneous region will be observed. If, on the other hand, non-linear viscoelasticity is involved, the region of homogeneity may either still be increased or diminished, depending on the relative position of the critical point and the composition at which the maximum energy is stored for a given shear rate or rate of elongation. The occurrence of shear-demixing, however, is in any case an unequivocal indication of non-linear viscoelastic behavior.

Acknowledgements: The author thanks the "Deutsche Forschungsgemeinschaft" for financial support and his coworkers, in particular H. Krämer-Lucas, H. Schenck and R. Horst, without their engaged collaboration this study would not have been possible.

REFERENCES

1. A. Silberberg, W. Kuhn; Nature **170**,450(1952)
2. A. Silberberg, W. Kuhn; J.Pol.Sci. **123**,21(1954)
3. J.W. Breitenbach, B.A. Wolf; Makromol.Chem. 117,163(1968)
4. G. Ver Strate, W. Philippoff; J.Pol.Sci., Polym. Lett. 12,267(1974)
5. J.R. Schmidt, B.A. Wolf; Coll. & Polym. Sci. **257**,1188(1979)
6. B.A. Wolf, H. Krämer; J.Polym.Sci., Polym. Lett., **18**,789(1980)
7. K.A. Mazich, S.H. Carr; J.Appl.Phys. **54**,5511(1983)
8. C. Rangel-Nafaile, A.B. Metzner, K.F. Wissbrun; Macromolecules 17,1187(1984)
9. H. Krämer, B.A. Wolf; Makromol.Chem., Rapid Commun. **6**,21(1985)
10. J.D. Katsaros, M.F. Malone, H.H. Winter; Polym. Bull. **16**,83(1986)
11. J. Lyngaae-Jorgenson, K. Sondergaard; Polym.Eng.Sci. **27**,344(1987) and **27**,351(1987)
12. H. Krämer-Lucas, H. Schenck, B.A. Wolf; Makromol.Chem. **189**,1613(1988) and **189**,1627(1988) and H. Krämer, Ph.D. Dissertation, Mainz 1983
13. L.P. Rector, K.A. Mazich, S.H. Carr; J.Macromol.Sci., Phys., **B27**,421(1988)
14. F.B. Cheikh Larbi, M.F. Malone, H.H. Winter, J.L. Halary, M.H. Leviet, L. Monnerie; Macromolecules 21,3532(1988)
15. J.D. Katsaros, M.F. Malone, H.H. Winter; Pol. Engin. and Sci. **29**,1434(1989)
16. T. Takebe, R. Sawaoka, T. Hashimoto; J.Chem.Phys. **91**,4369(1989)
17. H. Schenck; Ph.D. Dissertation, Mainz 1989
18. B.A. Wolf, M.C. Sezen; Macromolecules **10**,1010(1977)
19. B.A. Wolf, R. Jend; Macromolecules **12**,732(1979)
20. B.A. Wolf; Macromolecules **17**,615(1984)

21. E. Vrahopoulou-Gilbert, A.J. McHugh; Macromolecules 17,2657(1984)
22. B.A. Wolf; Makromol.Chem. Rapid Commun. 1,231(1980)
23. N. Pistoor, K. Binder; Colloid & Polymer Sci. 266,132(1988) and literature cited therein
24. S. Hess; Lecture Notes in Physics, present volume, Springer-Verlag Berlin, Heidelberg, New York, London, Paris, Tokyo
25. R. Koningsveld, L.A. Kleintjens; Macromolecules 4,637(1971)
26. G. Marucci; Trans.Soc.Rheolog. 16,321(1972)
27. G.V. Vinogradov, A. Ya. Malkin; "Rheology of Polymers" Springer Verlag, Berlin (1980)
28. W.W. Graessley; Adv. Poylmer Sci. 16,1(1974)
29. F.K. Herold; Ph.D. Dissertation, Mainz 1985
30. P.E. Rouse; J.Chem.Phys. 21,1272(1953)
31. R. Horst, B.A. Wolf; Macromolecules in press

Figs. 3, 4, 11, 13-15 are reprinted with permission from "Die Makromolekulare Chemie" Hüthig & Wepf Verlag, Basel, and Figs. 2, 7, 9, 10 and 12 with permission form "Macromolecules", American Chemical Society.

TOWARDS A UNIFIED
FORMULATION OF MICRORHEOLOGICAL MODELS

BY

R.J.J. JONGSCHAAP

Centre for Rheology, Department of Applied Physics
Twente University, PO B 217 7500 AE Enschede
The Netherlands

1. Introduction

In the theories about the rheological behaviour of materials their actual state is to represented by a model. Such a model may be considered an image of the system, simplified in such a way that it is tractable for mathematical analysis and still representing the features of the system that are expected to be important for its rheological behaviour. The development of models may be viewed as a process in which some kind of balance between physical reality, mathematical simplicity and usefulness is optimized.

In the present paper a thermodynamic approach is given which may be applied to many type of models in a unified and systematic way. Our results are similar to the ones obtained earlier by Lhuillier [1] and Maugin and Drout [2], but derived in a slightly different manner and generalized to a broader class of systems. There is also a close connection with the more general abstract formulation of Grmela [3] based upon a bracket formulation of diffusion and convection equations. In the work of Grmela however the general matrix representation (see eq. (2.7) below) which plays a central role in our formulation was not obtained. An important notion in our approach is the explicit distinguishment of different levels of description. This means that the representation of system by a model may be a more or less detailed description of the real microstructure. The following levels of description may be considered:

level 1: molecular dynamics

A molecular model representation very similar to the real microstucture and governed by classical mechanics. This level is used in computer simulations

level 2: **phase space**
Representation of the structure like level 1, but specification of the state by a distribution function in phase space.

level 3: **configuration space**
Representation of the structure like level 2 but preaveraged with respect to velocities. The state is specified by a configurational distribution function.

level 4: **structural variables**
The structure is represented by a (set of) scalar and/or tensorial variables.

level 5: **continuum**
The system is represented by a continuum, specified by (a) constitutive equation(s).

In the present paper we will consider only the levels 3, 4 and 5. Nevertheless there will be still many possible descriptions for each system. The reason for this is that, as we shall see in the next section, the model is based upon a division of the system in a subsystem and its environment, both of which may be described at different levels of description.

In section 2 the general theory -which, for reasons to be given in section 3.1, will be called "the triangle model" -will be presented and in section 3 the application to a number of rheological models will be discussed. These applications are merely to illustrate the method. No attempt has been made to be complete in some sense or to obtain new results. In section 4 we discuss some features of the present approach and some prospects for future investigations.

2. Theory

In any microrheological model it is possible to define a subsystem and a set of external stresses and forces are acting on it. This set will be denoted here by Σ, and the associated set of external flows and velocities by Γ. The power supply to the subsystem then becomes

$$W = \Sigma \cdot \dot{\Gamma} \tag{2.1}$$

Here, the dot denotes an inner product in the linear space to which Σ and $\dot{\Gamma}$ belong. If Σ and $\dot{\Gamma}$ are vectors or tensors the product will be a single or multiple contraction. Σ and $\dot{\Gamma}$ may however also denote spatial fields or other functions, in that case the inner product in (1) may contain an integration with respect to the space coordinates or other variables.

Besides the external variables Σ and $\dot{\Gamma}$ we introduce a set of state variables Φ, such that, at constant temperature, the free energy A of the system may be expressed as a function, or a functional of Φ:

$$A = A\ (\Phi) \tag{2.2}$$

The derivative of A with respect to Φ has the significance of a (set of) thermodynamic force(s):

$$\Pi = \frac{\delta A}{\delta \Phi} = \Pi\ (\Phi) \tag{2.3}$$

Here, and elsewhere in this paper, the kind of derivative is not explicitly specified, it depends upon the nature of the quantity Φ: if Φ is a set of scalar or tensorial variables (3) consists of a set of partial derivatives and in the case that Φ is a function, an apropriate functional derivative. The latter situation occurs, for instance in configurational-space molecular models, if the distribution function ψ is used as the state variable Φ.

We now consider the rate of dissipation Δ. Bij definition, this is the power supply (1) minus the rate of reversible storage of energy. In the isothermal case, the latter part equals the rate of change of the free energy $\dot{A} = \Pi * \dot{\Phi}$ (the * denotes an inner product in the space to which Π and Φ belong) so we have

$$\Delta = \Sigma \cdot \dot{\Gamma} - \Pi * \dot{\Phi} \tag{2.4}$$

We assume that the set of variables, introduced so far, is complete in the sense that all quantities in (4) may be considered as functions (or functionals) of $\dot{\Gamma}$ and Π:

$$\Sigma = \Sigma\ (\dot{\Gamma},\Pi) \qquad \dot{\Phi} = \dot{\Phi}\ (\dot{\Gamma},\ \Pi) \tag{2.5}$$

If from the three equations (4) and (5) the internal variables Π and Φ (including $\dot{\Phi}$) are eliminated a relation between Σ and $\dot{\Gamma}$ results. This is the microscopic constitutive equation of the model.

We will now impose some general restrictions upon the functionals (5). To that end we introduce the concept of a "macroscopic time reversal". With this we mean a reversal of the external flows and velocities. The term "macroscopic" is used, in order to distinguish this kind of time reversal from a real time reversal in which also on a microscopic scale all velocities and rates of change of state variables are reversed. A microscopic time reversal implies a macroscopic one but the opposite statement is not necessarily true. In the present formalism a macroscopic time reversal is a change of sign of the variable $\dot{\Gamma}$. Under such a transformation the variables Δ, Σ, π and $\dot{\Phi}$ will also change but some restrictions upon these changes are imposed by the laws of thermodynamics. In order to analyse this, we define for any functional $f\ (\dot{\Gamma})$ an even and an odd part with respect to macroscopic time reversal as

$$f^{+}\ (\dot{\Gamma}) + \frac{1}{2}\ (f\ (\dot{\Gamma}) + f\ (-\dot{\Gamma})) \quad \text{and} \quad f^{-}\ (\dot{\Gamma}) = \frac{1}{2}\ (f\ (\dot{\Gamma}) - f\ (-\dot{\Gamma}))$$

For the variables in (4) we then have: $\Delta = \Delta^{+}$, since the dissipation Δ has to be positive, by the second law of thermodynamics; $\dot{\Gamma} = \dot{\Gamma}^{-}$, by definition; and $\Pi = \Pi^{+}$, since according to (3) Π is a variable of state and not directly dependent upon $\dot{\Gamma}$. The quantities Σ and $\dot{\Phi}$ have no definite parity, so, in general: $\Sigma = \Sigma^{+} + \Sigma^{-}$ and $\dot{\Phi} = \dot{\Phi}^{-} + \dot{\Phi}^{+}$.

We now consider the even part of (4):

$$\Delta = \Sigma^{-} \cdot \dot{\Gamma} \qquad - \Pi * \dot{\Phi}^{+} \tag{2.6}$$

and the odd part:

$$0 = \Sigma^{+} \cdot \dot{\Gamma} - \Pi * \dot{\Phi}^{-} \tag{2.7}$$

From (6) we see that only the odd part of Σ and the even part of $\dot{\Phi}$ contribute to the dissipation Δ. Therefore we define the dissipative stress Σ^{D} as

$$\Sigma^{D} = \Sigma^{-} \tag{2.8}$$

The even part of Σ will considered as a reversible stress:

$$\Sigma^{R} = \Sigma^{+} \tag{2.9}$$

Similarly we define $\Phi = \dot{\Phi}^{D} + \dot{\Phi}^{R}$ with

$$\dot{\Phi}^{D} = \dot{\Phi}^{+} \tag{2.10}$$

and

$$\dot{\Phi}^{R} = \dot{\Phi}^{-} \tag{2.11}$$

We now will derive some more explicit results about the dependence of Σ and $\dot{\Phi}$ on $\dot{\Gamma}$ and Π. From the parity of $\dot{\Gamma}$, Π, Σ^{D} and $\dot{\Phi}^{D}$, as defined in (8) - (11) it can be seen that we may write:

$$\Sigma^{D} = \eta\ (\dot{\Gamma},\ \Pi) \cdot \dot{\Gamma} \tag{2.12}$$

and

$$\dot{\Phi}^{D} = -\beta\ (\dot{\Gamma},\ \Pi) * \Pi \tag{2.13}$$

in which the quantities η and β are even with respect to $\dot{\Gamma}$. The minus sign in (13) will be explained after eq. (17), below. For the reversible part of $\dot{\Phi}$ we may write

$$\dot{\Phi}^R = \Lambda\ (\dot{\Gamma},\ \Pi) \cdot \dot{\Gamma} \tag{2.14}$$

in which Λ is even with respect to $\dot{\Gamma}$ and Π. If (14) is substituted in (7), on using (9) and (11) we see that

$$\Sigma^R \cdot \dot{\Gamma} = \Pi * \Lambda\ (\dot{\Gamma},\ \Pi) \cdot \dot{\Gamma} \tag{2.15}$$

so

$$\Sigma^R = \Lambda^T\ (\dot{\Gamma},\ \Pi) * \Pi \tag{2.16}$$

The quantity Λ^T in this expression is the adjoint of the operator Λ, in the sense that $\Pi * \Lambda \cdot \dot{\Gamma} = \dot{\Gamma} \cdot \Lambda^T * \Pi$ for arbirary Π and $\dot{\Gamma}$. (In the case that Λ is a tensor, Λ^T denotes the transposed of Λ). The transposed of higher order tensors, defined in this way depends upon the number of contractions corresponding to the products "$*$" and "$\cdot$". If, for instance $\underline{\lambda}$ is a third order tensor $\underline{a} \cdot \underline{\lambda} : \underline{B} = \underline{B} : \underline{\lambda}^T \cdot \underline{a}$ implies that $\lambda_{ijk} = \lambda^T_{jki}$ whereas $\underline{A} : \underline{\lambda} \cdot \underline{b} = \underline{b} \cdot \underline{\lambda}^T : \underline{A}$ would imply $\lambda_{ijk} = \lambda^T_{kij}$. We will make no distinction in the notation of these type of transpositions since in our case their meaning can always be deduced easily from the expressions in which they are used.

We see that, as a direct result of eq. (7), which was obtained by considering the parity of the quantities in (4) with respect to macroscopic time reversal, obtain a close relationship between the evalution equation (14) of $\dot{\Phi}^R$ and the expression (16) of Σ^R.

The results (12), (13), (14) and (16) may be collected in a matrix expression, similar to the one obtained in classical treatment of non-equilibrium thermodynamics from the expression of the entropy production, (which, in the isothermal case considered here, is proportional to the dissipation (4)).

This matrix expression becomes:

$$\begin{bmatrix} \Sigma \\ \dot{\Phi} \end{bmatrix} \qquad \begin{bmatrix} \eta\ \cdot & -\Lambda^T * \\ \Lambda\ \cdot & \beta\ * \end{bmatrix} \quad \begin{bmatrix} \dot{\Gamma} \\ -\Pi \end{bmatrix} \tag{2.17}$$

The skew-semmetry of the off diagonal elements is in accordance with the Onsager-Casimir reciprocal relations. The minus sign in (17) was introduced, in order to use the "forces" and "fluxes" of the expression of the dissipation (4), written as $\Delta \dot{=} \Sigma \cdot \dot{\Gamma} + (-\Pi) * \dot{\Phi}$.

A derivation of a set of rate equations similar to our derivation of (17) by taking into account the time-reversal properties of dissipative effects was first applied to rheological models by Lhuillier [1], (see also the review article by Maugin and Drout [2]).

The main difference to the approach in these papers and the one presented here is that we do not consider the parity of variables with respect to ordinary time reversal but to what was called a macroscopic time reversal. This enabled us to split up the original expression (4) of the dissipation into the two parts (6) and (7). In this way the reversible and dissipative parts of the quantities $\dot{\Phi}$ amd Σ may be defined in an unambigious way. In the method based upon ordinary time reversal-symmetry a dissipation expression similar to (6) and an expression similar to (7) are introduced ad hoc.

It is interesting to note that since Λ is independent of $\dot{\Gamma}$, if follows from (17) $\Lambda = \delta\, \dot{\Phi}^R \,/\, \delta\, \dot{\Gamma}$ and so we obtain

$$\Sigma^R = \Pi \cdot \frac{\delta\, \dot{\Phi}^R}{\delta\, \dot{\Gamma}} \tag{2.18}$$

This result is in accordance with an expression first derived by Grmela [3] in a general theory based upon a bracket formulation of convection and diffusion equations. In [3] and also in subsequent publications by Grmela et al. [4-6] it was shown that the expression (18) is consistent with the stress-tensor expressions in many rheological models.
In the application to specific rheological models it is possible to make different choices for the variables Σ, $\dot{\Gamma}$, Π and Φ, depending upon the choice of the sub(system) and the level of description. A particular choice will be referred to as the "(Σ, $\dot{\Gamma}$, / Π, $\dot{\Phi}$) - level of description" with the choosen variables substituted for Σ, $\dot{\Gamma}$, Π and $\dot{\Phi}$ In section 3 we will discuss several rheological models at various levels of description.

3. Applications

The theory, described in section 2 will now be applied to a number of special rheological models. In section 3.1 we start with the treatment of spring dashpot models. This results in the construction of a special type of mechanical model, called "the triangle model" which represents the important features of the general theory in a schematic way. In section 3.2 the elastic dumbbell will be discussed at several levels of description and in section 3.3. the rigid dumbbell model as an example of a model with a constraint.

In section 3.4 it will be shown that it is also possible to describe reptation models in the present formalism and in section 3.5 we discuss the transient-network model as an example of a kinetic model. Finally in section 3.6 some examples of configurational tensor theories will be treated.

3.1 Mechanical models

In the treatment of linear viscoelasticity one often makes use of spring-dashpot models, as a phenomenological, representation of material behaviour. Although the theory presented in section 2 is not restricted to linear viscoelastic behaviour it is still very instructive to apply it to this type of models. The reason for this is that some of the basic elements of the present model: reversible storage of energy, dissipation and conpling of internal parts of the system to the environment are also the basic characteristics of spring-dashpot models.

The external variables Σ and $\dot{\Gamma}$ in (2.17) then become the external stress τ and the external rate of extension $\dot{\gamma}$, so we have:

$$\Sigma = \tau \qquad\qquad \dot{\Gamma} = \dot{\gamma} \qquad\qquad (3.1\text{-}1)$$

The mechanical energy, stored in the springs of the model corresponds to the free energy A (Φ) defined in (2.2). We will consider here models with one spring, with spring force σ and an extension ε, so we have:

$$\pi = \sigma \qquad\qquad \Phi = \varepsilon \qquad\qquad (3.1\text{-}2)$$

The dissipation expression (2.4) now becomes:

$$\Delta = \tau\,\dot{\gamma} \quad - \sigma\,\dot{\varepsilon} \qquad\qquad (3.1\text{-}3)$$

so we, similar to (2.17), we have

$$\begin{bmatrix} \tau \\ \dot{\varepsilon} \end{bmatrix} = \begin{bmatrix} \eta & -\lambda \\ \lambda & \beta \end{bmatrix} \begin{bmatrix} \dot{\gamma} \\ -\sigma \end{bmatrix} \qquad\qquad (3.1\text{-}4)$$

In table 1 the constans η, β and λ are given for a few well known mechanical models. We see that the structure of the matrix in (4) reflects the connection structure of the corresponding network:

Table I

Eq. (3.1-4) applied to some mechanical models

Model	Parameters in ( )		
	η	β	λ
	0	0	1
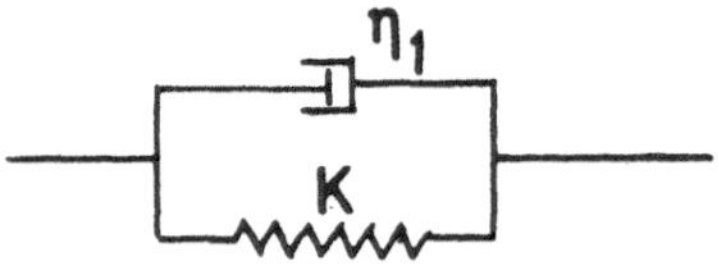	η_1	0	1
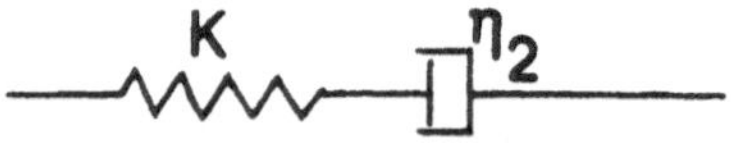	0	$1/\eta_2$	1
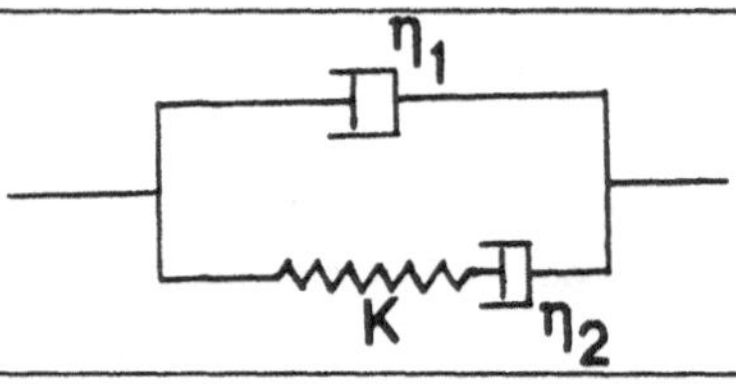	η_1	$1/\eta_2$	1

if the internal stress σ is directly coupled to the external stress τ (like in the Maxwell model) the parameter η vanishes. If on the other hand the rate of extension of the spring Σ is directly coupled to the external rate of extension $\dot{\gamma}$ (like in the Voigt-Kelvin model) the parameter β vanishes. the other two possibilities are combinations of these two cases. In all mechanical models discussed so far, we have $\lambda = 1$. The case $\lambda \neq 1$ corresponds to amplification or reduction of stresses and rates of extension in the connection between internal and external variables.

A model in which such is the case is presented in figure 1. This model, which -because of its characteristic shape- will be called "the triangle model", is described by the complete set of equations (4) with $\eta \neq 0$, $\beta \neq 0$ and $\lambda \neq 1$. Since (4) is similar to the general expression (2.17) the triangle model may be considered also as a symbolic representation of the general theory, described in section 2. For that reason, in the rest of this paper we will also refer to the model described in section 2 as the "triangle theory".

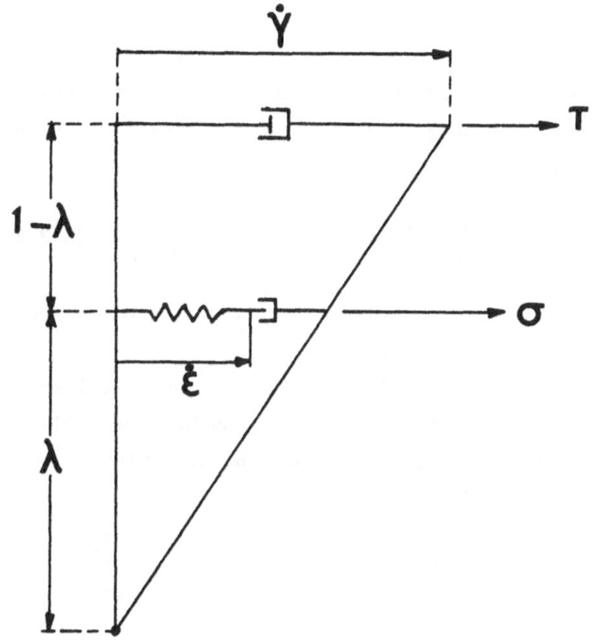

Fig. 1 **The triangle Model**

3.2 The elastic dumbbell model

As a first application of the triangle model to a microscopic model we consider the elastic dumbbell model. This model is used to describe approximately some of the rheological properties of dilute solutions of flexible macromolecules. For an extensive discussion we refer to chapter 13 of ref. [7]. Here we recall that a dumbbell consists of two beads on which hydrodynamic forces are acting, connected by a spring. The spring vector will be denoted by $\underline{q}$ and the configurational distribution function by $\psi(\underline{q}, t)$.

In order to apply the triangle model to the dumbbell model, we first have to define the variables Σ, $\dot{\Gamma}$, Π and Φ. This means that a level of description has to be specified. We start at the ($\underline{T}$, $\underline{L}$/ μ, $\dot{\psi}$)-level, so

$$\Sigma = \underline{T} \quad ; \qquad \dot{\Gamma} = \underline{\underline{L}}$$

$$\Phi = \psi \quad ; \qquad \Pi = \mu \underline{=} \frac{\delta A}{\delta \psi} \qquad (3.2\text{-}1)$$

Here $\underline{T}$ is the macroscopic stress tensor; $\underline{L}$ the velocity gradient tensor and $A = A\{\psi\}$ the free energy, considered as a functional of the distribution function $\psi \equiv \psi\ (\underline{q}, t)$. In the present case we have

$$A\{\psi\} = nkT \int \psi \ln \frac{\psi}{\psi_o} \, d^3\zeta \qquad (3.2\text{-}2)$$

in which n is the number density of dumbbells, k Boltzmanns's constant, T the temperature and ψ_o the equilibrium distribution.
This functional is (within an additative constant) the so called dynamical free energy, introduced by Doi [8]. (see also Sarti and Marrucci [9]).
The associated thermodynamic force, which has the significance of a chemical potential in configuration space becomes

$$\mu = \frac{\delta A}{\delta \psi} = nkT \left(1 + \ln \frac{\psi}{\psi_o}\right) \qquad (3.2\text{-}3)$$

The rate of reversible storage of energy due to a change of ψ may now be written as

$$\dot{\underline{A}} = \frac{\delta A}{\delta \psi} * \frac{\partial \psi}{\partial t} = \int \mu \frac{\partial \psi}{\partial t} \, d^3\underline{\underline{q}} \qquad (3.2\text{-}4)$$

On the other hand, the macroscopic power supply is given by the familiar expression W = $\underline{T}$: $\underline{L}$, so the dissipation becomes:

$$\Delta = \underline{T} : \underline{L} - \mu * \frac{\partial \psi}{\partial t} \qquad (3.2\text{-}5)$$

For the matrix espression (2.17) we also obtain

$$\begin{bmatrix} \underline{T} \\ \frac{\partial \psi}{\partial t} \end{bmatrix} = \begin{bmatrix} \underline{\underline{\eta}}: & -\underline{\Lambda}^{T}* \\ \underline{\Lambda} & \beta* \end{bmatrix} \begin{bmatrix} \underline{L} \\ -\mu \end{bmatrix} \qquad (3.2\text{-}6)$$

With $\underline{\mu}$ a fourth order tensor, $\underline{\Lambda}$ a second order tensor and β a scalar operator.

In order to determine the explicit forms of these quantities we have to consider the evolution equations and the stress-tensor espressions of the dumbbell model.
First, from the diffusion equation

$$\frac{\partial\psi}{\partial t} = - \frac{\partial}{\partial \underline{q}} \cdot (\psi \, \underline{L} \cdot \underline{q}) + \frac{2}{n\zeta} \frac{\partial}{\partial \underline{q}} \cdot (\psi \frac{\partial}{\partial \underline{q}} \mu) \qquad (3.2\text{-}7)$$

in which ζ is a friction parameter and μ the chemical potential defined by (1), on noting that the first term in the r.h.s of (7) is odd with respect to $\underline{L}$ and the second term is even, we see that the tensor $\underline{\Lambda}$ becomes

$$\underline{\Lambda} = - \frac{\delta}{\delta \underline{q}} (\psi \, \underline{q}) \qquad (3.2\text{-}8)$$

and that the operator β may be represented as

$$\beta = \frac{-2}{n\zeta} \frac{\delta}{\delta \underline{q}} \cdot \psi \frac{\delta}{\delta \underline{q}} \qquad (3.2\text{-}9)$$

It is possible, by using derivatives of the Dirac-delta function to express β in a form such that the last term in (7) becomes of the form $-\beta * \mu$ in which, like in (4) the symbol * denotes an integration in configuration space. We will not use this representation here, and represent β by the differential operator (9). We now consider the stress tensor expression. We first calculate the reversible part from (2) and and (6):

$$\begin{aligned}\underline{T}^R = & \quad \underline{\Lambda}^T * \mu = \mu * \underline{\Lambda} = \\ = & \quad nkT \int \left(1 - \ln \frac{\psi}{\psi_o}\right) \left(- \frac{\partial}{\partial \underline{q}} \psi \, \underline{q}\right) d^3\underline{q} \\ = & \quad nkT \int \left(\frac{\partial}{\partial \underline{q}} \ln \frac{\psi}{\psi_o}\right) (\psi \, \underline{q}) \, d^3\underline{q} \\ = & \quad nkT \, \underline{1} + n \langle \underline{f}^I \underline{q} \rangle \qquad (3.2\text{-}10)\end{aligned}$$

with $\underline{f}^I = -kT \frac{\partial}{\partial \underline{q}} \, n \, \psi_o$, the so called connector force. In this way the "Kramers form" of the (reversible part of the) stress tensor is obtained.
The total stress tensor expression for the dumbbell model is given by

$$\underline{T} = 2 \, \eta \, \underline{D} \quad - nkT \, \underline{1} + n \langle \underline{f}^I \, \underline{q} \rangle \qquad (3.2\text{-}11)$$

in which $\underline{D} = \frac{1}{2} (\underline{L} + \underline{L}^T)$. The first term of the r.h.s. is the solvent contribution to the stress tensor. Being odd in $\underline{L}$ this term is the dissipative part $\underline{T}^D$ of $\underline{T}$. So, form (6) and (11) we obtain for the tensor $\underline{\underline{\eta}}$:

$$\underline{\underline{\eta}} = 2\eta \, \underline{\underline{I}} \qquad (3.2\text{-}12)$$

in which $\underline{\underline{I}}$ is a fourth order tensor, defined by

$$I_{ijkm} = \frac{1}{2} (\delta_{ik} \delta_{jm} + \delta_{im} \delta_{jk}) \qquad (3.2\text{-}13)$$

The results (8). (9), (10) and (11) may be collected now in the following matrixs expression:

$$\begin{bmatrix} \underline{\underline{T}} \\ \frac{\partial \psi}{\partial t} \end{bmatrix} = \begin{bmatrix} 2 \eta \underline{\underline{I}} : & \frac{\partial}{\partial \underline{q}} (\psi \underline{q}) * \\ -\frac{\partial}{\partial \underline{q}} (\psi \underline{q}) : & \frac{-2}{n\zeta} \frac{\partial}{\partial \underline{q}} \cdot \psi \frac{\delta}{\delta \underline{q}} \end{bmatrix} \begin{bmatrix} \underline{\underline{L}} \\ -\mu \end{bmatrix} \qquad (3.2\text{-}14)$$

This result, together with the free energy expression(2) contains all relevant information of the elastic dumbbell model at the present level of description.

We now change to a second level of description. Here we do not take the whole distribution function ψ but the configuration vector $\underline{q}$ as the variable of state. This will be done however in an average sense, namely such that $\dot{\underline{q}}$ is not the actual rapidly fluctuation rate of change of $\underline{q}$ of an individual dumbbell due to the thermal motion, but a flux, proportional to the diffusive flow in configuration space. It is the flux which is also present in the equation of continuity is configuration space:

$$\frac{\partial \psi}{\partial t} = - \frac{\partial}{\partial \underline{q}} \cdot (\psi \dot{\underline{q}}) \qquad (3.2\text{-}15)$$

The reversible force, associated with $\underline{q}$ may be obtained by substitution of (15) in (14) and ingretation by parts:

$$\dot{A} = -\int \mu \frac{\partial}{\partial \underline{q}} \cdot (\psi \dot{\underline{q}})\, d^3\underline{q} = \int \psi \frac{\partial \mu}{\partial \underline{q}} \cdot \dot{\underline{q}}\, d^3\underline{q} \equiv n < \underline{m} \cdot \dot{\underline{q}}> \qquad (3.2\text{-}16)$$

$$= n <\dot{a}>$$

In this expression $\underline{m}$ is defined as

$$\underline{m} = \frac{1}{n} \frac{\partial \mu}{\partial \underline{q}} = kT \frac{\partial}{\partial \underline{q}} \ln \frac{\psi}{\psi_o}$$

and

$$\dot{a} = \underline{m} \cdot \dot{\underline{q}}$$

The vector $\underline{m}$ is the thermodynamic force associated with the gradient of the chemical potential μ in configuration space. It may also be interpreted as the resultant of minus the Brownian force $\underline{f}^B = -kT \frac{\partial}{\partial \underline{q}} \ln \psi$ and the connector force $\underline{f}^I = - kT \frac{\delta}{\delta \xi} \ln \psi_o$.

The quantity $\dot{a}$ in (16) may be considered as a density in configuration space of the rate of change of the free energy per dumbbell, associated with the flux $\dot{\underline{q}}$ discussed above. This makes it possible to apply the triangle model now to one dumbbell. In that case we have

$$\Pi = \underline{m} \quad , \quad \Phi = \underline{q} \tag{3.2-17}$$

And instead of the total stress tensor $\underline{T}$ we now use the corresponding tensor $\underline{\tau}$, defined by

$$\underline{T} = n \int \psi \, \underline{\tau} \, d^3 \; = n \, \langle \underline{\tau} \rangle \tag{3.2-18}$$

The quantity $\underline{\tau}$ will be called a stresslet. The total power supply may be written now as

$$W = \underline{T} : \underline{L} = n \, \langle \underline{\tau} \rangle : \underline{L} \quad = \; n \, \langle w \rangle \tag{3.2-19}$$

with

$$w = \underline{\tau} : \underline{L} \tag{3.2-20}$$

the local density in configuration space, corresponding to the total power W. From (20) we see that in the present description the quantities Σ and $\dot{\Gamma}$ defined in (2.1) become:

$$\Sigma = \underline{\tau} \qquad , \qquad \dot{\Gamma} = \underline{L} \tag{3.2-21}$$

The matrix representation analogous to (2.17) becomes:

$$\begin{bmatrix} \underline{\tau} \\ \dot{\underline{q}} \end{bmatrix} = \begin{bmatrix} \underline{\underline{\eta}}: & -\underline{\underline{\Lambda}}^T \cdot \\ \underline{\underline{\Lambda}}: & \underline{\underline{\beta}} \cdot \end{bmatrix} \begin{bmatrix} \underline{L} \\ -\underline{m} \end{bmatrix} \tag{3.2-22}$$

In this case $\underline{\eta}$ is a fourth order tensor Λ a third order tensor and $\underline{\beta}$ a second order tensor. The explicit form for these tensors may be obtained from some further properties of the model. First, from the equation of motion

$$\dot{\underline{q}} = \underline{L} \cdot \underline{q} - \frac{2}{\zeta} m \tag{3.2-23}$$

in which ζ is a friction coefficient, one obtains

$$\dot{\underline{q}}^R = \underline{L} \cdot \underline{q} = \underline{\underline{\Lambda}} : \underline{L} \tag{3.2-24}$$

so

$$\underline{\underline{\Lambda}} = \underline{1} \, \underline{q} \tag{3.2-25}$$

and

$$\dot{\underline{q}}^{D} = - \frac{2}{\zeta} \underline{m} = \underline{\beta} \cdot (-\underline{m}) \qquad (3.2\text{-}26)$$

so

$$\underline{\beta} = \frac{2}{\zeta} \underline{1} \qquad (3.2\text{-}27)$$

From (22) and (25) we obtain for the reversible part of the stresslet

$$\underline{\underline{\tau}}^{R} = \underline{\underline{\Lambda}}^{T} \cdot \underline{m} = \underline{m} \cdot \underline{\underline{\Lambda}} = \underline{m}\,\underline{q} \qquad (3.2\text{-}28)$$

In accordance with the Kramers form (see form ref. [7])

$$\underline{T}^{R} = n < \underline{mq} > = n\,k\,T \int \left(\frac{\partial}{\partial \underline{q}} \ln \frac{\psi}{\psi_o} \right) \psi\,\underline{q}\;\; d^3\underline{q} \qquad (3.2\text{-}29)$$

The dissipative part of $\underline{\tau}$ is similar to the first term *m* (11) given by

$$\underline{\tau}^{D} = \frac{2}{n} \eta\, \underline{D} \equiv \underline{\underline{\eta}} : \underline{\underline{L}} \qquad (3.2\text{-}30)$$

so

$$\underline{\underline{\eta}} = 2 \frac{\eta}{n} \underline{\underline{I}} \qquad (3.2\text{-}31)$$

with $\underline{I}$, given by (13)

The results (25), (26), (28) and (30), collected in matrix form, become:

$$\begin{bmatrix} \underline{\tau} \\ \dot{\underline{q}} \end{bmatrix} = \begin{bmatrix} \eta\, \underline{\underline{I}} : & -\,(\underline{1}\underline{q})^{T} \cdot \\ \underline{1}\,\underline{q} : & \frac{2}{\zeta} \underline{1} \cdot \end{bmatrix} \begin{bmatrix} \underline{L} \\ -\underline{m} \end{bmatrix} \qquad (3.2\text{-}32)$$

Instead of taking $\underline{\tau}$ and $\underline{L}$ as the macroscopic variables in which the external power supply is expressed one also may write

$$W = \underline{f} \cdot \dot{\underline{d}} \qquad (3.2\text{-}33)$$

in which $\underline{f}$ is the hydrodynamic (external) force acting upon a dumbbell and $\dot{\underline{d}} = \underline{L} \cdot \underline{q}$ the relative velocity of the fluid flow at its end points.

The matrix representation (32) then becomes:

$$\begin{bmatrix} \underline{f} \\ \dot{\underline{q}} \end{bmatrix} = \begin{bmatrix} \underline{0} & -\underline{1} \\ \underline{1} & \frac{2}{\zeta} \underline{1} \end{bmatrix} \begin{bmatrix} \dot{\underline{d}} \\ -\underline{m} \end{bmatrix} \qquad (3.2\text{-}34)$$

This form expresse the equilibrium of forces : $\underline{f} = \underline{m}$ and the Stokes law $\underline{f} = \frac{\zeta}{2} (\dot{\underline{d}} - \dot{\underline{q}})$. We see that again, the skew symmetry of the off-diagonal elements of the matrix is obeyed. A final description that we will cosider in connection with the elastic dumbbell model is one in terms of the configuration tensor

$$\underline{\underline{S}} = < \underline{q}\,\underline{q} > \qquad (3.2\text{-}35)$$

as a variable of state. A closed description at that level is possible if the free energy A can be expressed explicitely as a function of $\underline{\underline{S}}$. In the case of linear dumbbells such is the case and we have

$$A = - \frac{1}{2} n\,k\,T \log\;\; (\det \underline{\underline{S}}) + \frac{1}{2} n\,\kappa\, \mathrm{tr}\, \underline{\underline{S}} \qquad (3.2\text{-}36)$$

with k = the spring modulus of the dumbbells. The expression (36), first obtained by Grmela and Carreau [6], is ensistent with the Booy-Wiegen [10] expression for the configuration distribution function in the expression (2) of the free energy. The thermodynamic force conjugate with $\underline{S}$ is obtained by differentation of (36).

In this way we get the variable Π at the configuration-tensor level of description:

$$\Pi = \underline{M} = \frac{\partial A}{\partial \underline{S}} = \frac{1}{2} n \kappa (\underline{1} - \frac{kT}{\kappa} \underline{S}^{-1}) \qquad (3.2\text{-}37)$$

Again, we will show that a matrix representation of the form

$$\begin{bmatrix} \underline{T} \\ \dot{\underline{S}} \end{bmatrix} = \begin{bmatrix} \underline{\underline{\eta}} : & -\underline{\underline{\Lambda}}^T : \\ \underline{\underline{\Lambda}} : & \underline{\underline{\beta}} : \end{bmatrix} \begin{bmatrix} \underline{L} \\ -\underline{M} \end{bmatrix} \qquad (3.2\text{-}38)$$

applies and specify the matrix elements $\underline{\underline{\eta}}$, $\underline{\underline{\Lambda}}$ and $\underline{\underline{\beta}}$.
To this end an expression for $\dot{\underline{S}}$ is needed. This may be derived by noting that $\dot{S} = \langle \dot{\underline{q}}\, \underline{q} + \underline{q}\, \dot{\underline{q}} \rangle$ in which $\dot{\underline{q}}$ is given by (23).
If subsequently the result (21) for $\underline{m}$ is used one obtains for the case of linear springs.

$$\dot{\underline{S}} = \underline{L} \cdot \underline{S} + \underline{S} \cdot \underline{L}^T + \frac{4kT}{\zeta} (\underline{1} - \frac{\kappa}{kT} \underline{S}) \qquad (3.2\text{-}39)$$

From (38) and (39) it follows that

$$\Lambda_{ijkm} = \delta_{ik} S_{mj} + S_{im} \delta_{jk} \qquad (3.2\text{-}40)$$

and

$$\underline{\underline{\beta}} = \frac{4}{n\zeta} \underline{\underline{\Lambda}} \qquad (3.2\text{-}41)$$

The dissipative part of the stress tensor is given by $\underline{T}^D = 2\eta\, \underline{D}$, so analogous to (30) we have $\underline{T}^D = \underline{\underline{\eta}} : \underline{L}$ with

$$\underline{\underline{\eta}} = 2\eta\, \underline{\underline{I}} \qquad (3.2\text{-}42)$$

The reversible part is given by the Kramers form (29) with, in the present case $\underline{f}^I = \kappa\, \underline{q}$, so

$$\underline{T}^R = nkT\, (\frac{\kappa}{kT} \underline{S} - \underline{1}) \qquad (3.2\text{-}43)$$

and we see from (38), (40) and (43) that indeed

$$\underline{T}^R = \underline{\underline{\Lambda}}^T : \underline{M} \qquad (3.2\text{-}44)$$

It is important to note that the consistency of the evolution equation (39) and the stress tensor expression (43) is only obtained if the correct expression (36) of the free energy is used. This point (see also Maugin and Drauot [2]) was overlooked by Lhuillier [1] who used a quadratic form for the free energy function A($\underline{S}$) and arrived at the conclusion that a convection laws similar to (39) (i.e. based upon the Oldroyd upper convective time derivative $\overset{\nabla}{\underline{S}} = \dot{\underline{S}} - \underline{L} \cdot \underline{S} - \underline{S} \cdot \underline{L}^T$) are incompatible with the Kramers expression of the stress tensor. Instead, Lhuilier obtains $\underline{T}^R = 2 \frac{\delta A}{\delta \underline{S}} \cdot \underline{S}$, the so called Eringen "thermodynamic microstress tensor" as the correct stress tensor expression in this case. We have now seen, however that a consistent structure tensor formulation of the elastic dummbbell model is possible with as well an Oldroyd upper convected derivative in the evolution equation as a Kramers expression for the stress tensor.

3.3 **The rigid dumbbell model**

It is interesting to see how constraints may be incorporated in the present formalism. This will be illustrated now for the rigid dumbbell model. This model is very similar to the elastic dumbbell described in the previous section. The difference is the rigidity constraint: $|\underline{q}| = q$ = constant.
As a consequence the connector force $\underline{f}^I$ is no longer a function of $\underline{q}$, but a constraining force, determined by the equilibrium of forces in the $\underline{q}$-direction. It has been shown [11] that the treatment of the rigidity constraint is facilated, by using the projection operator

$$\underline{P} = \underline{1} - \underline{e}\,\underline{e} \tag{3.3-1}$$

with $\underline{e} = \underline{q}/|\underline{q}|$ a unit vector in the direction of $\underline{q}$. We shall see that this operator also plays a prominent role in the present treatment.
We start with configuration-space level of description. Similar to (2) the free energy then becomes:

$$A\{\psi\} = nkT \int \psi \ln \psi \, d^2\underline{e} \tag{3.3-2}$$

and the corresponding chemical potential:

$$\mu = \frac{\partial A}{\partial \psi} = nkT \, (1 + \ln \psi) \tag{3.3-3}$$

The diffusion equation may analogous to (3.2-7) be expressed as

$$\frac{\partial \psi}{\partial t} = \frac{\partial}{\partial \underline{e}} \cdot (\psi \, \underline{P} \cdot \underline{L} \cdot \underline{e}) + \frac{2}{n \zeta q^2} \, \frac{\partial}{\partial \underline{e}} \cdot \left(\psi \frac{\partial \mu}{\partial \underline{e}} \right) \tag{3.3-4}$$

If a matrix expression similar to (3.2-6) is defined one obtains from (4):

$$\underline{\Lambda} = - \frac{\partial}{\partial \underline{e}} \cdot (\psi \, \underline{P} \, \underline{e}) \tag{3.3-5}$$

and

$$\beta = \frac{2}{n \zeta q^2} \frac{\partial}{\partial \underline{e}} \cdot \psi \frac{\partial}{\partial \underline{e}} \qquad (3.3\text{-}6)$$

The stress tensor for the rigid dumbbell is given by

$$\underline{T} = 2\eta \underline{D} + \frac{1}{2} n \psi q^2 \langle \underline{e}\, \underline{e}\, \underline{e}\, \underline{e} \rangle : \underline{D} + nkT \langle 3\, \underline{e}\, \underline{e} - \underline{1} \rangle \qquad (3.3\text{-}7)$$

The first two terms of the r.h.s. consitutes the dissipative stress $\underline{T}^D$, so we have

$$\eta = 2\eta \underline{\underline{I}} + \frac{1}{2} n \zeta \underline{q}^2 \langle \underline{e}\, \underline{e}\, \underline{e}\, \underline{e} \rangle \qquad (3.3\text{-}8)$$

The last term in (7) is the reversible stress $\underline{T}^R$. Similar to (3.2-10) it may be proved that also in this case we have

$$\underline{T}^R = \mu * \Lambda \equiv \int \mu \Lambda\, d^2 \underline{e} \qquad (3.3\text{-}9)$$

So, the stress tensor expression (7) is compatible with the formalism of section 2. Analogous to (3.2-14) the matrix formulation of the rigid dumbbell model at the present level of description becomes:

$$\begin{bmatrix} \underline{T} \\ \frac{\partial \psi}{\partial t} \end{bmatrix} = \begin{bmatrix} 2\eta \underline{\underline{I}} : & \frac{\partial}{\partial \underline{e}} \cdot (\psi \underline{P}\, \underline{e}) * \\ -\frac{\partial}{\partial e} (\psi \underline{P}\, \underline{e}) : & \frac{-2}{n \zeta \underline{q}^2} \frac{\partial}{\partial \underline{e}} \cdot \psi \frac{\partial}{\partial \underline{e}} \end{bmatrix} \begin{bmatrix} \underline{L} \\ \mu \end{bmatrix} \qquad (3.3\text{-}10)$$

At the $(\underline{T}, \underline{L} \,/\, \underline{m}, \dot{\underline{q}})$ - level of description the thermodynamic force becomes

$$\underline{m} = \frac{\partial \mu}{\partial \underline{e}} = k\, T \frac{\partial}{\partial \underline{e}} \ln \psi \qquad (3.3\text{-}11)$$

and the equation of motion is given by

$$\dot{\underline{e}} = \underline{P} \cdot \underline{L} \cdot \underline{e} - \frac{2}{\zeta \underline{q}^2} \underline{m} \qquad (3.3\text{-}12)$$

If, again, we define the stresslet $\underline{\tau}$ by $\underline{T} = n \langle \underline{\tau} \rangle$, from (10) and the stress tensor expression (7) the following matrix representation may be obtained:

$$\begin{bmatrix} \underline{\tau} \\ \dot{\underline{e}} \end{bmatrix} = \begin{bmatrix} (\frac{2\eta}{n} \underline{I} + \frac{\zeta \underline{q}^2}{2} \underline{e}\, \underline{e}\, \underline{e}\, \underline{e}): & -(\underline{P}\, \underline{e})^T \cdot \\ \underline{P}\, \underline{e} : & \frac{2}{\zeta \underline{q}^2} \underline{1} \cdot \end{bmatrix} \begin{bmatrix} \underline{L} \\ -\underline{m} \end{bmatrix} \qquad (3.3\text{-}13)$$

In verifying the equality $\underline{T}^R = n \langle \underline{\tau}^R \rangle = n \langle \underline{m} \cdot \underline{P}\, \underline{e} \rangle$ it should be noted that

$$\int \frac{\partial \psi}{\partial \underline{e}} \cdot (\underline{1} - \underline{e}\, \underline{e})\, \underline{e}\, d^2 \underline{e} = -\int \psi \frac{\partial}{\partial \underline{e}} \cdot [(\underline{1} - \underline{e}\, \underline{e})\underline{e}]\, d^2 \underline{e} = \int \psi\, (3\, \underline{e}\, \underline{e} - \underline{1})\, d^2 \underline{e}$$

The result (13) may be brought in a form which is more similar to the elastic dumbbell result (3.2-32) by defining a thermodynamic force

$$\underline{m}_q = \frac{\partial \mu}{\partial \underline{q}} \qquad (3.3\text{-}14)$$

in which $\eta = \eta$ ($\underline{e}$) is an arbitrary extension of the function $\eta(\underline{e})$ to the whole $\underline{q}$ space. Then we have:

$$\underline{m} = q \, \underline{P} \cdot \underline{m}_q \qquad (3.3\text{-}15)$$

If we also use that $\dot{q} = \underline{q} \cdot \underline{e}$, (13) may be written as

$$\begin{bmatrix} \underline{\tau} \\ \dot{\underline{g}} \end{bmatrix} \begin{bmatrix} (\frac{2\eta}{n} \underline{I} + \frac{\zeta q^2}{2} \; \underline{e}\,\underline{e}\,\underline{e}\,\underline{e}): & -(\underline{P}\,\underline{q})^{T} \cdot \\ \underline{P}\,\underline{q} : & \frac{2}{\zeta} \; \underline{P} \cdot \end{bmatrix} \begin{bmatrix} \underline{L} \\ -\underline{m}_q \end{bmatrix} \qquad (3.3\text{-}16)$$

From (16) the description on the ($\underline{\tau}$, $\underline{L}$ / $\underline{m}_q$, $\underline{q}$)-level is readily obtained by writing

$$\underline{\tau} = \underline{f} \; \underline{g} \qquad (3.3\text{-}17)$$

which follows from the Kirkwood-Kramers expression of the stress tensor [7], and using the fact that $\dot{\underline{d}} = \underline{L} \cdot \underline{g}$.

The result is:

$$\begin{bmatrix} \underline{f} \\ \dot{q} \end{bmatrix} = \begin{bmatrix} \frac{\zeta q}{2} (\underline{1} - \underline{P}) & -\underline{P} \\ \underline{P} & \frac{2}{\zeta} \; \underline{P} \end{bmatrix} \cdot \begin{bmatrix} \dot{\underline{d}} \\ -\underline{m}_q \end{bmatrix} \qquad (3.3\text{-}18)$$

Note that the corresponding equation (3.2-34) for the case of elastic dumbbels is obtained if we take $\underline{P} = \underline{1}$

The theory of Doi [12] for nematic liquid crystals may be formulated very similar to equation (16). In that case the viscous stresses are neglected, so the (1,1)-element of the matrix in (16) becomes zero. Furthermore a mean field potential $\Phi(\underline{e})$ is introduced, in order to describe the tendency of the rods to assign in preferred directions. Instead of (11) we then have

$$\underline{m} = kT \frac{\partial}{\partial \underline{e}} \ln \psi + \frac{\partial}{\partial \underline{e}} \Phi \qquad (3.3\text{-}19)$$

Usually Φ is taken to be the Maier-Saupe potential:

$$\Phi = \text{const} - \frac{3}{2} NkT \; \underline{e}\,\underline{e} : \underline{S} \qquad (3.3\text{-}20)$$

in which $\underline{S}$ is the structure tensor, defined as

$$\underline{S} = \langle \underline{e}\,\underline{e} - \frac{1}{3} \underline{1} \rangle \qquad (3.3\text{-}21)$$

The evolution equation for the rotary motion of the rods is given by [12]

$$\frac{\partial\psi}{\partial t} = - \frac{\partial}{\partial \underline{e}} \cdot (\psi\ \underline{P} \cdot \underline{L} \cdot \underline{e}) + \bar{D}_r \frac{\partial}{\partial \underline{e}} \cdot (\psi \frac{\partial \ln \psi}{\partial \underline{e}} + \psi \frac{\partial}{\partial \underline{e}} \frac{\Phi}{kT}) \qquad (3.3\text{-}22)$$

in which D_r is an average rotary diffusivity.

From this expression we see that

$$\dot{\underline{e}} = \underline{P} \cdot \underline{L} \cdot \underline{e} - \frac{D_r}{kT} \underline{m} \qquad (3.3\text{-}23)$$

which is indeed similar to the expression of $\dot{\underline{e}}$. So we expect that in this case

$$\begin{bmatrix} \underline{\tau} \\ \dot{\underline{e}} \end{bmatrix} = \begin{bmatrix} \underline{\underline{0}} & : & -(\underline{P}\ \underline{e})^{T}\cdot \\ (\underline{P}\ \underline{e}) & : & \frac{\bar{D}_r}{kT} \underline{1} \cdot \end{bmatrix} \begin{bmatrix} \underline{L} \\ -\underline{m} \end{bmatrix} \qquad (3.3\text{-}24)$$

In order to prove this, we still have to verify the expression for $\underline{\tau}$, implied by (24). To this end we calculate

$\underline{T} = n <\underline{\tau}> = n <kT \frac{\partial}{\partial \underline{e}} \ln \psi \cdot (\underline{1} - \underline{e}\ \underline{e})\underline{e}> + n <\frac{\partial \Phi}{\partial \underline{e}} \cdot (\underline{1} - \underline{e}\ \underline{e})\ \underline{e}>.$

The first term delivers the dilute-solution result nkT $<3\ \underline{e}\ \underline{e} - 1> = 3$ nkT $\underline{S}$.
For the second term, by using the Maier Saupe potential (20) we obtain $-3n$ UkT $(\underline{S} \cdot \underline{S} - \underline{S}: <\underline{e}\ \underline{e}\ \underline{e}\ \underline{e}>)$, so

$$\underline{T} = 3\ nkT\ \underline{S} - 3n\ UkT\ (\underline{S} \cdot \underline{S} - \underline{S}: <\underline{e}\ \underline{e}\ \underline{e}\ \underline{e}> \qquad (3.2\text{-}25)$$

This result is indeed in accordance with the stress tensor expression obtained by Doi. So we see that the matrix representation (24) is consistent with the theory of Doi.

In the theory of nematics one often employs the preaveraging assumption.

$$<\underline{e}\ \underline{e}\ \underline{e}\ \underline{e}> = <\underline{e}\ \underline{e}> <\underline{e}\ \underline{e}> \qquad (3.3\text{-}26)$$

In that case a closed description at the $(\underline{T}, \underline{L}/\underline{S}, \underline{M})$ level is possible. Noting that the free energy function corresponding to the Maier-Saupe potential is given by $A_\Phi = -\frac{3}{4}n$ UkT $\underline{S}:\underline{S}$ we obtain for the thermodynamic force $\underline{M}$:

$$\underline{M} = \frac{\partial A_\Phi}{\partial \underline{S}} = -\frac{3}{2}n\ UkT \qquad (3.3\text{-}27)$$

It can be seen that the equation for $\dot{\underline{S}}$, see ref. [13], obtained from (22) may be represented as

$$\dot{\underline{\underline{S}}} = \underline{\underline{\Lambda}} : L \; - \; \underline{\underline{\beta}} : \underline{\underline{M}} \tag{3.3-28}$$

with,

$$\underline{\underline{\Lambda}} : \underline{\underline{L}} = \frac{1}{3} (\underline{\underline{L}} + \underline{\underline{L}}^T) + \underline{\underline{L}} \cdot \underline{\underline{S}} + \underline{\underline{S}} \cdot \underline{\underline{L}}^T - \frac{1}{3}(\underline{\underline{L}} + \underline{\underline{L}}^T) : \underline{\underline{S}}\, \underline{\underline{1}} - (\underline{\underline{L}} + \underline{\underline{L}}^T) : \underline{\underline{S}}\, \underline{\underline{S}} \tag{3.3-29}$$

and,

$$\underline{\underline{\beta}} : \underline{\underline{M}} = \frac{4\, D_r}{n\, UkT} \left[(1 - \frac{U}{3}) \underline{\underline{M}} - U\, (\underline{\underline{S}} \cdot \underline{\underline{M}} - \frac{1}{3}\underline{\underline{S}} : \underline{\underline{M}}\, \underline{\underline{1}}) + U\, \underline{\underline{S}}\, \underline{\underline{S}} : \underline{\underline{M}}\right. \tag{3.3-30}$$

From (29) we calculate

$$\underline{\underline{\Lambda}}^T : \underline{\underline{M}} = \frac{1}{3} (\underline{\underline{M}} + \underline{\underline{M}}^T) + \underline{\underline{M}} \cdot \underline{\underline{S}} + \underline{\underline{S}} \cdot \underline{\underline{M}} - \frac{1}{3} (\underline{\underline{M}} + \underline{\underline{M}}^T) : \underline{\underline{1}}\, \underline{\underline{S}} - (\underline{\underline{M}} + \underline{\underline{M}}^T) : \underline{\underline{S}}\, \underline{\underline{S}} \tag{3.3-31}$$

or by eliminating $\underline{\underline{M}}$ with (27)

$$\underline{\underline{\Lambda}}^T : \underline{\underline{M}} = 3\, n\, kT\, (\underline{\underline{S}} - U\, \underline{\underline{S}} \cdot \underline{\underline{S}} + U\, \underline{\underline{S}} : \underline{\underline{S}}\, \underline{\underline{S}}) \tag{3.3-32}$$

Within an isotropic term this is indeed the stress tensor expression, obtained by Doi [13]. So, again the formalism of our triangle model applies. The matrix representation in this case becomes of the form(3.2-38)with $\eta = 0$, $\underline{\underline{\Lambda}}$ and $\underline{\underline{\beta}}$ defined by (29) and (30) and a thermodynamic force $\underline{\underline{M}}$, given by (27).

3.4 **Reptation models**

The concept of reptation, was proposed originally by de Gennes [14] and used in a rheological model by Doi and Edwards [15] and by Curtiss and Bird [16] in a different way.

We will follow here the approach of Doi and Edwards in which the polymer molecule is treated as a chain, confined in a tube. The tube consists of N segments of a length a. The average contour of the molecule coincides with the center line of the tube and is called the "primitive chain". Due to the thermal motion at the molecule the primitiva chain performs a diffusive motion along its own contour (reptation) and tube elements are created and destructed at the endpoints of the primitive chain.

The motion of a primitive chain segment consists of two parts : a convective part and a diffusive part.

The convective part is fully determined by the motion of the tube segments and causes a rate of rotation.

$$\dot{\underline{e}} = \underline{L} \cdot \underline{e} - \underline{e} \cdot \underline{D} \cdot \underline{e} = \underline{P}\, \underline{e} : \underline{L} \qquad (3.4\text{-}1)$$

of the segment-orientation vectors $\underline{e}$. In the second expression (1) we have used the projection operator defined by (3.3-1). The diffusive part of the motion of the segments determines the motion of the chain along the tube. This motion is governed by the diffusion law

$$\dot{s} = - D \frac{\partial \ln \psi}{\partial s} \qquad (3.4\text{-}2)$$

in which D is a diffusion constant, s the curve linear position along the tube and $\psi = \psi(\underline{e}, s, t)$ a probability density for one segment of being at a position s with an orientation $\underline{e}$ at time t. The function ψ may be represented [15] by the integral expression:

$$\psi(\underline{e}, s, t) = \int_{-a}^{t}\int \kappa\,(t - t', s)\; \delta(\underline{e} - \bar{\underline{e}}(e', t', t)\; \hat{\psi}(e')\; d\, \overset{2}{e}'\; dt' \qquad (3.4\text{-}3)$$

in which $\kappa(t - t', s)$ a memory function determined by the diffusion process along the rope, $\bar{\underline{e}}(\underline{e}', t', t) = \underline{F}_t(t') \cdot \underline{e}' / |(\underline{F}_t(t') \cdot \underline{e}'|$ (with $\underline{F}_t(t')$ the relative deformation gradient) a function which determines the relation between the orientation ($\underline{e}'$) of a tube element at the tense of creation (t′) and its orientation at the present tense (t) and $\hat{\psi}(\underline{e}')$ is the orientation distribution function of tube elements at the constant of creation.

We also will use the averages of κ along the tube.:

$$\phi(t - t') = \frac{1}{L} \int \kappa(t - t', s)\; ds \qquad (3.4\text{-}4)$$

The calculation of the stress tensor in the Doi and Edwards theory is based upon the expression

$$\underline{T} = n \int_{0}^{L} \langle \sigma\; \underline{e}\; \underline{e} \rangle\; ds \qquad (3.4\text{-}5)$$

in which σ is the tension in the chain. From the theory of rubber elasticity one obtains:

$$\sigma = \frac{3\; kT}{a} \qquad (3.4\text{-}6)$$

If (3) is used to perform the averaging in (5) and (6) is substituted we obtain

$$\underline{T} = 3\; n\; N\; kT \int_{\infty}^{t} \phi(t - t')\; \langle \underline{e}\; \underline{e} \rangle'\; dt' \qquad (3.4\text{-}7)$$

in which < >' denotes an average with respect to the creation distribution function $\psi(e')$. In the derivation of (7) we also have used the definition (4) of ϕ.

We will show now in which way it is possible to formulate the Doi and Edwards theory in the framework outlined in section 2. To that end we have to define for this case the tensor $\underline{\tau}$, the vector $\underline{m}$ and the matrix elements in the expression

$$\begin{bmatrix} \underline{\tau} \\ \dot{\underline{e}} \end{bmatrix} = \begin{bmatrix} \underline{\underline{\eta}} & -\underline{\underline{\Lambda}}^T \\ \underline{\underline{\Lambda}} & \underline{\beta} \end{bmatrix} \begin{bmatrix} \underline{\underline{L}} \\ -\underline{m} \end{bmatrix} \tag{3.4-8}$$

From (1) we see that

$$\underline{\underline{\Lambda}} = \underline{P}\ \underline{e} \tag{3.4-9}$$

and

$$\underline{\beta} = \underline{0} \tag{3.4-10}$$

Since in the Doi and Edwards theory no dissipative stresses occur, we also have

$$\underline{\eta} = \underline{0} \tag{3.4-11}$$

and remaining problem is to verify that $\underline{\tau} = \underline{\underline{\Lambda}}^T \cdot \underline{m}$ is consistent with the stress tensor expression of Doi and Edwards. To this end, like in our previous examples we use as the thermodynamic force

$$\underline{m} = kT \frac{\partial}{\partial \underline{e}} \ln \psi \tag{3.4-12}$$

With the form (10) of $\underline{\underline{\Lambda}}$ we then have

$$\underline{\tau} = (\underline{P}\ \underline{e})^T \cdot \underline{m} = \underline{m} \cdot \underline{P}\ \underline{e} = \underline{m}\ \underline{e} \tag{3.4-13}$$

Here we have used the expressions (3.3-1) and (13) of $\underline{P}$ and $\underline{m}$ respectively. By performing the average $<\underline{\tau}>$ with the distribution function (3) and comparing the result with the stress tensor expression (7) is may be shown, [note that $\int (\frac{\partial}{\partial \underline{e}} \phi)\ \underline{e}\ d^2\underline{e} = \int (3\ \underline{e}\ \underline{e} - \underline{1})\ \phi\ d^2\ \underline{e})$], that

$$\underline{T} = \frac{m\ N}{L} \int_0^L <\underline{\tau}>\ ds + \text{isotropic term} \tag{3.4-14}$$

So we see that the quantity $\underline{\tau}$ introduced above is indeed the appropriate variable to be used in the matrix formulation. We also obtain the following result

$$\begin{bmatrix} \underline{\tau} \\ \dot{\underline{e}} \end{bmatrix} = \begin{bmatrix} \underline{0} & : & -(\underline{P}\ \underline{e})^T \cdot \\ (\underline{P}\ \underline{e}) : & & \underline{0}\ \cdot \end{bmatrix} \begin{bmatrix} \underline{\underline{L}} \\ -\underline{m} \end{bmatrix} \tag{3.4-15}$$

It is possible to include the variable s in the the theory. To this end we introduce a thermodynamic force associated with the diffusive motion of the chain. This force is the thermodynamic potential

$$\mu = kT \frac{\partial \ln \psi}{\partial s} \qquad (3.4\text{-}16)$$

The equation (2) may be written then as s = - (D/kt) η and the matrix form (15) becomes:

$$\begin{bmatrix} \underline{\tau} \\ \dot{\underline{e}} \\ \dot{\underline{s}} \end{bmatrix} = \begin{bmatrix} \underline{0} & : & -(\underline{P}\ \underline{e})^{T} & \underline{0}\ \cdot \\ (\underline{P}\ \underline{e}): & & \underline{0}\ \cdot & \underline{0}\ \cdot \\ \underline{0} & : & \underline{0}\ \cdot & \frac{D}{kt} \end{bmatrix} \begin{bmatrix} \underline{L} \\ -\underline{m} \\ -\mu \end{bmatrix} \qquad (3.4\text{-}17)$$

We see that μ only contributes to the flux s and not to the tensor $\underline{\tau}$.

A generalisation of the Doi and Edwards theory in which frictional forces between the chain and the tube wall were taken into account was presented first by Jongschaap [17] (see also Geurts and Jongschaap [18]). This model, which was called the "Reptating Rope Model" was shown to be equivalent to the theory of Curtiss and Bird. The main difference with the Doi and Edwards theory is an extra term in the stress tensor of the form.

$$\underline{T}^{D} = n\, N \frac{\xi}{2} \int_{-\infty}^{t} \int_{0}^{t} s(L - s)\, \kappa(t - t', s) \langle \underline{e}\ \underline{e}\ \underline{e}\ \underline{e} \rangle : \underline{D}\ ds\ dt' \qquad (3.4\text{-}18)$$

Comparing this with (14) and (17) we see that the matrix representation of Reptating Rope model becomes:

$$\begin{bmatrix} \underline{\tau} \\ \dot{\underline{e}} \\ \dot{\underline{s}} \end{bmatrix} = \begin{bmatrix} \frac{1}{2}\xi\ s(L-s)\ \underline{e}\ \underline{e}\ \underline{e}\ \underline{e} & : & -(\underline{P}\ \dot{\underline{e}}^{T}) & \cdot & \underline{0}\ \cdot \\ \underline{P}\ \underline{e} & : & \underline{0}\ . & & \underline{0}\ \cdot \\ \underline{0} & : & \underline{0}\ \cdot & & \frac{D}{kt} \end{bmatrix} \begin{bmatrix} \underline{L} \\ -\underline{m} \\ -\underline{\mu} \end{bmatrix} \qquad (3.4\text{-}19)$$

Although we now have seen how the triangle model may be applied to reptation models some point remain to be clarified. First we see that a front-factor 3 in the stress tensor expression is obtained. This factor arises from the averaging in orientation space and is similar to the same factor in reversible part of the stress tensor (3.3-13) of the rigid dumbbell model. The factor 3 is also present in the theory of Doi and Edwards but seems to have a different origin there. It does nog appear in the Curtiss Bird theory or in the Reptating Rope Model. The main difference is that in our present formulation the thermodynamic force $\underline{m}$ associated with the orientation distribution of the segments plays a prominent role whereas in the original reptation theories it is the tensile force along the chain which contributes to the stress.

3.5 Transient-Network models

The Transient-Network Model, originally developed by Green and Toblsky [18] Lodge [19] and Yamamoto [20]is used to describe the rheological behaviour of polymer melts and concentrated polymer solutions. In the model, the system of highly entangled polymers is represented by a rubberlike network of segments. The network is not permanent since the segments are created and annihilated at specified rates.

Before describing it in our present formalism we will first briefly summarise some basic notions of the Transient Network Model. The number density $\Psi(\underline{q},t)$ of segments with a specified configuration may, analogous to (3.4-3) be expressed as

$$\Psi = \int_{-\infty}^{t} \int \tilde{n}\ (t,t') \quad \delta(\underline{q}-\bar{\underline{q}}(\underline{q}',t't))\ \hat{\psi}(\underline{q}')\ d^3\underline{q}'\ dt' \qquad (3.5\text{-}1)$$

In this expression $\tilde{n}(t,t')$ is the number of segments created per unit time at time t' and still present at time t,

$$\bar{\underline{q}} = \underline{F}_t^{-1}(t') \cdot \underline{q}' \qquad (3.5\text{-}2)$$

A function specifying the motion of the segments and $\hat{\psi}(\underline{q}')$ the creation distribution function of a segment which usually is assumed to be afunctional. The equation (2) specifies the assumption of affine motion of the segments. This may also be expressed as

$$\dot{\underline{q}} = \underline{L} \cdot \underline{q} \qquad (3.5\text{-}3)$$

The kinetics of loss and creation of segments may be specified by the properties of the function $\tilde{n}(t,t')$. At constant t', the change of $\tilde{n}$ with t is due to loss. This is expressed by the loss function h(t), defined by

$$\frac{\partial\ \tilde{n}(t,t')}{\partial\ t} = -\ h(t)\ \tilde{n}(t,t') \qquad (3.5\text{-}4)$$

On the other hand $\tilde{n}$ (t',t') dt' is the number of segments created in the time interval [t',t' + dt'], therefore a creation function g(t') is defined by

$$\tilde{n}\ (t',t') = n_o\ g(t') \qquad (3.5\text{-}5)$$

In this expression n_o denotes the equilibrium value of the number density of segments n. The relation between n and $\tilde{n}$ is:

$$n(t) = \int_{-\infty}^{t} \tilde{n}(t,t')\ dt' \qquad (3.5\text{-}6)$$

From this expression and (4) and (5) the following rate equation for n is obtained:

$$\frac{dn}{dt} = g\, n_o - h\, n \qquad (3.5\text{-}7)$$

The solution of (7) with the initial condition (5) may be expressed as

$$\tilde{n}(t,t') = n_o\, g(t,t')\, e^{-\int_{t'}^{t} h(t'')\, dt''} \qquad (3.5\text{-}8)$$

We will now apply the triangle theory to the transient network model like in some of the previous applications we take the one-segment contribution τ to the stress tensor as the stress variable, so

$$\underline{T} = \int \Psi(\underline{q},t)\ \underline{\tau}\ d^3\underline{q} \qquad (3.5\text{-}9)$$

The matrix representation then becomes:

$$\begin{bmatrix} \underline{\tau} \\ \dot{\underline{q}} \\ \dot{n} \end{bmatrix} = \begin{bmatrix} \underline{\underline{0}} & -(\underline{1}\ \underline{q})^T & \underline{0} \\ \underline{1}\ \underline{q} & \underline{0} & \underline{0} \\ \underline{0} & \underline{0} & \frac{g}{kT} \end{bmatrix} \begin{bmatrix} -\underline{\underline{L}} \\ -\underline{m} \\ -\mu \end{bmatrix} \qquad (3.5\text{-}10)$$

with

$$\underline{m} = k\ \underline{q} \qquad (3.5\text{-}11)$$

the spring-force in a segment and

$$\mu = kT\ (\frac{h}{g}\, n - n_o) \qquad (3.5\text{-}12)$$

a chemical potential associated with the change in the free energy of the network due to a change of n. In principle this quantity could be derived form details of the entanglement-desentanglement previous regarded as a chemical reaction, but, following Ajji at al [33], we use the form (12)suggested by the rate equation (7) of the transient network model. The expression for $\underline{\tau}$ obtained form (10) and (11) is in accordance with the usual stress tensor expression $\underline{T} = k \int \Psi\ \underline{q}\ \underline{q}\ d^3\underline{q}$. Substitution of Ψ as given by (1) and making use of (2), and (8) results in the constitutive equation

$$\underline{T} = n_o\, kT \int^{t} g(t')\, e^{-\int_{t'}^{t} h(t'')\, dt''}\ \underline{c}_t^{-1}(t')\ dt' \qquad (3.5\text{-}13)$$

in which $\underline{c}_t(\tau) = \underline{F}_t(\tau) \cdot \underline{F}_t\ (\tau)^{-1}$

Returning to the matrix expression (10) we will consider now some possible modifications of the model. First like has been done in the case of the elastic dumbbell model (section 3.1) one may change the level of description. For instance by using the distribution function $\Psi(\underline{e},t)$ or alternatively a structure tensor. $\underline{S} = \langle \underline{q}\ \underline{q} \rangle$ as a state variable. We will not discuss these modifications here. An interesting change at the present level of description however is a change of the convection law. A well known alternative is the so called Gordon and Schowalter [22] non affine convection law, also used in the Phan Thien-Tanner [23] formulation of the Transient Network Model. In that case instead of (3) we have

$$\dot{\underline{q}} = \underline{\underline{L}} \cdot \underline{q} - \xi \, \underline{\underline{D}} \cdot \underline{q} \qquad (3.5\text{-}14)$$

in which $\underline{\underline{D}} = \frac{1}{2}(\underline{\underline{L}} + \underline{\underline{L}}^T)$, the rate of strain tensor. This may be written as,

$$\dot{\underline{q}} = \left[(1 - \frac{\xi}{2}) \, \underline{\underline{1}} \, \underline{q} - \frac{\xi}{2} \, (\underline{\underline{1}} \, \underline{q})^T\right] : \underline{\underline{L}} \qquad (3.5\text{-}15)$$

The skew symmetry of the matrix in (2.17) then implies that

$$\underline{\underline{\tau}} = \left[(1 - \frac{\xi}{2}) \, (\underline{\underline{1}} \, \underline{q})^T - \frac{\xi}{2} \, (\underline{\underline{1}} \, \underline{q})^T\right] \cdot \underline{m}$$

$$= \ (1 - \frac{\xi}{2}) \, \underline{m} \, \underline{q} + (\text{isotropic term}) \qquad (3.5\text{-}16)$$

This implies that also in the stress tensor expression a factor $(1 - \frac{\xi}{2})$ should be included. The point that in the case of non-affine motion the stress tensor expression should be modified has been discussed in an other context by Larson [24] and also by Maugin and Drout [2] and by Grmela [3]. As pointed out by Larson [24], the physical reason for the extra factor in the stress tensor is some kind of "slip" of the network strands. We will illustrate this now for the case of slip in entanglement. In figure 2 a piece of a polymer chain between two entanglements is shown.

If the forces $\underline{f}$ and $\underline{m}$, and the vectors $\underline{q}$ and $\underline{s}$ are defined as indicated in figure 2. Conservation of energy requires (in the case of no friction in the entanglements) that

$$\underline{f} \cdot \dot{\underline{q}} = \underline{m} \cdot \dot{\underline{s}} \qquad (3.5\text{-}17)$$

So, if we introduce a slip factor a by requiring

$$\dot{\underline{s}} = a \, \dot{\underline{q}} \qquad (3.5\text{-}18)$$

we have,

$$\underline{m} = \frac{1}{a} \, \underline{f} \qquad (3.5\text{-}19)$$

(In the case represented by Figure 2b we would have a = 2.) The stress tensor in a network corresponding to the entanglement structure of Figure 2a would become. $\underline{\underline{T}} = n\langle \underline{f} \, \underline{q} \rangle = n \, a \langle \underline{m} \, \underline{q} \rangle$. If in this expression we take $\underline{m} = \kappa \, \underline{s}$ with $\underline{s} = \underline{q}$ (since at any instant the total part of the chain between two entanglements contributes to the elastic stress), we obtain,

$$\underline{\underline{T}} = n\kappa \, a \langle \underline{q} \, \underline{q} \rangle \qquad (3.5\text{-}20)$$

in accordance with our previous result (16).

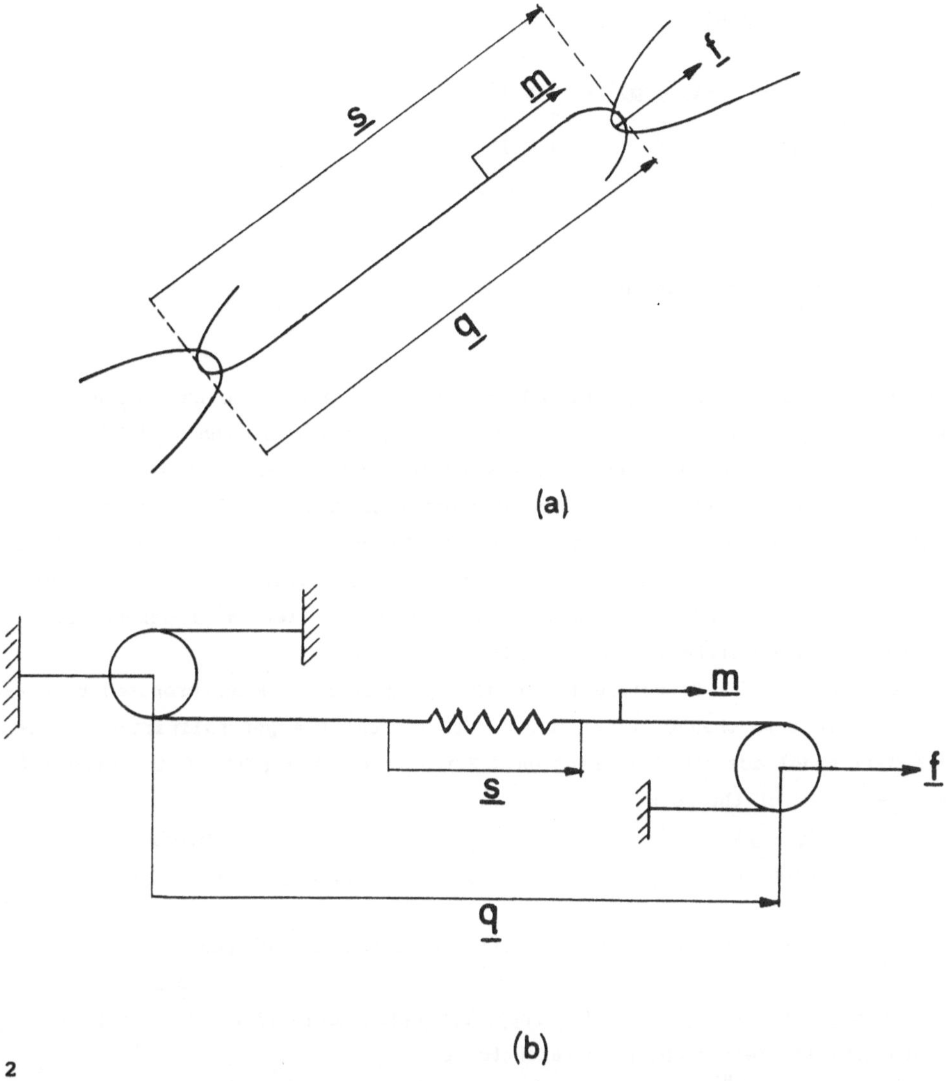

Figure 2

Non affine motion due to slip in entanglements : schematic picture of the entanglement structure (a) and mechanical analogue (b). The chain vector $\underline{s}$ of a part of the chain changes at rate that differs from the rate of change of the vector $\underline{q}$ between two entanglements. The elastic force $\underline{m}$ in the chain also differs from the force $\underline{f}$, acting upon the entanglements.

Also in this case the situation may be clearly summarised in a matrix representation. At the $(\underline{f},\dot{\underline{q}}/\underline{m},\dot{\underline{s}})$ - level we have

$$\begin{bmatrix} \underline{f} \\ \dot{\underline{s}} \end{bmatrix} = \begin{bmatrix} \underline{0} & -a \\ a & \underline{0} \end{bmatrix} \cdot \begin{bmatrix} \dot{\underline{q}} \\ -\underline{m} \end{bmatrix} \qquad (3.5\text{-}21)$$

and at the $(\underline{\tau}, \underline{L}/\dot{\underline{s}}, \underline{m})$ - level :with $\underline{\tau} = \underline{f}\,\underline{q}$

$$\begin{bmatrix} \underline{\tau} \\ \dot{\underline{s}} \end{bmatrix} = \begin{bmatrix} \underline{\underline{0}} & -(a\,\underline{\underline{1}}\,\underline{q})^T \\ a\,\underline{\underline{1}}\,\underline{q} & \underline{\underline{0}} \end{bmatrix} \begin{bmatrix} \underline{\underline{L}} \\ \underline{m} \end{bmatrix} \qquad (3.5\text{-}22)$$

3.6 Configuration tensor models

In the case of the elastic dumbbell model (section 3.1) we have seen an example in which a closed formulation of the theory is possible at the $(\underline{\underline{T}}, \underline{\underline{L}}/\underline{\underline{M}}, \underline{\underline{S}})$ - level. In that case the structure tensor $\underline{\underline{S}}$ was defined by (3.2-35) in terms of microscopic variables and its properties were derived from the underlying configuration-space theory. In some cases, it may be usefull to introduce the properties of the structure tensors without an explicit reference to molecular theories. This possibility - in fact - is the main advantage of the use of different levels of description in combination with a consistent thermodynamic formulation.

A nice example of a structure tensor theory which is a model proposed by Giesekus [27,28]. In this theory, which may be considered as a generalisation of the elastic dumbbell model described in section 3.2.the reversible part of the stress tensor is assumed to be of the form

$$\underline{\underline{T}}^R = \mu(\underline{\underline{C}} - \underline{\underline{1}}) \qquad (3.6\text{-}1)$$

This expression is consistent with our previous result (3.2-43) if we take $\mu = nkT$ and $\underline{\underline{C}} = \frac{\kappa}{kT}\underline{\underline{S}}$.

For the tensor $\underline{\underline{C}}$, the following evolution equation is proposed:

$$\overset{\nabla}{\underline{\underline{C}}} = - \underline{\underline{B}} \cdot \underline{\underline{T}}^R \qquad (3.6\text{-}2)$$

with $\overset{\nabla}{\underline{\underline{C}}} = \dot{\underline{\underline{C}}} - \underline{\underline{L}} \cdot \underline{\underline{C}} - \underline{\underline{C}} \cdot \underline{\underline{L}}^T$, the upper convected derivative, and $\underline{\underline{B}}$ a kind of generalized mobility tensor which is taken to be

$$\underline{\underline{B}} = \beta(\underline{\underline{1}} + a\,\underline{\underline{T}}^R) \qquad (3.6\text{-}3)$$

In order to compare this with our previous results we note that from

$$\dot{\underline{\underline{C}}}^R = \underline{\underline{L}} \cdot \underline{\underline{C}} + \underline{\underline{C}} \cdot \underline{\underline{L}}^T = \underline{\underline{\Lambda}} : \underline{\underline{C}} \qquad (3.6\text{-}4)$$

in which $\underline{\underline{\Lambda}}$ is of the same form (3.2-40) as in the dumbbell model. With this expression for $\underline{\underline{\Lambda}}$, on using the result $T^R = \underline{\underline{\Lambda}}^T : \underline{\underline{M}}$ we obtain from (1) that

$$\underline{\underline{M}} = \frac{1}{2}\mu\,(\underline{\underline{1}} - \underline{\underline{C}}^{-1}) \qquad (3.6\text{-}5)$$

which is indeed in accordance with (3.2-37)

The tensor $\underline{\underline{\beta}}$ in the expression $\dot{\underline{\underline{C}}}^B = - \underline{\underline{\beta}} : \underline{\underline{M}}$ may be obtained now by using the result $\dot{\underline{\underline{C}}}^D = - \underline{\underline{B}} \cdot \underline{\underline{T}}^R$, which follows from (2) and the expressions (1), (3) and (5) for $\underline{\underline{T}}^R$, $\underline{\underline{B}}$ and $\underline{\underline{M}}$. The result is

$$\underline{\underline{\beta}} = - \underline{\underline{B}} \cdot \underline{\underline{\Lambda}} \qquad (3.6\text{-}6)$$

If this is compared with the elastic dumbbell result (3.2-41) one sees that the tensor $\underline{\underline{B}}$ generalizes the mobility factor $4/n\ \xi$. In the case that $a = 0$ the elastic dumbbell result is reobtained. The viscous stress in the Giesekus model has the usual form $\underline{T}^D = \underline{\underline{\eta}} : \underline{L}$ with $\underline{\underline{\eta}}$ given by (3.2-42), so the model may be summarizes now as follows:

$$\begin{bmatrix} \underline{T} \\ \dot{\underline{C}} \end{bmatrix} = \begin{bmatrix} 2\eta\ \underline{\underline{I}} & -\underline{\underline{\Lambda}}^T \\ \underline{\underline{\Lambda}} & \underline{\underline{B}} \cdot \underline{\underline{\Lambda}} \end{bmatrix} : \begin{bmatrix} \underline{L} \\ -\underline{M} \end{bmatrix} \qquad (3.6\text{-}7)$$

with $\underline{\underline{\Lambda}}$, $\underline{\underline{B}}$ and $\underline{M}$ given by (3.2-40), (3) and (5) respectively. By elimination of $\underline{M}$, $\underline{C}$ and $\dot{\underline{C}}$ the following constitutive equations may be obtained:

$$\underline{T}^D = 2\ \eta\ \underline{D} \qquad (3.6\text{-}8)$$

$$\underline{T}^R + \lambda\ \overset{\nabla}{\underline{T}}{}^R + a\ \underline{T}^R \cdot \underline{T}^R = 2\ \beta^{-1}\ \underline{D} \qquad (3.6\text{-}9)$$

with $\lambda = \dfrac{1}{\beta\ \eta}$.

The Giesekus model is a good illustration of how a slight modification of the dumbbell model, expressed at the configuration-tensor level of description may cause signigicant changes in the constitutive equations.

In the Giesekus model, the treatment was partially based upon the underlying molecular description. In other models at the configuration- tensor-level the approach is entirely at a macroscopic level. An example is the theory of Leonov [25] based upon the concept of a recoverable strain and theories based upon Eckarts [29] concept of a variable relaxed state. For further information about the latter class of theories we refer to a paper by Stickforth [30]. For our present discussion it is sufficient to know that in those theories the stress and the velocity gradient are decomposed in a reversible (elastic) and a dissipative (inelastic) part.

So we have like before : $\underline{T} = \underline{T}^R + \underline{T}^D$ but also : $\underline{L} = \underline{L}^R + \underline{L}^D$. In order to compare this with our formalism we have to define $\underline{L}^R$ and $\underline{L}^D$ in a consistent way. to this end we rewrite the dissipation form as follows

$$\begin{aligned} \Delta &= \underline{T} : \underline{L} - \underline{M} : \dot{\underline{S}} \\ &= \underline{T} : \underline{L} - \underline{M} : \underline{\underline{\Lambda}} : \underline{\underline{\Lambda}}^{-1} : \dot{\underline{S}} \\ &= \underline{T} : \underline{L} - \underline{T}^R : \underline{L}^R \end{aligned} \qquad (3.6\text{-}10)$$

in which (3.2-44) has been used and

$$\underline{L}^R = \underline{\underline{\Lambda}}^{-1} : \dot{\underline{S}} \qquad (3.6\text{-}11)$$

Instead of (10) one may also write

$$\Delta = \underline{T}^D : \underline{L} + \underline{T}^R : \underline{L}^D \qquad (3.6\text{-}12)$$

The phenomenological relations according to non-equilibrium thermodynamics then become :

$$\begin{bmatrix} \underline{T}^D \\ \underline{L}^D \end{bmatrix} = \begin{bmatrix} \underline{\underline{\eta}} & -\underline{\underline{\Phi}}^T \\ \underline{\underline{\Phi}} & \underline{\underline{\upsilon}} \end{bmatrix} : \begin{bmatrix} \underline{L} \\ -\underline{T}^R \end{bmatrix} \qquad (3.6\text{-}13)$$

This, indeed, is the form obtained by Leonov [25], Stickforth [27] and others. In our present approach, however, it is possible to go one step further. If we start from our matrix expression (3.2-38)and eliminate $\underline{M}$, $\underline{S}$ and $\underline{T}$ in the same way as in deriving(13), we obtain :

$$\begin{bmatrix} \underline{T}^D \\ \underline{L}^D \end{bmatrix} = \begin{bmatrix} \underline{\underline{\eta}} & \underline{\underline{0}} \\ \underline{\underline{0}} & \underline{\underline{\upsilon}} \end{bmatrix} : \begin{bmatrix} \underline{L} \\ -\underline{T}^R \end{bmatrix} \qquad (3.6\text{-}14)$$

with $\underline{\underline{\upsilon}} = \underline{\Lambda}^{-1} : \beta : \underline{\Lambda}^{-T}$ which is of the same form as (13) but with one important difference, namely that the matrix now has the diagonal form. The meaning of the variables $\underline{L}^R$ and $\underline{L}^D$ becomes more clear in a schematic representation similar to the triangle model given in Figure 3. We see that $\underline{L}^R$ and $\underline{L}^D$ correspond to a mapping of the spring and dashpot inside the system. In the case that $\underline{\underline{\Lambda}} = \underline{1}$, $\underline{L}^R$ coincides with $\dot{\underline{S}}$ and $\underline{T}^R$ with $\underline{M}$. In other cases these quantities will differ from each other.
From figure 3 it can also be seen that the strain corresponding to $\underline{L}^R$ is indeed the so called recoverable strain, used in the Leonov Model: it is an elastic recovery measured at the part of the system where the external variables $\underline{T}$ and $\underline{L}$ apply.

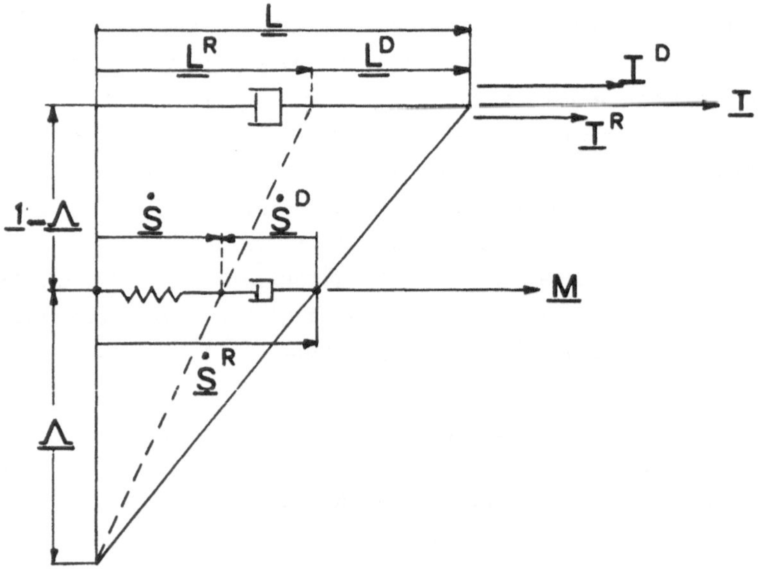

Figure 3
Representation of the model described by the equations (3.2-38) and (3.6-14) by the triangle model (see also Figure 1)

4. Discussion

In this paper a theory was presented by which a unified treatment of various rheological models is possible. The theory may be considered as an extension of earlier work by Grmela [3] and others [1,2]. In our treatment the concept of a so called macroscopic time reversal played a central role. On the basis of this, an unambigious distinguishment between reversible and dissipative variables was possible and a universal matrix representation (2.17) could be given. The skew symmetry of this matrix is in accordance with the Onsager-Casimir symmetry relations.

In the applications, discussed in section 3, we have seen that this skew-symmetry may be used as a check on the consistency of the stress tensor expression with the evolution equation of a model. In the network models with non-affine motion (sec. 3-6), for instance, it implies the necessity of an extra factor in the stress tensor expression. In configuration tensor models the skew symmetry of the matrix in (2.17) may be used to construct the stress tensor expression, from a given evolution equation. In section 3 some examples were given of application of our formalism to existing models. Our purpose there was not to obtain new results in the sense of new constitutive equations, but merely to demonstrate the capability and flexibility of the new approach. Only in the treatment of the dumbbell models (section 3.2 - 3.3) the level of description with functions as state variables and functional dependences was employed. In the other applications a tensor-formulation was sufficient. Nevertheless one should keep in mind that in future applications a functional formulation might be necessary. For instance in cases where nonlocal phenomena are important. The description could be based upon global fields instead of local variables of state ; this makes a functional formulation necessary. In this context the problem of the effects of domains on orientation and stress liquid crystals might be of interest. In the examples of section 3, in most cases only a reformulation of existing theories was obtained. The main advantage of the present formulation, however is it shows in which directions modifications of the model are possible and also that those modifications fall into a few categories. Referring to the general equation (2.17) we first have : a change in the free energy functional $A(\Phi)$ and so of the expression for the thermodynamic stress : $M(\Phi)$, seemed of the viscous effects described by the quantities η and β and third a change of the coupling effect, expressed by the quantity Λ. It has been shown that the latter 4.1 changes significant in the case of constraints and in the case of non-affine motion. So we see that our theory offers the possibility of analysing the implications of particular modifications of a model in a systematic and consistent manner.

In applying the theory to existing models sometimes new insight is obtained. For instance, in the case of the Doi and Edwards model, where we haven seen that the stress is due to the thermodynamic force associated with the orientation of the tube segments, rather than a tensile force along the tube. We also have clarified the origin of the extra factor in the stress tensor expression due to non-affine motion. Finally, in section 3.7 we arrived at the result that the matrix (3.6-13) in the theory of Leonov and in similar theories should be diagonal.

The use of different levels of description was only fully employed in the case of the elastic dumbbell model. In some other cases this could also be done, but we did not make an attempt to be complete here. Still it should be stressed that a change of level of description result in a considerable simplification of the problem. An obvious case is of course the change from the configuration space level to the structure tensor level of description. The opposite change from the configuration space level to the level of forces and deformations of individual particles, however, is in many cases even more usefull. At this level, in fact, many of the examples that we have considered were formulated. In general one could say that our triangle modelcan be applied in any case in which a thermodynamic subsystem can be defined for which the free energy may be expressed as a function of functional of some state variables, a set of external variables by which the exchange of power with the environment is described and an evolution equation of the microstructure.

References

1. Lhuillur, D. and A. Quibrahum
   J. Mecanique 19, 1-17 (1980)
2. Maugin, G.A. and R. Drout
   Int. J. Eng. Sci. 21, 705-24 (1983)
3. Grmela, M. Phys. Lett 111A, 41-4 (1985)
4. Grmela, M. Physica 21D, 179-212 (1986)
5. Grmela, M. J. Rheol. 30, 707-728 (1986)
6. Grmela, M. and P.J. Carreau
   J. Non-Newtonian Fluid Mechanics 4, 269-75 (1987)
7. Bird, R.B. et.al. Dynamics of Polymeric
   Liquids (2nd edn) vol 2 Kinetic Theory
   (New York: Wiley) (1987)
8. Doi, M. J. Chem. Phys. 79, 5080 (1983)
9. Sarti, G.C. and G. Marrucci
   Chem. Eng. Sci 28, 1053-1059 (1973)

10. H.C. Booij and P.N. van Wiechen
J. Chem. Phys. 52, 5056-5068 (1970)
11. Jongschaap, R.J.J. Rheol. Acta 26, 328-335 (1987)
12. Doi. M. J. Polym. Sci Phys. Ed. 19, 229 (1981)
13. Doi, M. and S.F. Edwards, The Theory of
Polymer Dynamics, Oxford Press, New York (1986)
14. Gennes, P.J. de, J. Chem. Phys. 55, 572-579 (1971)
15. Doi. M. and Edwards, S.F. J. Chem. Soc.
Trans II 74, 1789-1832 (1978); 75, 373-382 (1979)
16. Curtiss C.F. and R.D. Bird
J. Chem. Phys. 75, 2016-33 (1981)
17. Jongschaap, R.J.J. Rheol. Acta 26, 99-102 (1988)
18. Geurts, B.G. and Jongschaap, R.J.J.
J. Rheol. 32, 353-365 (1988)
19. Green, M.S. and Tobolsky, A.V.
J. Chem. Phys. 15, 651 (1947)
20. Lodge, A.S. Trans. Farnaday Soc. 52, 120-130 (1956)
21. Yamamoto, M. J. Phys. Soc. Japan 11, 413 (1956)
22. Ajji, A, et.al. J. Rheol. 33, 401-420 (1989)
23. Gordon, R.J. and W.R. Schowalter
Trans. Soc. Rheol. 16, 79-97 (1972)
24. Phar Thien, N. and R.I. Tanner
J. Non-Newtonian Fluid Mech 2, 235-265 (1977)
25. Larson, R.G. J. Non-Newtonian Fluid Mech. 8, 271 (1981)
26. Leonov, 15, 85-98 (1976)
27 Giesekus, H. Rheol. Acta 2 , 50-62 (1962)
28. Giesekus, H. Rheol. Acta 5, 29-35 (1966)
29. Eckart, C. Phys., Rev. 73, 373-82 (1948)
30. Stickforth J. Rheol. Acta 25, 447-458 (1966)

ADHESION AND RHEOLOGY

P. G. de Gennes
Collège de France
75231 Paris Cedex 05

(Abstract): In most adhesive polymers, the energy G per unit area of peeled surface, is related to strong (non linear) dissipative processes in a thin ribbon ahead of the crack tip: we call this the junction region. The field variables inside the junction are the applied stress $\sigma(x)$, and the rate of opening $\dot{h}$ of the junction width (h). For glassy polymers, the junction is often a "craze" with a complex set of fibrils linking the two sides[1][2]. For rubbery polymers, crazes are not observed, and the microscopic processes involved may possibly be described in terms of a simple "pull out" of polymer chains, after a certain level of chemical rupture[3]. Various forms of the constitutive law $\sigma(\dot{h})$ have been proposed for these two distinct situations[4][5], and will be presented here. They are essentially characterized by three parameters.

a) A threshold stress (σ_c), which corresponds to plastic flow and crazing in glassy polymers, and to chemical scission processes in rubbers.

b) A terminal value of the junction opening (h_f), which can be estimated simply in models of homogeneous "pull out", but which is less well understood for crazes.

c) A friction coefficient $Q = (d\sigma/d\dot{h})$ in the high stress regime $(\sigma > \sigma_c)$.

Having specified the relation $\sigma(\dot{h})$, the theoretical analysis of fracture (at a velocity V) incorporates another (integral) relation between σ and $h/2$, which represents normal stresses and displacements at the surface of an elastic half space. These two coupled equations can be solved analytically in terms of hypergeometric functions[6][5], when $\sigma(\dot{h})$ is a piecewise linear function. For glassy polymers, where $\sigma(\dot{h})$ is often expected to be logarithmic, numerical studies are required[6][7]. In the piecewise linear models, an important parameter is the characteristic velocity $V^* = \mu/Q$ (where μ is the shear modulus of the elastic matrix surrounding the junction). At $V < V^*$, the stress $\sigma(x)$ is nearly constant and equal to the threshold value σ_c in all the junction. The resulting adhesive energy $G(V \to 0)$ is simply $\sigma_c h_f$. At $V > V^*$, most the stress sources are concentrated near the tip (where $h = h_f$), and in the linear models, one predicts $G(V) \sim \sigma_c h_f V/V^*$[3]. For the glassy systems, on the other hand, one expects $G(V) \sim \ln V$ as observed.

This general discussion will then be supplemented by a description of two especial points:

1) In weakly cross linked rubbers, the adhesive energy $G(V)$ is known -from work by Gent and Petrick[9]- to increase steeply with V, and to show a sharp peak. An interpretation of these features has been constructed in terms of viscoelastic losses far from the junction region[10].

2) Smectic A liquid crystals (composed of stacked liquid layers), usually respond to tensile stresses by classical undulation instability[11][12]. But "semi-smectics", made with an alternation of liquid/glassy layers, should (at least in thin samples) show a remarkable form of "lenticular fracture" with a cusp at the fracture tip. The features may be of interest for the mechanical properties of stratified coextrudates[13].

References

[1] E. Kramer, Adv. Polymer Science, 52/53, p. 1-56, (H. Kausch ed.), Springer (1983).

[2] E. Kramer, L. Berger, Adv. Polymer Science, to be published.

[3] P. G. de Gennes, J. de Physique, 50, 2551- 2562, (1989).

[4] E. Kramer, C. Y. Hui, to be published.

[5] S. Troian, P. G. de Gennes, C. R. Acad. Sci. (Paris), 311 II, 389-392, (1990).

[6] C. Y. Hui, to be published.

[7] P. G. de Gennes, H. Hervet, unpublished.

[8] For a simple discussion of the singularities in the linear models, see P. G. de Gennes, C. R. Acad. Sci. (Paris), 309 II, 1125-1128, (1989).

[9] A. Gent, R. Petick, Proc. Roy. Soc., A 310, 433-450, (1969).

[10] P. G. de Gennes, C. R. Acad. Sci. (Paris), 307 II, 1949-1953, (1988).

[11] M. Delayde, G. Durand, R. Ribotta, Phys. Lett., A 44, 139-143, (1973).

[12] P. G. de Gennes, "The physics of liquid crystals", Oxford U. Press, 2nd prining, (1983), chap. 7.

[13] P. G. de Gennes, submitted to Physics Letters, (1990).

RHEOLOGY OF HARD SPHERE SUSPENSIONS

B.U. Felderhof
Institut für Theoretische Physik A
RWTH Aachen
Templergraben 55, D -5100 Aachen

A suspension of rigid hard spheres in a viscous incompressible liquid constitutes an attractively simple rheological model system. Such a suspension may be realized experimentally as a collection of silica spheres immersed in an organic solvent. The spheres are neutral and can be made monodisperse with a radius of about 1000 Å. In suspensions of polystyrene spheres in water the spheres are charged and surrounded by a Debye cloud of small ions. The static structure factor of such a suspension is well approximated by that of a system of hard spheres with an effective radius equal to the actual sphere radius plus the Debye length.

On the time scale seen in a dynamic light scattering experiment one may visualize the system as a collection of interacting Brownian particles. The spheres diffuse with a bare diffusion coefficient given by the Stokes-Einstein expression $D_o = k_BT/6\pi\eta a$, where T is absolute temperature, η is the shear viscosity of the solvent, and a is the sphere radius. For water at room temperature and for a sphere radius $a = 10^{-5}$ cm this amounts to $D_o \approx 10^{-8}$ cm²/sec. The corresponding diffusion time scale $\tau_o = a^2/D_o$ is about 10^{-2} sec. This should be compared with the Brownian time scale $\tau_B = m/\zeta$, which is the time in which a particle of mass m and friction coefficient $\zeta = 6\pi\eta a$ loses its momentum. This time is of the order 10^{-8} sec , much shorter than the diffusion time scale. As a consequence, on the time scale of order 10^{-4} sec seen in light scattering experiments [1] one may disregard momentum relaxation and view the system as a collection of diffusing, interacting Brownian particles, with a momentum distribution close to a Maxwellian.

We consider N spheres in a volume Ω with instantaneous configuration $\underset{\sim}{X} = (\underset{\sim}{R}_1,\ldots,\underset{\sim}{R}_N)$. The configurations are random, so that we must envisage a probability distribution $P(\underset{\sim}{X},t)$. On the diffusion time scale the evolution of the distribution function is described by the generalized Smoluchowski equation [1]

$$\frac{\partial P(\underset{\sim}{X},t)}{\partial t} = \underset{\sim}{\nabla} \cdot \underset{\approx}{D}(\underset{\sim}{X}) \cdot [\underset{\sim}{\nabla} P + \beta(\underset{\sim}{\nabla}\Phi)P] \quad . \tag{1}$$

The diffusion is influenced by hydrodynamic interactions, as incorporated in the 3Nx3N diffusion matrix $\underset{\approx}{D}(\underset{\sim}{X})$, and by direct interactions incorporated in the potential $\Phi(\underset{\sim}{X})$. The latter has also a contribution from the wall potential. Finally $\beta = 1/k_BT$. The diffusion matrix is given by the generalized Einstein relation

$$\underset{\approx}{D}(\underset{\sim}{X}) = k_BT \; \underset{\approx}{\mu}(\underset{\sim}{X}) \quad , \tag{2}$$

where $\underset{\approx}{\mu}(\underset{\sim}{X})$ is the 3Nx3N mobility matrix, which may in principle be found from the solution of the linearized Navier-Stokes equations.

The Smoluchowski equation (1) has the equilibrium solution

$$P_{eq}(\underset{\sim}{X}) = \exp\,(-\beta\Phi(\underset{\sim}{X}))/Z(\beta) \quad , \tag{3}$$

where $Z(\beta)$ is the normalization integral. The equation describes how an arbitrary initial distribution $P(\underset{\sim}{X},0)$ tends to equilibrium in the course of time. In rheology one perturbs the distribution by imposing a flow on the solvent. Then Eq. (1) must be supplemented with a convection term proportional to the imposed flow velocity $\underset{\sim}{v}_0(\underset{\sim}{r},t)$.

We consider in particular the oscillatory linear flow

$$\underset{\sim}{v}_0(\underset{\sim}{r},t) = \underset{\approx}{E}_\omega \cdot \underset{\sim}{r}\, e^{-i\omega t} \quad , \tag{4}$$

where $\underset{\approx}{E}_\omega$ is a traceless tensor. The resulting average stress in the suspension may be evaluated by linear response theory [2] and leads to the frequency-dependent shear viscosity $\eta(\omega)$. To second order in the density one may write

$$\eta(\omega) = \eta_\infty + \alpha_V(\omega)\phi^2\eta + O(\phi^3) \quad , \tag{5}$$

where $\phi = (4\pi/3)na^3$ is the volume fraction at density $n = N/\Omega$. Here η_∞ is the high-frequency limit which itself has the density expansion

$$\eta_\infty = \eta\left[1 + \frac{5}{2}\phi + \frac{25}{4}\,k_H\phi^2 + O(\phi^3)\right] \quad . \tag{6}$$

The coefficient $\frac{5}{2}$ for hard spheres with stick boundary conditions was evaluated by Einstein [3], and the Huggins coefficient k_H for such a system [4,5] is given by $k_H = 0.800$. The high-frequency value η_∞ may be evaluated from the equilibrium distribution $P_{eq}(\underset{\sim}{X})$. In order to find the frequency-dependent viscosity $\eta(\omega)$ one must study the nonequilibrium change of distribution. The frequency-dependence describes the viscoelasticity of the system.

Here we are interested in the coefficient $\alpha_V(\omega)$ in (5). In order to calculate this coefficient it suffices to consider the Smoluchowski equation for a pair of particles. The zero-frequency value $\alpha_V(0)$ was first evaluated by Batchelor [6], who found $\alpha_V(0) = 0.97$. A more precise calculation [7] yields $\alpha_V(0) = 0.913$. By definition the coefficient $\alpha_V(\omega)$ tends to zero at high frequency. Recently we have found the complete frequency dependence for hard spheres with neglect of hydrodynamic interactions [8].

In linear response theory the change of the pair distribution function from its equilibrium form is given by

$$\delta P(\underset{\sim}{R},\omega) = -3\pi\beta\eta a^3 n^2 \; \underset{\approx}{E}_{\omega}:\underset{\sim}{\hat{R}}\underset{\sim}{\hat{R}} \; g(R)f(x,\omega) \quad , \tag{7}$$

where $\underset{\sim}{R}$ is the relative distance vector of a pair of particles, $g(R)$ is the equilibrium radial distribution function, $x = R/2a$ is the dimensionless distance and $f(x,\omega)$ is a frequency-dependent radial function. To lowest order in the density the radial distribution function is given by $g(\mathbf{R}) = \theta(2a-R)$, where $\theta(r)$ is the step-function. With neglect of hydrodynamic interactions the radial function $f(x,\omega)$ satisfies the equation

$$\frac{d}{dx}\left(x^2 \frac{df}{dx}\right) - 6f - \alpha^2 x^2 f = 0 \quad , \tag{8}$$

where α is given by

$$\alpha = (1-i)\sqrt{\omega a^2/D_0} \qquad \text{for } \omega > 0 \quad . \tag{9}$$

Of course, the function $f(x,\omega)$ tends to zero at large x. It may be shown [9] that at touching it must satisfy the boundary condition $f'(1,\omega) = -4$. The solution of (8) with the above boundary conditions is simple and given by

$$f(x,\omega) = -\frac{4}{\alpha k_2'(\alpha)}\, k_2(\alpha x) \quad , \tag{10}$$

where $k_2(z)$ is a modified spherical Bessel function. In particular in the steady state

$$f(x,0) = 4/(3x^3) \quad . \tag{11}$$

It may be shown [9] that the coefficient $\alpha_V(\omega)$ in (5) is given by the value at touching according to

$$\alpha_V(\omega) = \frac{9}{5} f(1,\omega) \quad . \tag{12}$$

From (10) one finds the explicit result

$$\alpha_V(\omega) = \frac{36}{5} \frac{\alpha^2+4\alpha+3}{\alpha^3+4\alpha^2+9\alpha+9} \quad . \tag{13}$$

The steady state value $\alpha_V(0) = 12/5$ was found earlier by Russel [10]. It differs significantly from the value $\alpha_V(0) = 0.913$ mentioned above. This suggests that hydrodynamic interactions make an important difference. In Fig. 1 we plot the dimensionless functions

$$R(\omega) = \frac{\eta'(\omega)-\eta_\infty}{\eta(0)-\eta_\infty} \ , \qquad I(\omega) = \frac{\eta''(\omega)}{\eta(0)-\eta_\infty} \quad , \tag{14}$$

as given by (5) and (13).

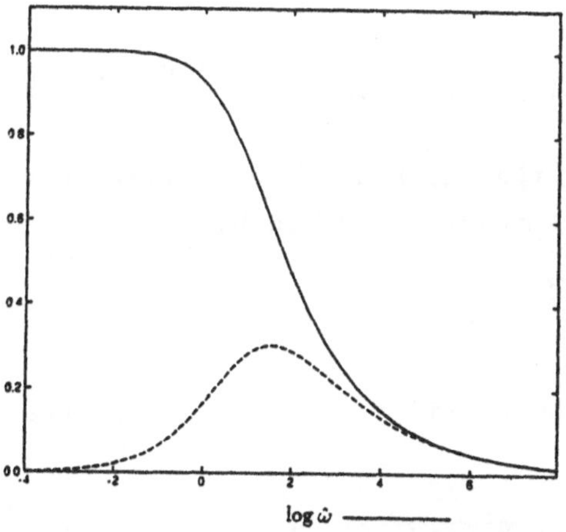

Figure 1: $R(\omega)$ (drawn curve), $I(\omega)$ (dashed curve), $\hat{\omega} = \omega\tau_0$.

The frequency scale is logarithmic, so that in fact the functions vary over a wide range of frequency.

The high frequency behavior of the real and imaginary parts of the viscosity is given by

$$\eta'(\omega) \approx \eta_\infty + \frac{18}{5} \frac{\phi^2}{\sqrt{\omega\tau_0}} \eta \quad ,$$
$$\eta''(\omega) \approx \frac{18}{5} \frac{\phi^2}{\sqrt{\omega\tau_0}} \eta \quad , \qquad \text{as} \quad \omega \to \infty \quad . \tag{15}$$

This is reminiscent of similar power law behavior of the dynamic viscosity of dilute polymer solutions [11,12]. We can write

$$\eta(\omega) = \eta_\infty + G_V \tau_0 \int_0^\infty \frac{p(u)}{u-z}\, du \quad , \tag{16}$$

where $z = i\omega\tau_0$ and the relaxation strength G_V is fixed by the specification of the coefficient in the asymptotic behavior of the spectral density $p(u)$. In the theory of polymer solutions one encounters the asymptotic behavior

$$p(u) \approx \frac{1}{\mu}\, u^{-1+1/\mu} \qquad \text{as} \qquad u \to \infty \quad , \tag{17}$$

with a characteristic exponent μ. The corresponding high frequency behavior of the dynamic viscosity is given by

$$\eta'(\omega) \approx \eta_\infty + \frac{\pi}{2\mu\cos(\pi/2\mu)}\, G_V\, \tau_0 (\omega\tau_0)^{-1+1/\mu} \quad ,$$

$$\eta''(\omega) \approx \frac{\pi}{2\mu\sin(\pi/2\mu)}\, G_V\, \tau_0 (\omega\tau_0)^{-1+1/\mu} \quad \text{as} \quad \omega \to \infty \quad . \tag{18}$$

In the present case the exponent is $\mu=2$, like in the Rouse model, and the relaxation strength is given by

$$G_V = \frac{36\sqrt{2}}{5\pi}\, \phi^2\, \frac{\eta}{\tau_0} \quad . \tag{19}$$

One can also find the explicit form of the spectral density in (16). From (13) one finds

$$p(u) = \frac{4u^{5/2}}{8u^3-8u^2+18u+81} \quad . \tag{20}$$

In Fig. 2 we show a doubly-logarithmic plot of the spectral density.

Figure 2:

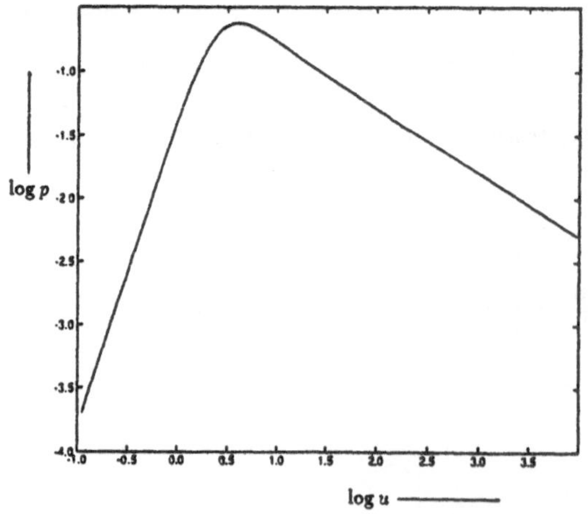

The dynamic viscosity may be expressed as a one-sided Fourier transform

$$\eta(\omega) = \eta_\infty + \eta \int_0^\infty e^{i\omega t}\, \Psi(t)dt \quad , \qquad (21)$$

where $\Psi(t)$ is the dimensionless stress relaxation function. It is related to the spectral density in (16) by

$$\Psi(t) = \frac{G_V \tau_0}{\eta} \int_0^\infty p(u) e^{-ut/\tau_0}\, du \quad . \qquad (22)$$

From (20) one finds the long-time behavior

$$\Psi(t) \approx \frac{2}{3}\sqrt{\frac{2}{\pi}}\, \phi^2 \left(\frac{\tau_0}{t}\right)^{7/2} \quad \text{as} \quad t \to \infty \qquad (23)$$

and the short-time behavior

$$\Psi(t) \approx \frac{18}{5}\sqrt{\frac{2}{\pi}}\, \phi^2 \left(\frac{\tau_0}{t}\right)^{1/2} \quad \text{as} \quad t \to 0 \quad . \qquad (24)$$

One may also find an explicit expression valid for all times [8]. We plot the relaxation function in Fig. 3. Here $s = t/\tau_0$ and $\psi(s)$ is defined by

$$\Psi(t) = \frac{18\sqrt{2}}{5\pi}\, \phi^2\, \psi(t/\tau_0) \quad . \qquad (25)$$

Figure 3:

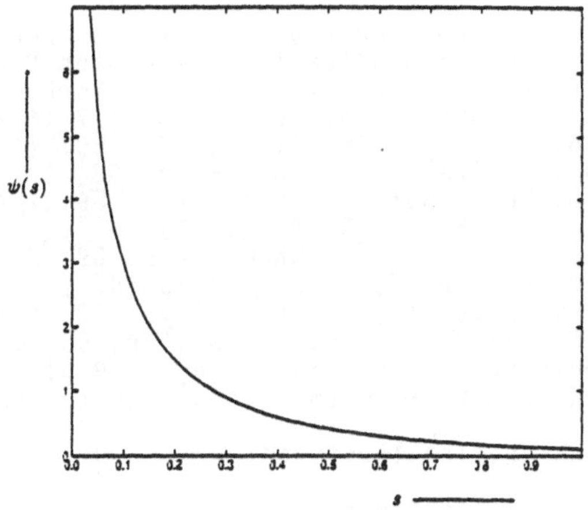

The coefficient $\alpha_V(\omega)$ in (5) may also be evaluated explicitly for a slightly different model, in which again the spheres diffuse with diffusion coefficient D_0 in the absence of hydrodynamic interactions, but cannot approach each other closer than 2b, where the radius b is larger than a. The function $\alpha_V(\omega)$ is related to the expression (13) by a simple scaling. This model may be a good approximation for suspensions of charged polystyrene spheres. For such systems the hydrodynamic interactions are relatively unimportant.

Behavior of the dynamic viscosity as shown in Fig. 1 has been found experimentally by van der Werff et al. [13] in suspensions of silica spheres. In such suspensions hydrodynamic interactions cannot be neglected. The experiments were carried out at volume fractions ϕ between 0.3 and 0.5, so that the low density theory presented above must be extended. The experiments show the asymptotic behavior

$$\eta'(\omega) \approx \eta_\infty + \frac{\pi}{2\sqrt{2}} G_1 \tau_1^{1/2} \omega^{-1/2} ,$$

$$\eta''(\omega) \approx \frac{\pi}{2\sqrt{2}} G_1 \tau_1^{1/2} \omega^{-1/2} , \tag{26}$$

similar to (15). The relaxation strength G_1 and the time scale τ_1 must be fitted to the experimental data. The agreement with Fig. 1 suggests that the diffusion mechanism discussed above is responsible for the $\omega^{-1/2}$-behavior seen experimentally. The influence of hydrodynamic interactions and higher order density corrections remain to be investigated theoretically.

References

1. See the review by P.N. Pusey and R.J.A. Tough, in Dynamic Light Scattering and Velocimetry: Applications of Photon Correlation Spectroscopy, edited by R. Pecora (Plenum, New York, 1981).
2. B.U. Felderhof and R.B. Jones, Physica A 146, 417 (1987).
3. A. Einstein, Ann. d. Phys. 19, 289 (1906); 34, 591 (1911).
4. G.K. Batchelor and J.T. Green, J. Fluid Mech. 56, 401 (1972).
5. B. Cichocki and B.U. Felderhof, J. Chem. Phys. 89, 1049 (1988).
6. G.K. Batchelor, J. Fluid Mech. 83, 97 (1976).
7. B. Cichocki and B.U. Felderhof, J. Chem. Phys. 89, 3705 (1988).
8. B. Cichocki and B.U. Felderhof, to be published.
9. B. Cichocki and B.U. Felderhof, J. Chem. Phys. 93, 4427 (1990).
10. W.B. Russel, D.A. Saville, and W.R. Schowalter, Colloidal Dispersions (Cambridge University Press, Cambridge, 1989).
11. H. Yamakawa, Modern Theory of Polymer Solutions (Harper and Row, New York, 1971).
12. M. Doi and S.F. Edwards, The Theory of Polymer Dynamics (Clarendon, Oxford, 1986).
13. J.C. van der Werff, C.G. de Kruif, C. Blom, and J. Mellema, Phys. Rev. A39, 795 (1989).

EXTENDED IRREVERSIBLE THERMODYNAMICS VERSUS RHEOLOGY

G. LEBON
Institut de Physique, Université de Liège, B5 Sart Tilman, B-4000 Liège, Belgique.

D. JOU and J. CASAS-VAZQUEZ
Departament de Fisica, Universitat Autonoma de Barcelona, 08193, Bellaterra, Catalonia, España.

INTRODUCTION

The purpose of the present work is to examine to which extent non-equilibrium thermodynamics is useful for establishing rheological equations. Several works have been devoted to the description of rheological bodies within the framework of thermodynamics. The originality of the present paper lies in the fact that the analysis rests on a recent version of non-equilibrium thermodynamics, referred to as Extended Irreversible Thermodynamics (EIT) [1,2]. This new theory has fuelled much interest during the last decade and has been applied with success to heat conductivity [3], deformable solids [4] and viscous fluids [5,6]. Since the contents and scope of EIT may not be familiar to the audience, we have prefaced the description of rheological materials by a brief summary of EIT in section 1. In sections 2 and 3, EIT is applied to derive the constitutive equations of rheological materials. In section 2 it is shown that linear viscoelasticity is easily interpreted within the framework of EIT and that the classical models of Maxwell, Kelvin-Voigt, Poynting-Thomson and Jeffreys are recovered as particular cases of the formalism. In section 3, a non-linear analysis is proposed to describe second order non-Newtonian fluids like these of Reiner-Rivlin, Rivlin-Ericksen or Giesekus. As particular results, it is shown that EIT leads to the correct signs of the normal stress coefficients in the Rivlin-Ericksen model.

1. EXTENDED IRREVERSIBLE THERMODYNAMICS

To make clear and explicit the structure and the main hypotheses underlying Extended Irreversible Thermodynamics (EIT), we consider a very simple system consisting of an incompressible viscous fluid at uniform temperature. The differences with Classical Irreversible Thermodynamics (CIT) [7,8] will be emphasized.

The basic problem in fluid mechanics is to determine the behaviour of the velocity field $v_i(x_i, t)$ as a function of position x_i and time t. The evolution of v_i in the course of time is governed by the momentum balance

$$\rho d_t v_i = -p_{,i} + P^v_{ij,j} + \rho f_i \quad , \tag{1.1}$$

supplemented by the incompressiblility condition $v_{i,i} = 0$. In (1.1) d_t stands for the material time derivative and a comma for the derivation with respect to the spatial coordinates, f_i designates the body force per unit mass, ρ the density , p the hydrostatic pressure, P^v_{ij} the viscous pressure tensor related to the total pressure tensor by

$$P_{ij} = p\delta_{ij} + P^v_{ij} \quad ; \tag{1.2}$$

like P_{ij} the quantity P^v_{ij} is assumed to be symmetric; moreover as bulk viscosity effects are ignored, it is also traceless. Summation convention on repeated indices is used throughout this work.

In CIT [7,8], the viscous pressure P^v_{ij} is given by the constitutive relation

$$P^v_{ij} = -2\eta V_{ij} \quad , \quad \left[V_{ij} = \frac{1}{2}(v_{i,j} + v_{j,i}) \right] \quad . \tag{1.3}$$

Equation (1.3) is Newton's law and η is the shear viscosity, assumed to be independent of the velocity gradient. Substitution of Newton's law in the momentum balance (1.1) results in the Navier-Stokes equation; it is a parabolic partial differential equation from which follows that velocity disturbances will be felt instantaneously everywhere within the system, at variance with the causality principle requiring that the effect will be felt after application of the cause.

Classical irreversible thermodynamics is based on the so-called local equilibrium hypothesis stating that the specific entropy s depends on the same variables as in equilibrium. Although the local equilibrium hypothesis is satisfactory for solving a wide variety of problems in continuum mechanics, it is not appropriate for describing materials with memory (like polymers) and high frequencies or short waves processes (like ultrasound propagation, light and neutron scattering by fluids). EIT was developed to provide a thermodynamic framework for the description of systems and phenomena not covered by CIT.

1.1. The basic statements of EIT

Extended irreversible thermodynamics was born out of the necessity to go beyond the local equilibrium hypothesis and to remove the unpleasant physical property of propagation of disturbances with an infinite velocity. These requirements are achieved by making the three following statements.

1 - It is assumed that there exists a generalized entropy function s with the following properties :

i. it is additive,
ii. it is a convex function of the whole set of variables,
iii. its rate of production is positive definite.

The importance of these properties cannot be assessed a priori but will become clear from their consequences, to be analyzed subsequently. In view of proposal (iii) the evolution of s in the course of time takes the form

$$\rho d_t s = -J^s_{i,i} + \sigma^s \tag{1.4}$$

with

$$\sigma^s \geq 0 \quad . \tag{1.5}$$

2 - The space of the basic variables used in CIT is enlarged to include non-conserved fast variables. The latter take usually the form of thermodynamic fluxes as the heat flux, the diffusion flux of matter, and the viscous pressure tensor. In contrast with the classical variables, like mass, energy, momentum which are slow and conserved quantities, the new variables are usually quickly decreasing in time and are not obeying conservation laws, moreover they have to be zero at equilibrium. The space V of the state variables can be considered as the union of the space C of the classical variables and the space F of the fast thermodynamic fluxes :

$$V = C \cup F \quad . \tag{1.6}$$

3 - The extra variables F are assumed to satisfy evolution equations of the general form

$$d_t F = -J^F_{i,i} + S^F \tag{1.7}$$

where J^F_i is the flux related to the variable F and S^F the corresponding source term. Both J^F_i and S^F must be determined by means of constitutive equations. Restrictions on the possible forms of the constitutive relations will be placed by the second law and the objectivity criterion, requiring invariance with respect to reference systems in motion [9].

1.2. The evolution equation of the pressure tensor

In the aforementioned example of a one-component ordinary viscous fluid, the set of classical variables is complemented by the viscous pressure tensor P^v_{ij} supposed to satisfy an evolution equation of the form

$$d_t P^v_{(ij)} = -J^v_{(ij)k,k} + S^v_{(ij)} \quad , \tag{1.8}$$

round brackets mean traceless symmetrization. The general form of $J^{v}_{(ij)k}$ (a third rank tensor) and $S^{v}_{(ij)}$ (a second rank symmetric tensor) are derived from the usual representation theorems of tensors. By restricting the analysis to the linear approximation, $J^{v}_{(ij)k}$ and $S^{v}_{(ij)}$ are simply given by

$$J^{v}_{(ij)k} = A\left[\frac{1}{2}(v_i\delta_{jk} + v_j\delta_{ik}) - \frac{1}{3}v_k\delta_{ij}\right] \quad , \quad S^{v}_{(ij)} = -BP^{v}_{ij} \tag{1.9}$$

where A and B are constant scalars. After substitution of (1.9) in (1.8), one obtains

$$\tau d_t P^{V}_{ij} = -2\eta V_{ij} - P^{v}_{ij} \tag{1.10}$$

wherein τ and η have been defined as

$$\tau = \frac{1}{B} \quad , \quad 2\eta = \frac{A}{B} \quad . \tag{1.11}$$

It is easily checked that η has the dimension of a viscosity and τ the dimension of a time : it is the relaxation time of the "flux" P^{v}_{ij}. By setting in (1.10) $\tau = 0$, one recovers Newton's law. It is worth stressing that the steady constitutive Newton equation of CIT is replaced in EIT by unsteady evolution equations of the Maxwell type.

1.3. Restriction imposed by the second law

Informations about the signs of the coefficients τ and η are provided by the second law of thermodynamics requiring that the entropy production σ^s is positive definite. The quantity σ^s is calculated from the balance law (1.4) wherein the entropy flux J^s_i and s will be given by constitutive equations. Up to second order terms, it can be shown [2] that σ^s is given by

$$\sigma^s = \frac{1}{2\eta T} P^v_{ij} P^v_{ij} \geq 0 \quad , \tag{1.12}$$

wherein T is the positive absolute temperature. The positiveness of σ^s is guaranteed at the condition that

$$\eta > 0 \quad .$$

Another interesting consequence from the calculation of σ^s is that it leads to the explicit expression of the Gibbs equation [6]. The latter plays a dominant role in *CIT wherein it is postulated* from the outset. In *EIT, Gibbs equation is derived* and found to be given by [2,6]

$$ds = \frac{1}{T} du - \frac{\tau}{2\rho\eta T} P^v_{ij} dP^v_{ij} \quad , \tag{1.13}$$

u is the specific internal energy. The first term in the r.h.s. of (1.13) is classical while the second term is typical of EIT.

1.4. Restriction placed by the convexity of entropy

Expanding entropy around equilibrium yields

$$s = s_{eq} - \frac{\tau}{4\rho\eta T} P^v_{ij} P^v_{ij} \tag{1.14}$$

Convexity of s around equilibrium implies that its second derivatives with respect to P^v_{ij} is negative. Since it has been shown earlier that $\eta > 0$, it follows from (1.14) that

$$\tau > 0.$$

This result is important as it ensures that the evolution equations are hyperbolic, allowing the disturbances to propagate with finite velocity even at infinite frequencies.

The behaviour of the fluid is completely described by means of the momentum balance (1.1) and the evolution equation (1.10). The latter together with the constraints $\eta > 0$ and $\tau > 0$ are the essential features of the EIT model. Boundary and initial conditions will be provided by experimental observations.

2. LINEAR VISCOELASTICITY

In this section, the classical theory of linear viscoelasticity is revisited. It is seen that EIT provides a simple and systematic way to derive the main results of linear viscoelasticity.

2.1. Constitutive and evolution equations

We take for granted the following hypotheses :

i. the deformations are infinitesimally small,
ii. the body is at uniform temperature and heat effects are neglected,
iii. the material is isotropic.

The choice of the variables is inspired by the results of section 1 derived for viscous fluids . Let us recall that in the latter case, the basic parameters are the velocity v_i, the internal energy u and the viscous pressure P^v_{ij}. By analogy with the decomposition (1.2) we shall decompose the pressure tensor, assumed symmetric, into an elastic part P'_{ij} and an inelastic part P''_{ij} :

$$P_{ij} = P'_{ij} + P''_{ij} \quad , \tag{2.1}$$

P'_{ij} obeys Hooke's law

$$P'_{ij} = -2G\varepsilon_{ij} \tag{2.2}$$

with G the Lamé coefficients and ε_{ij} the symmetric strain tensor, defined in terms of the deformation vector u_i as

$$\varepsilon_{ij} = 1/2(u_{i,j} + u_{j,i}) \quad , \tag{2.3}$$

recall that bulk effects are ignored ($\varepsilon_{kk} = 0$).

In parallel to the treatment of a viscous fluid [1,2,5,6], we choose as basic variables the rate of deformation $\delta_t u_i$, the internal energy u and the inelastic pressure P''_{ij}. The behaviour of the classical variable u_i and u is governed by the usual balance laws of momentum and energy :

$$\rho\partial_t^2 u_i = -P_{ij,j} + \rho f_i \quad , \tag{2.4}$$

$$\rho\partial_t u = -P_{ij}\partial_t\varepsilon_{ij} \quad , \tag{2.5}$$

Where in ∂_t is the partial time derivative. By strict analogy with the classical balance laws (2.4) and (2.5), it is assumed that the supplementary variable P''_{ij} satisfies a balance equation of the general form

$$\partial_t P''_{ij} = -J_{(ij)k,k} + S_{(ij)} \quad . \tag{2.6}$$

In order that the description be complete, it remains to express the flux J_{ijk} and source S_{ij} as functions of the basic set of variables by means of constitutive equations :

$$J_{ijk} = J_{ijk}(u, \partial_t u_i, P''_{ij}) \quad , \tag{2.7}$$

$$S_{ij} = S_{ij}(u, \partial_t u_i, P''_{ij}) \quad , \tag{2.8}$$

the most general relations compatible with a linear analysis are

$$J_{(ij)k} = \frac{\eta}{\tau_\varepsilon}\left[\frac{1}{2}(\partial_t u_i \delta_{jk} + \partial_t u_j \delta_{ik}) - \frac{2}{3}\partial_t u_k \delta_{ij}\right] \quad , \tag{2.9}$$

$$S_{(ij)} = -\frac{1}{\tau_\varepsilon} P''_{ij} \quad , \tag{2.10}$$

where τ_ε and η are undetermined coefficients. Substitution of (2.9) and (2.10) in (2.6) results in the following field equation for the extra variable

$$\tau_\varepsilon \partial_t P''_{ij} = -P''_{ij} - 2\eta \partial_t \varepsilon_{ij} \quad . \tag{2.11}$$

Equation (2.11) is the required linear evolution equation of the new variables P''_{ij}. It is possible to obtain evolution equations for the total pressure components by elimination of P''_{ij} between (2.11) and (2.1). This operations leads to the following results

$$\tau_\varepsilon \partial_t P_{ij} + P_{ij} = -2G(\varepsilon_{ij} + \tau_\sigma \partial_t \varepsilon_{ij}) \quad , \tag{2.12}$$

wherein τ_σ stands for

$$\tau_\sigma = \frac{\eta}{G} + \tau_\varepsilon \quad . \tag{2.13}$$

It is worth noticing that relation (2.12) is the constitutive equation for a Poynting-Thomson body and that this relation arises naturally from extended irreversible thermodynamics.

The following particular cases are also of interest. By setting $\tau_\varepsilon = 0$ in (2.12), one obtains

$$P_{ij} = -2G\varepsilon_{ij} - 2\eta \partial_t \varepsilon_{ij} \quad . \tag{2.14}$$

Equation (2.14) is representative of a Kelvin-Voigt body. If one assumes that in equation (2.12) $G = 0$ which means that the total stress has only an inelastic contribution, one recovers the basic equation of Maxwell's model, namely,

$$\tau_\varepsilon \partial_t P_{ij} + P_{ij} = -2\eta \partial_t \varepsilon_{ij} \quad . \tag{2.15}$$

An interesting generalization is provided by the following model which is a coupling of Newton's viscous fluid and the material described by equation (2.12). Let us write for the stress tensor a relation of the form

$$P_{ij} = P^s_{ij} + P^p_{ij} \quad , \tag{2.16}$$

P^s_{ij} is the viscous pressure tensor whose traceless part is given by

$$P^s_{ij} = -2\eta^s \partial_t \varepsilon_{ij} \quad , \quad (Newton's \quad law) \tag{2.17}$$

with η^s the shear viscosity, while P^p_{ij} is supposed to be formed of an elastic and an inelastic part :

$$P^p_{ij} = -2G\varepsilon_{ij} + P''_{ij} \quad . \tag{2.18}$$

Repeating the reasoning leading to (2.11), with P''_{ij} selected as basic variables, we are led to the relaxational equation

$$\tau \partial_t P''_{ij} = -P''_{ij} - 2\eta^p \partial_t \varepsilon_{ij} \quad . \tag{2.19}$$

By taking the time derivative of equation (2.16) and eliminating P''_{ij} by means of (2.19), one obtains

$$\tau\partial_t P_{ij} + P_{ij} = -2G\varepsilon_{ij} - 2G\left(\tau + \frac{\eta^p + \eta^s}{G}\right)\partial_t\varepsilon_{ij} + 2\tau\eta^s\partial_t^2\varepsilon_{ij} \quad . \tag{2.20}$$

Setting $G = 0$ in (2.20) yields

$$\tau\partial_t P_{ij} + P_{ij} = -2(\eta^p + \eta^s)\partial_t\varepsilon_{ij} + 2\tau\eta^s\partial_t^2\varepsilon_{ij} \quad , \tag{2.21}$$

which is nothing but Jeffreys' model while by letting $\eta^s = 0$, one recovers the previous model (2.12).

The material coefficients appearing in the aforementioned evolution equations for the pressure tensor components are subjected to some constraints imposed by the second law of thermodynamics, the criterion of objectivity, and the condition that entropy is a convex function at equilibrium. These restrictions are examined in the next subsection.

2.2. Restrictions imposed by the second law of thermodynamics. objectivity. and the convexity requirement

We postulate the existence of a regular and continuous function, the specific entropy, given by the constitutive relation

$$s = s(u, \varepsilon_{ij}, P''_{ij}) \quad , \tag{2.22}$$

and obeying a balance equation of the form (1.4). In classical theories, s depends only on u and ε_{ij}, here it depends in addition on P''_{ij}. Objectivity prevents the entropy to depend on the velocity because the latter is a non-objective quantity. In the absence of heat flux, the entropy flux J_i^s is zero so that (1.4) reduces to

$$\sigma^s = \rho\partial_t s \quad . \tag{2.23}$$

Using the chain differentiation rule to calculate $\delta_t s$, one obtains

$$\sigma^s = \rho\left(\frac{\partial s}{\partial u}\partial_t u + \frac{\partial s}{\partial \varepsilon_{ij}}\partial_t \varepsilon_{ij} + \frac{\partial s}{\partial P''_{ij}}\partial_t P''_{ij}\right) \geq 0 \quad . \tag{2.24}$$

Define as usual the absolute temperature T and the components of the elastic pressure P'_{ij} tensor by

$$\frac{\partial s}{\partial u} = \frac{1}{T} \quad , \quad \frac{\partial s}{\partial \varepsilon_{ij}} = \frac{1}{\rho T}P'_{ij} \quad . \tag{2.25}$$

Making use of the energy balance (2.5) and the evolution equation (2.11) for P''_{ij}, the entropy inequality (2.24) reads as

$$\sigma^s = -\frac{\rho}{T}P''_{ij}\partial_t \varepsilon_{ij} - \frac{\rho}{\tau_\varepsilon}(P''_{ij} + 2\eta\partial_t \varepsilon_{ij})\frac{\partial s}{\partial P''_{ij}} \geq 0 \quad . \tag{2.26}$$

For isotropic systems, the most general form for the derivatives of s with respect to P''_{ij} is given, in the linear approximation, by

$$\frac{\partial s}{\partial P''_{ij}} = -dP''_{ij} - \Lambda \varepsilon_{ij} \quad , \tag{2.27}$$

By taking the mixed derivative of (2.27) with respect to ε_{ij}, respectively and comparing with the mixed derivative of (2.25b) with respect to P''_{ij}, it is found that the coefficient Λ is zero. Substitution of equation (2.27) in (2.26) results then in

$$\sigma^s = \rho\frac{d}{\tau_\varepsilon}P''_{ij}P''_{ij} - P''_{ij}\partial_t \varepsilon_{ij}\left(\frac{1}{T} - 2\frac{d\eta}{\tau_\varepsilon}\right) \geq 0 \quad . \tag{2.28}$$

Positiveness of expression (2.28) demands that

$$\frac{d}{\tau_\varepsilon} > 0 \quad , \quad 2\eta = \frac{\tau_\varepsilon}{Td} > 0 \quad . \tag{2.29}$$

We turn now our attention to the consequences stemming from the convexity requirement of entropy. Expanding s around equilibrium, for fixed values of energy and strain, one obtains

$$s = s_{eq} - \frac{1}{2} dP''_{ij} P''_{ij} + O(3) \quad . \tag{2.30}$$

and, since s is maximum at equilibrium,

$$d > 0 \quad .$$

By combining this result with inequality (2.29a), it is found that

$$\tau_\varepsilon > 0 \quad .$$

The requirement that the entropy production is positive definite has led to the important result that the viscosity coefficient η is positive while from the convexity property of entropy it is concluded that the relaxation time τ_ε is positive.

It may also be asked what are the consequences of introducing a whole spectrum of relaxational modes for the pressure tensor instead of working with one single mode. Such a behaviour is typical of the Rouse and Zimm [12] models. These molecular models are very useful for describing dilute polymer solutions. It was recently shown that the Rouse-Zimm models can easily be incorporated into EIT description. For details, the reader is referred to reference [13].

3. NON-NEWTONIAN FLUIDS

It is shown that EIT leads to a simple and coherent modelling of non-linear systems, like non-Newtonian fluids. The basic hypothesis is still to raise the viscous pressure tensor to the status of independent variable. Applying the technique developed previously, constitutive equations for a wide class of second-order non-Newtonian fluids are established.

3.1. Extended thermodynamics of second-order non-Newtonian fluids

The basic variables are selected to be the same as in section 2, namely

$$v_i \quad , \quad u \quad , \quad P^v_{ij} \quad , \tag{3.1}$$

where P^v_{ij} satisfies a balance relation of the form

$$d_t P^v_{ij} = -J_{(ij)k,k} + S_{(ij)} \quad . \tag{3.2}$$

The quantities J_{ijk} and S_{ij} will be expressed by means of constitutive relation :

$$J_{ijk} = J_{ijk}(v_i, u, P^v_{ij}) \quad , \quad S_{ij} = S_{ij}(v_i, u, P^v_{ij}) \quad . \tag{3.3}$$

To comply with the criterion of objectivity, the material time derivative will be replaced by an objective time derivative D, for instance, Jaumann's derivative. The source term in (3.2) must be a traceless symmetric objective quantity that can be cast in the general form

$$S_{(ij)} = -AP^v_{ij} - B(P^v_{ik}P^v_{kj} - \pi^v\delta_{ij}) \quad , \quad \pi^v = \frac{1}{3}P^v_{kl}P^v_{kl} \quad . \tag{3.4}$$

The most general form for the flux J_{ijk}, symmetric with respect to the indices i and j, is given by

$$J_{(ij)k} = C[\frac{1}{2}(v_i\delta_{jk} + v_j\delta_{ik}) - \frac{1}{3}v_k\delta_{ij}] + E(v_iP^v_{jk} + v_jP^v_{ik}) \quad . \tag{3.5}$$

A, B, C, E are arbitrary coefficients which may depend on the invariants of P^v_{ij}. Substitution of (3.4) and (3.5) in (3.2) yields the following evolution equation for P^v_{ik} :

$$DP^v_{ij} = -CV_{ij} - AP^v_{ij} - B(P^v_{ik}P^v_{kj} - \pi^v\delta_{ij}) \quad . \tag{3.6}$$

In order to fulfil the objectivity requirement, the coefficient C has to be taken constant while E must be zero : otherwise $J_{ijk,k}$ and consequently the r.h.s. of (3.6) would contain undesirable non-objective terms in v_i. For further purpose, it is interesting to put $1/A = \tau$, $C/A = 2\eta$, $B/A = a$ so that (3.6) takes the more familiar form

$$\tau DP^v_{ij} = -2\eta V_{ij} - P^v_{ij} - a(P^v_{ik}P^v_{kj} - \pi^v\delta_{ij}) \quad , \tag{3.7}$$

wherein τ has the dimension of a time and η the dimension of a viscosity. The relation (3.7) is the keystone of the model. After exploring the restrictions placed by the second law and the convexity property of s it is concluded that $\tau > 0$ and $\eta > 0$ indicating that P^v_{ij} satisfies an evolution equation with a positive relaxation time and a positive viscosity. The model (3.7) has been shown to account in a fairly good way for the steady and oscillatory shearing flows if it is admitted that τ, η and a are power-laws of the invariants of P^v_{ij} [14]. Comparison between experimental data and the present model is reported on figures 1 and 2 for a 2.5% solution of polyacrylamide in a 50% water and 50% glycerine solution (PAA). Figure 1 shows the shear rate dependence of the classical viscometric functions, the first and second normal stress coefficients Ψ_1, Ψ_2 and the apparent viscosity η. In figure 2, the material functions η' (dynamic viscosity) and G' (storage modulus) are represented as a function of the oscillations frequency ω. The solid lines represent the theoretical predictions; results corresponding to other materials can be found in [14].

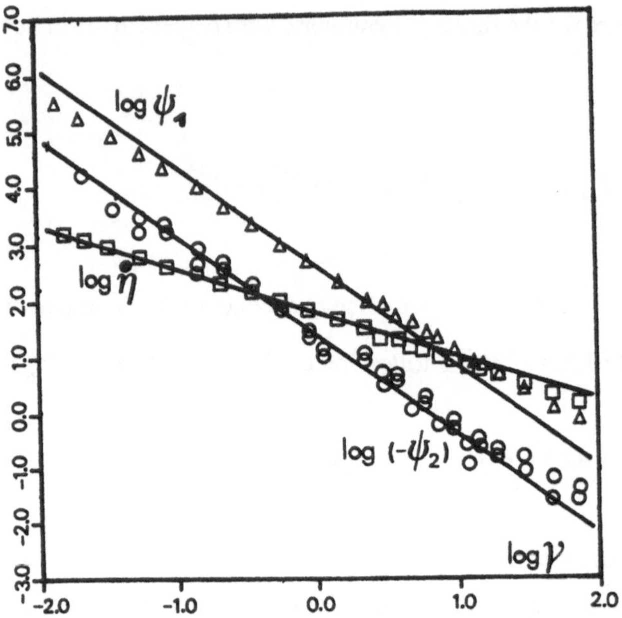

Figure 1 : Dependence of the viscometric fuctions on the shear rate for PAA.

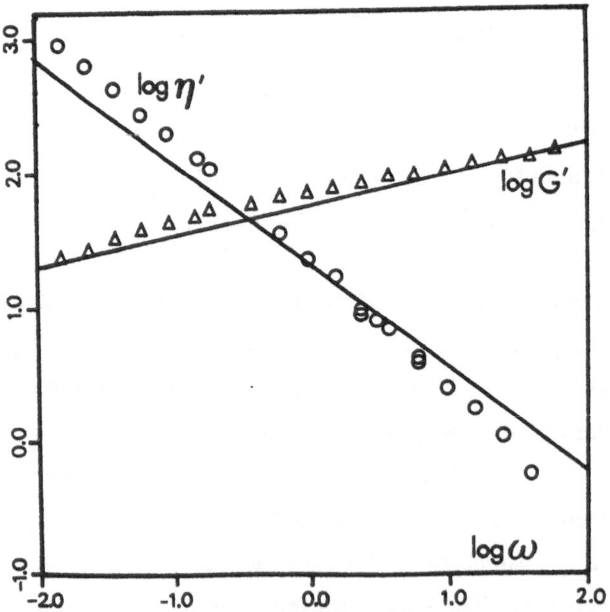

Figure 2 : Dynamic viscosity and storage modulus versus frequency for PAA.

3.2. Comparison with the Reiner-Rivlin and Rivlin-Ericksen models

At this point of the analysis, it is interesting to investigate the transition from a rate-type equation like (3.7) to constitutive relations like the Reiner-Rivlin or the Rivlin-Ericksen equations which are expressed by [15] :

$$P_{ij} = p\delta_{ij} - 2\eta V_{ij} - 4\alpha_1 V_{ik} V_{kj} \quad , \qquad (Reiner-Rivlin) \quad (3.8)$$

$$P_{ij} = p\delta_{ij} - 2\eta V_{ij} - 2\alpha_2 V_{ij}^{(2)} - 4\alpha_1 V_{ik} V_{kj} \;, \qquad (Rivlin-Ericksen) \qquad (3.9)$$

η, α_1 and α_2 are material coefficients depending in general on the principal invariants of V_{ij} while $V_{ij}^{(2)}$ is the Rivlin-Ericksen time-derivative of order two defined by

$$V_{ij}^{(2)} = d_t V_{ij} + v_{k,i} V_{kj} + V_{ik} v_{k,j} \quad . \qquad (3.10)$$

To recover (3.8) and (3.9) we rewrite equation (3.7) in term of Rivlin-Ericksen's time derivative $P_{ij}^{(2)}$; one obtains

$$P_{ij}^{v} = -2\eta V_{ij} - \tau P_{ij}^{(2)} - a(P_{ik}^{v} P_{kj}^{v} - \pi^{v} \delta_{ij})$$

$$+F\tau(P_{ik}^{v} V_{kj} + V_{ik} P_{kj}^{v}) \quad . \qquad (3.11)$$

For $F = 0, 1$ and 2 , one recovers respectively the lower convected $D_{\downarrow}$, the Jaumann D and the upper convected time derivative $D_{\uparrow}$.

At the first order approximation and in the limit $\tau = 0$, (3.11) reduces to Newton's law. The second order approximation with $\tau = 0$ is obtained by substituting in the right hand side of (3.11) P_{ij}^{v} by Newton's law, this operation leads to the Reiner-Rivlin relation

$$P_{ij}^{v} = -2\eta V_{ij} - 4\eta^2 a V_{ik} V_{kj} \quad . \tag{3.12}$$

If the relaxation time τ is not zero, one obtains from (3.11)

$$P_{ij}^{v} = -2\eta V_{ij} + 2\tau\eta V_{ij}^{(2)} - 4\eta(\eta a + F\tau) V_{ik} V_{kj} \quad . \tag{3.13}$$

Comparison between (3.13) and the Rivlin-Ericksen equation (3.9) allows to express the coefficients α_1 and α_2 in terms of the parameters η, τ and a, namely

$$\alpha_1 = F\eta\tau + \eta^2 a \quad , \quad \alpha_2 = -\eta\tau < 0 \quad .$$

Recalling that τ and η are positive, it is clear that α_2 is a negative quantity in accordance with experiments. Moreover, addition of α_1 and α_2 yields

$$\alpha_1 + \alpha_2 = \mu\tau(F - 1) + \eta^2 a \quad ,$$

showing that the sum of the coefficients α_1 and α_2 is generally non-zero. In particular, when Jaumann's derivative is used, the sum α_1 and α_2 is simply equal to $\eta^2 a$. Let us recall that Dunn and Fosdick [16], working in the framework of rational thermodynamics, have found $\alpha_2 > 0$ and $\alpha_1 + \alpha_2 = 0$. It can thus be concluded that by considering the Rivlin-Ericksen equation as an approximation of the more general rate-type model (3.7), one avoids the contradictions raised by Dunn and Fosdick's work. The essential conclusion is that $\alpha_2 < 0$ is not in contradiction with thermodynamics.

3.3. More general models.

The model (3.7) contains only three parameters τ, η and a. A more complicated model involving more parameters may be easily generated if it is admitted as in section 2 that the total viscous pressure tensor P_{ij}^{v} is the sum of contributions P_{ij}^{s} from the solvent and P_{ij}^{p} from the polymer chain. It is assumed that the solvent is an incompressible Newtonian fluid and that the space of the independent variables is formed by v_i, u and P^p_{ij}. Repeating the procedure of section 3.1 results in the next evolution equation for P^p_{ij}

$$\tau^p D P^p_{ij} = -2\eta^p V_{ij} - P^p_{ij} - a^p (P^p_{ik} P^p_{kj} - \pi^p \delta_{ij}) \quad , \quad \left(\pi^p = \frac{1}{3} P^p_{ij} P^p_{ij} \right) \tag{3.14}$$

which contains three parameters τ^p, η^p and a^p, D is an objective time derivative. After elimination of P^p_{ij} between (3.14) and (2.16), one obtains the following rheological equation for the total pressure tensor :

$$\tau^p D P^v_{ij} + P^v_{ij} - \frac{a\tau^p}{\eta} (P^v_{ik} P^v_{kj} - \pi^v \delta_{ij}) - a\lambda (V_{ik} P^v_{kj} + P^v_{ik} V_{kj} - \bar{\pi} \delta_{ij})$$

$$= -2\eta \left[V_{ij} + \lambda D V_{ij} - \frac{2a\lambda^2}{\tau^p} \left(V_{ik} V_{kj} - \bar{\bar{\pi}} \delta_{ij} \right) \right] \tag{3.15}$$

where the non-identified coefficients η, a, λ, $\bar{\pi}$ and $\bar{\bar{\pi}}$ stand for

$$\eta = \eta^s + \eta^p \quad , \quad a = \frac{-\eta a^p}{\tau^p} \quad , \quad \lambda = \frac{\eta^s \tau^p}{\eta}$$

$$\bar{\pi} = \frac{2}{3} P^v_{ij} V_{ij} \quad , \quad \bar{\bar{\pi}} = \frac{1}{3} V_{ij} V_{ij} \quad .$$

Expression (3.15) is the Giesekus constitutive equation [17] except for the terms in π^v, $\bar{\pi}$ and π which appear as a consequence of the no-bulk viscous pressure assumption. Equation (3.15) contains four independent parameters that can be indifferently chosen as τ^p, a^p, η^s and η^p or τ^p, a, η and λ. The above result is particularly promising as it allows to derive the Giesekus model from very simple macroscopic considerations.

It is worth repeating that the above results were obtained from very simple considerations : it was only required that P_{ij}^{v} be selected as independent variable and that it obeys a second-order non-linear evolution equation of the relaxation type. Of course, it is still possible to improve the quality of the model by introducing supplementary variables, for instance, the conformation tensor. This results in a more realistic description of the geometrical configuration of the polymer chains and generates general nonlinear constitutive equations encompassing the most usual rheological models [18]. Other problems remain open like the role played by the temperature and the polymer concentrations. These effects will be examined in future works.

REFERENCES

[1] G. Lebon, Bull. Acad. Roy. Sc. Belgique, 64, 456 (1978).

[2] D. Jou, J. Casas-Vazquez and G. Lebon, Rep. Prog. Phys., 51, 1105 (1988).

[3] G. Lebon, Int. J. Engng. Sc., 18, 727 (1980).

[4] G. Lebon, J. Techn. Phys., 23, 37 (1982).

[5] G. Lebon and D. Jou, J. Chem. Phys., 77, 970 (1982).

[6] G. Lebon and M.S. Boukary, Int. J. Engng. Sc., 26, 471 (1988).

[7] I. Prigogine, Introduction to Thermodynamics of Irreversible Processes, New-York, Interscience, 1961.

[8] S.R. de Groot and P. Mazur, Non-equilibrium Thermodynamics. Amsterdam, North-Holland Publ., 1962.

[9] C. Truesdell and W. Noll, Hd der Physik, vol 3/3, Berlin, Springer (1965).

[10] G. Lebon, C. Perez-Garcia and J. Casas-Vazquez, J. Chem. Phys., 88, 5068 (1988).

[11] G. Lebon and A. Cloot, J. Non Newt. Fluid Mech., 28, 61 (1988).

[12] R. Bird, C.F. Curtiss, R.C. Armstrong and O. Hassager, Dynamics of Polymeric Liquids. Vol. I and II, 2nd ed., New York, Wiley, 1987.

[13] C. Perez-Garcia, J. Casas-Vazquez and G. Lebon, J. Polym. Sci., B-Polym. Phys., 27, 1807 (1989).

[14] G. Lebon, P. Dauby, G. Palumbo and A. Valenti, Rheol. Acta, 19, 127 (1990).

[15] R. Rivlin and J.L. Ericksen, J. Rat. Mech. Anal., 4, 323 (1955).

[16] J. Dunn and R.L. Fosdick, Arch. Rat. Mech. Anal., 56, 191 (1974).
[17] H. Giesekus, J. Non Newt. Fluid Mech., 11, 69 (1982).
[18] P. Dauby and G. Lebon, Applied. Math. Lett., 3, 45 (1990).

Objectivity and the Extended Thermodynamic Description of Rheology

P.C. DAUBY,
Institut de Physique, B5,
Université de Liège, Sart Tilman,
B-4000 Liège, Belgium.

1 Introduction

Extended Irreversible Thermodynamics (EIT) [1-4] has been developed to enlarge the domain of applicability of classical irreversible thermodynamics [5-7] which is limited to steady linear constitutive equations.

Futhermore it is known from continuum mechanics that the principle of objectivity [8,9] is a usefull tool to impose restrictions on the possible forms of the constitutive equations. In short one can say that this principle imposes the constitutive equations to be form invariant for a change of observer.

The purpose of this paper is to show how the principle of objectivity can be introduced in a natural way in EIT.

As a preliminary, the Gibbs description of EIT is briefly recalled. This description highlights the role of the history of the viscous pressure tensor in the constitutive equations. Since history of a tensor is not a univocal notion, it is studied in some details in sections 3 and 4. Using an intrinsic definition of the history of tensors, it is then shown that the principle of objectivity appears naturally within EIT.

2 Gibbs description of Extended Irreversible Thermodynamics [3]

The fundamental hypothesis in Extended Irreversible Thermodynamics is to consider the thermodynamic fluxes as independent variables. The space of state variables of ordinary thermodynamics is enlarged by means of thermodynamic fluxes which vanish at equilibrium. For simplicity, temperature effects are not considered here : the heat flux and heat supply are supposed to be zero.

In EIT it is assumed that specific entropy s is an analytic function not only of classical variables u (specific internal energy) and v (specific volume) but also of the viscous pressure tensor. This tensor is usually split up into its trace p^v and its deviatoric part $\mathbf{P}^v$. If developments are limited to second order terms in the flux variables, one has :

$$s = s(u,v) + \frac{v}{6T}\,\alpha\, p^v \mathbf{U} : p^v \mathbf{U} + \frac{v}{2T}\,\beta\, \mathbf{P}^v : \mathbf{P}^v , \tag{1}$$

where α and β are phenomenological functions of u and v ; T is the temperature, U the unit tensor and the colon denotes double contraction.

After taking the material time derivative of this equation and using the mass and energy conservation laws, one can calculate the entropy production per unit volume σ^s. Since the heat flux vanishes, the entropy flux is zero and one obtains :

$$T\,\sigma^s = -\,p^v\,(\nabla . v + \alpha\, \dot{p}^v) - P^v_{ij}\,(\hat{V}_{ij} + \beta\, \dot{P}^v_{ij}). \tag{2}$$

In this expression, v is the velocity field and $\hat{V}_{ij}$ the symmetric traceless part of the velocity gradient tensor. A upper dot stands for the material time derivative.

This expression can also be written as a bilinear form in the thermodynamic fluxes p^v and P^v_{ij} and in the conjugate expressions x and X_{ij} called thermodynamic forces :

$$T\,\sigma^s = -\,p^v\, x - P^v_{ij}\, X_{ij} , \tag{3}$$

where x and X_{ij} are defined by comparison of (2) and (3).

Then, by analogy with Classical Thermodynamics [5-7], the constitutive equations are obtained by expressing the thermodynamic forces as functions of the basic variables. If developments are supposed to be linear, one can write :

$$X_{ij} = B\, P^v_{ij} \text{ and } x = A\, p^v . \tag{4}$$

The constitutive equations obtained by comparing (2), (3) and (4) are the well-known Maxwell equations for a viscoelastic material [10] :

$$\lambda_2 \dot{\overset{v}{P}}_{ij} + \overset{v}{P}_{ij} = - 2 \eta_2 \widehat{V}_{ij} , \tag{5.a}$$

$$\lambda_0 \dot{\overset{v}{p}} + \overset{v}{p} = - 2 \eta_0 \nabla . v , \tag{5.b}$$

where

$$\lambda_0 = - \frac{\alpha}{A} , \eta_0 = - \frac{1}{2 A} , \lambda_2 = - \frac{\beta}{B} , \eta_2 = - \frac{1}{2 B} . \tag{5.c}$$

From the second law of thermodynamics one can infer the positiveness of the viscosities η_2 and η_0 [4]. Moreover, if the entropy defined in (1) is assumed to be a convex function of the viscous pressure tensor -this property reflects the stability of equilibrium-, the relaxation times λ_2 and λ_0 are also positive [4]. As a consequence, the momentum disturbances propagate at finite speed (hyperbolic partial differential equations).

In the constitutive equations (5), it is important to point out the appearance of the time derivative of the viscous pressure tensor. Actually this time derivative accounts for a part of the "history" of the viscous pressure tensor, that is for a part of its evolution in the course of time.

Unfortunately, the history of a vector or a tensor is not a clear notion. The history of a tensor is often identified with the history of its components in some reference basis. The major difficulty arises then from the fact that the basis chosen by an observer can be time dependent for another one, due to the fact that different observers can be in movement with respect to one another. For this reason, it is sometimes uneasy to distinguish the variations of the components of a tensor due to the movement of the basis from the intrinsic variations of this tensor.

For instance, the material time derivative of a tensor is not an intrinsic quantity : for a change of observer, the transformation law for the material time derivative of a tensor is not the usual "tensorial law". One usually says that such a quantity is not objective or that it is frame dependent [8,9]. The material time derivative is thus ambiguous as it does not represent the same thing for different observers.

For this reason equations (5) can not be considered as real constitutive equations describing intrinsic properties of materials.

In the remainder of this paper it is shown how to build up rheological models in the framework of Extended Irreversible Thermodynamics which are intrinsic and which do not present such ambiguity.

3 Definition of an intrinsic history for tensors

3.1 Definition of an observer

Before defining an intrinsic history for tensors it is necessary to give a mathematical definition of an observer.

We define an observer O as a class of (spatial) coordinate systems whose elements are "fixed" with respect to each other. If x^i and $x^{i'}$ represent respectively the coordinates of a point in a system S and in a system S' which define the same observer, one has

$$x^i = x^i(x^{k'}) \quad \text{and} \quad x^{i'} = x^{i'}(x^k), \tag{6.a}$$

with no dependence on the time t. The coordinates are completely general (curvilinear and non-orthogonal coordinates) [11,12].

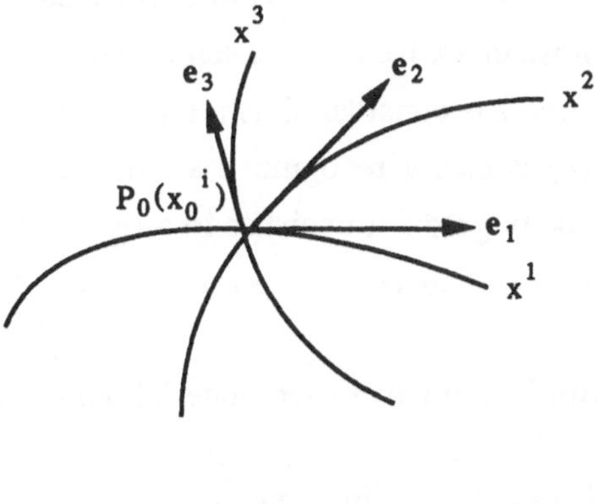

Fig. 1 Coordinate system S and natural basis at point P_0 whose coordinates are $x_0{}^i$

The metric tensor **G** is also independent of the time with

$$g_{ij} = g_{ij}\,(x^k) \;\text{ and }\; g_{i'j'} = g_{i'j'}(x^{k'}), \tag{6.b}$$

where g_{ij} represent the "natural" components of the metric, i.e. the components of the tensor **G** in a natural basis $\mathbf{e}_i$ whose vectors are the vectors tangent to the coordinates lines (see Fig. 1). Througout this paper indices like *i,j,k,i',j'*,... will always represent components of tensors in this natural basis of vectors $\mathbf{e}_i$.

The change of observer is characterized by a change of coordinate system which can depend on the time but in such a way that the metric remains independent of the time :

$$x^i = x^i(x^{k'},t) \quad \text{and} \quad x^{i'} = x^{i'}(x^k,t) \tag{7.a}$$

and

$$g_{ij} = g_{ij}\,(x^k) \;\text{ and }\; g_{i'j'} = g_{i'j'}(x^{k'}). \tag{7.b}$$

Using the Ricci identity which expresses that the covariant derivatives of the metric tensor vanish, one can prove that the independence of **G** with respect to the time is equivalent to the following condition :

$$\overset{e}{v}_{k;l} + \overset{e}{v}_{l;k} = 0 = \overset{e}{v'}_{k';l'} + \overset{e}{v'}_{l';k'}\,, \tag{8}$$

where $\overset{e}{v}_l = g_{lk}\,(\partial x^k(x^{i'},t)/\partial t)$ (or $\overset{e}{v'}_{l'} = g_{l'k'}\,(\partial x^{k'}(x^i,t)/\partial t)$) represents the components of the "entrainment velocity", i.e. the components of the velocity of one observer with respect to the other. A semi-colon indicates a covariant derivative with respect to the variable whose index follows the semi-colon.

If the change of observer is considered as a movement of an observer with respect to the other, condition (8) is easily recognized as the condition of rigid motion : the symmetric part of the velocity gradient vanishes in a rigid motion.

So a change of observer can be seen as a rigid and time dependent change of coordinate system.

Consider then a tensor **T** defined at each material point P and at each time t :

$$\mathbf{T} = \mathbf{T}(P,t)$$

At time t, the history of this tensor for an observer O and for the material point P is the following function of t':

$$t_{ij}(P, t') \quad \text{for} \quad -\infty < t' \leq t,$$

where t_{ij} denotes the components of **T** in the natural basis located where the material point P is at time t.

3.2 Corotational basis and intrinsic history for tensors

We know that the history of a tensor is ambiguous because of the movements of observers with respect to each other.

To avoid this ambiguity, one can attach rigidly a "privileged" observer to each material point of a continuous media and define the intrinsic history of a tensor as the history of this tensor with respect to this privileged observer.

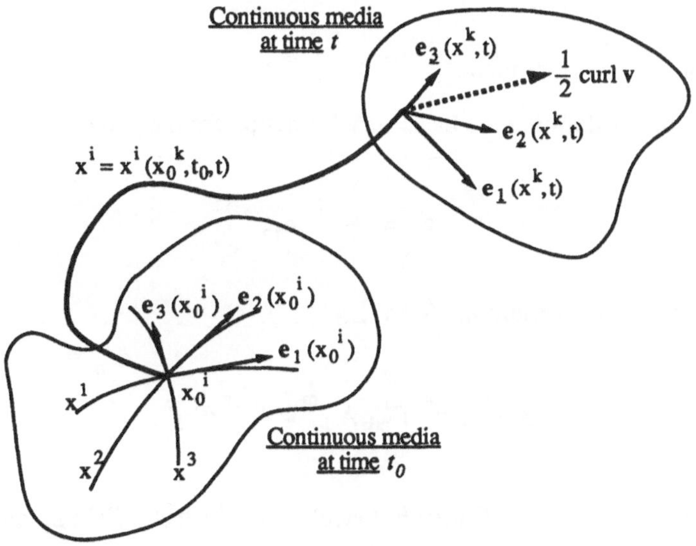

Fig. 2 Corotational basis rotating with angular velocity 1/2 curl v

Let us see more precisely how this privileged observer can be attached to each material point.

One can associate an observer with each material point, that is a coordinate system and the corresponding natural bases field. In fact, since different privileged observers are used to describe the histories of tensors for different material points, it is enough to explain how one can attach rigidly to each material point a basis of vectors in which the components of tensors are considered.

Consider an observer or a space coordinate system and a continuous media at time t_0 (see Fig. 2). To each material point P_0 whose coordinates are x_0^i corresponds the natural basis $\mathbf{e}_i$. Basis $\mathbf{e}_{\underline{i}}$ -the so-called corotational basis- attached rigidly to P_0 is the same as basis $\mathbf{e}_i$ at time t_0 and rotates with the angular velocity of P_0, i.e. with angular velocity 1/2 curl $\mathbf{v}$.

The rotation of the corotational basis is defined by the following equations :

$$\frac{\partial \mathbf{e}_{\underline{i}}(x_0^i, t_0, t)}{\partial t} = \frac{d\mathbf{e}_{\underline{i}}(x^i, t)}{dt} = \frac{1}{2}\,\mathrm{curl}\,\mathbf{v} \times \mathbf{e}_{\underline{i}}\,, \tag{9.a}$$

$$\mathbf{e}_{\underline{i}}(x_0^k, t_0, t_0) = \mathbf{e}_i(x_0^k), \tag{9.b}$$

where $\times$ denotes the cross-product.

If $A_{\underline{i}}^{\ j}$ is the matrix defining the basis transformation from $\mathbf{e}_i$ to $\mathbf{e}_{\underline{i}}$:

$$\mathbf{e}_{\underline{i}} = A_{\underline{i}}^{\ j}\, \mathbf{e}_j,$$

equation (9.a) can be recast into the form

$$\dot{A}_{\underline{i}}^{\ j} + v^k\, \Gamma^j_{mk} A_{\underline{i}}^{\ m} = A_{\underline{i}}^{\ r} W_r^{\ j}. \tag{10}$$

In this equation the Γ^j_{mk} are the Christoffel symbols of the bases field $\mathbf{e}_i$ and $W_r^{\ j}$ is the spin tensor.

So, for a general observer, the components in a natural basis of the "intrinsic" history of a tensor **T** at time t for a material point P (located at $x_0^{\;k}$ at time t_0) is given by

$$A_i^{\;\underline{i}}(x_0^{\;k},t)\,A_l^{\;\underline{l}}(x_0^{\;k},t)\,t_{\underline{i}\underline{l}}(x^m,t') \quad \text{for} \quad -\infty < t' \leq t. \tag{11}$$

In this equation, $A_i^{\;\underline{i}}$ is the inverse matrix of $A_{\underline{i}}^{\;i}$ and $x^m = x^m(x_0^{\;k},t_0,t')$.

3.3 "Objectivity" of the corotational basis

It is important to prove that the history defined by (11) is actually independent of the observer.

Consider two observers O and O' who build their own corotational bases $\mathbf{e}_{\underline{i}}$ and $\mathbf{e}_{\underline{i'}}$ from their own natural bases $\mathbf{e}_i$ and $\mathbf{e}_{i'}$.

First one has to know the transformation law for the vorticity tensor $W_r^{\;j}$ for a change of observer.

Using the rigidity condition (8), it is easy to check that the transformation law for $W_r^{\;j}$ reads :

$$W_{i'j'} = x^i_{\;,i'}\,x^j_{\;j'}\,W_{ij} - v^e_{\;i';j'}. \tag{12}$$

In (12) a comma denotes a partial derivation with respect to the variable whose index follows. So $x^i_{\;,i'}$ is the jacobian matrix of the coordinate transformation corresponding to the change from observer O to observer O'.

Thus W_{ij} does not transform like a tensor for a change of observer because of the presence of the second term which arises from the movement of one observer with respect to the other : the vorticity tensor is not objective.

In contrast, one can prove that the rate of strain tensor V_{ij} (the symmetric part of the velocity gradient tensor) follows the usual tensorial law :

$$V_{i'j'} = x^{i}_{,i'}\, x^{j}_{,j'}\, V_{ij}$$

and is thus objective.

The vectors $\mathbf{e}_i$, $\mathbf{e}_{i'}$, $\mathbf{e}_{\underline{i}}$ and $\mathbf{e}_{\underline{i}'}$ can be expressed in terms of each other by means of :

$$\mathbf{e}_{\underline{i}} = A_{\underline{i}}^{\ j}\, \mathbf{e}_j ,$$

$$\mathbf{e}_{\underline{i}'} = A_{\underline{i}'}^{\ j'}\, \mathbf{e}_{j'} ,$$

$$\mathbf{e}_{j'} = x^{i}_{\ ,j'}\, \mathbf{e}_i .$$

One can thus deduce :

$$\mathbf{e}_{\underline{i}} = A_{\underline{i}}^{\ i}\, x^{i'}_{\ ,i}\, A_{i'}^{\ \underline{i}'}\, \mathbf{e}_{\underline{i}'} = X_{\underline{i}}^{\ \underline{i}'}\, \mathbf{e}_{\underline{i}'} . \tag{13}$$

Using the evolution equations (10) for the matrixes $A_{\underline{i}}^{\ j}$ and $A_{\underline{i}'}^{\ j'}$ as well as the transformation law (12) for the vorticity tensor $W_{r}^{\ j}$, it can be shown that the coefficients $X_{\underline{i}}^{\ \underline{i}'}$ defined in (13) are constant in the course of time :

$$\frac{dX_{\underline{i}}^{\ \underline{i}'}}{dt} = 0$$

So, the corotational bases $\mathbf{e}_{\underline{i}}$ and $\mathbf{e}_{\underline{i}'}$ associated with the observers O and O' are fixed with respect to each other for each material point. Therefore the intrinsic history (11) for tensors can be defined in any of these bases without giving rise to any ambiguity even if the observers are in movement with respect to one another.

This property can also be seen as a consequence of the property of additivity of angular velocities.

4 Definition of an intrinsic time derivative : the Jaumann or corotational time derivative

The corotational or Jaumann time derivative [8] of a tensor is defined by deriving the components of this tensor in a corotational basis and bringing the result back to the natural basis of an observer.

Consider for instance a second order tensor **T**. If t_{ik} denotes the components of the tensor in the natural basis of an observer and $t_{\underline{ik}}$ its components in a corotational basis, those quantities are related by the tensorial law :

$$t_{\underline{ik}} = A_{\underline{i}}{}^{i} A_{\underline{k}}{}^{k} t_{ik} .$$

After derivation of the two members of this equality with respect to the time and use of equation (10), one obtains the following expression which defines the Jaumann derivative of the tensor **T** :

$$\frac{D_{Jaum} t_{ik}}{Dt} = A_{i}{}^{\underline{i}} A_{k}{}^{\underline{k}} \dot{t}_{\underline{ik}} = \frac{D t_{ik}}{Dt} + W_{i}{}^{j} t_{jk} + W_{k}{}^{j} t_{ij} . \qquad (14)$$

The notation D/Dt denotes the material time derivative in which the ordinary spatial derivatives are replaced by covariant ones :

$$\frac{D t_{ij}}{D t} = \frac{\partial t_{ij}}{\partial t} + v^{k} t_{ij;k} = \dot{t}_{ij} - v^{k} (\Gamma_{ki}^{l} t_{lj} + \Gamma_{kj}^{l} t_{il}). \qquad (15)$$

The Jaumann derivative consists of the material time derivative of the tensor plus two terms accounting for the rotation of the corotational basis.

Due to its intrinsic definition, the Jaumann derivative is independent of the observer. However this independence can also be checked explicitly by proving that this derivative follows the usual tensorial law for a change of observer.

It is also important to point out that the Jaumann derivative of the metric tensor vanish, so that raising or lowering indices commute with the corotational time derivation :

$$\frac{D_{Jaum}\, g_{ij}}{Dt} = \frac{D_{Jaum}\, g^{ij}}{Dt} = 0\,. \tag{16}$$

5 Extended Irreversible Thermodynamics and rheological modelling

We are now in a position to build up non ambiguous rheological models in the framework of Extended Irreversible Thermodynamics. For simplicity we shall restrict the present analysis to simple linear models.

The viscous pressure tensor is split up into its trace and its deviatoric part :

$$\Pi^{v} = p^{v}\,\mathbf{G} + \mathbf{P}^{v}, \tag{17}$$

where **G** is the metric tensor.

In a natural basis, this relation can be written in the form :

$$\Pi^{v}_{ik} = p^{v}\, g_{ik} + P^{v}_{ik}\,. \tag{18}$$

Using the components in a corotational basis, one has also :

$$\Pi^{v}_{\underline{ik}} = p^{v}\, g_{\underline{ik}} + P^{v}_{\underline{ik}}\,. \tag{19}$$

Following the usual procedure of Extended Thermodynamics, we express the entropy as an analytic function of the classical variables and the viscous pressure tensor :

$$s = s(u,v) + \frac{v}{6T}\,\alpha\, p^{v} g_{ik}\, p^{v} g^{ik} + \frac{v}{2T}\,\beta\, P^{v}_{ik}\, P^{v\,ik}\,. \tag{20}$$

This relation can be written, using the corotational components of the tensors :

$$s = s(u,v) + \frac{v}{6T}\alpha \overset{v}{p}\, g_{ik}\overset{v}{p}\, g^{ik} + \frac{v}{2T}\beta \overset{v}{P}_{ik}\overset{v}{P}^{ik}. \tag{21}$$

Using the mass and energy conservation laws, the entropy production is calculated. One obtains :

$$T\sigma^{s} = -\overset{v}{p}\,(v_{i;i} + \alpha \dot{\overset{v}{p}}) - \overset{v}{P}_{ik}(\widehat{V}^{ik} + \beta \dot{\overset{v}{P}}{}^{ik}). \tag{22}$$

It is important to stress that, to derive this expression, we have used equation (16).

Now that the pressure tensor history has been intrinsically introduced, it is easy to come back to the usual components of the tensors in a natural basis and rewrite the entropy production as :

$$T\sigma^{s} = -\overset{v}{p}\,(v_{i;i} + \alpha \dot{\overset{v}{p}}) - \overset{v}{P}_{ik}(\widehat{V}^{ik} + \beta \frac{D_{Jaum}\overset{v}{P}{}^{ik}}{Dt}). \tag{23}$$

The constitutive equations are obtained by expressing the thermodynamic forces as functions of the basic variables. If the relations fluxes-forces are assumed to be linear, one obtains a non-ambiguous version of Maxwell equations for a viscoelastic material, namely

$$\lambda_2 \frac{D_{Jaum}\overset{v}{P}_{ij}}{Dt} + \overset{v}{P}_{ij} = -2\eta_2 \widehat{V}_{ij}, \tag{24.a}$$

$$\lambda_0 \dot{\overset{v}{p}} + \overset{v}{p} = -2\eta_0 \nabla . v . \tag{24.b}$$

The positiveness of the entropy production implies that the viscosities η_2 and η_0 are positive quantities. Moreover, if the entropy is assumed to be a convex function, the relaxation times λ_2 and λ_0 are also positive.

Equations (24) are linear with respect to the pressure tensor. They can thus only describe situations where this tensor and the deformations remain rather small. However, it is no longer necessary to assume that the displacement gradients due for instance to superposed rigid rotations are small as it was actually the case with the first "ambiguous" version of Maxwell equations (5) [10].

6 Conclusion

We have thus been able to introduce in EIT an objective time derivative -the Jaumann or corotational derivative- in a rather natural way, using no other assumptions than the usual ones framing this theory. The objective time derivative appears as a consequence the necessity to make a choice and even a "good choice" for the description of the history of a tensor. As a matter of fact we have supposed that the history of a tensor at a given material point can be described at best by a privileged observer who is the material point itself, i.e. by an hypothetical observer attached rigidly to this material point.

Finally, let us point out that what has been said here about objectivity does not validate or invalidate the famous "Principle of Frame Indifference" stating that constitutive equations should be frame independent [8,9]. In this work we have not introduced dependences on external forces in the relations fluxes-forces but it is not forbidden *a priori* in EIT. This point as well as the possibility to describe other objective derivatives and non linear models in EIT will be studied in details in a forthcoming paper.

Acknowledgements

Frequent and very fruitful discussions with Prof. G. Lebon are cordially acknowledged.

References

1. Lebon G, Jou D and Casas-Vázquez J, J. Phys. A **13**, 275 (1980)
2. Casas-Vázquez J, Jou D and Lebon G eds, *Recent Developments in Nonequilibrium Thermo-dynamics*, Lect. Notes in Phys. **199**, Springer, Berlin (1984)
3. Jou D, Casas-Vázquez J and Lebon G, Rep. Prog. in Phys. **51**, 1105 (1988)
4. Lebon G, Jou D and Casas-Vázquez J , this volume (1990)
5. Onsager L, Phys. Rev. **37**, 405 (1931)
6. Prigogine I, *Introduction to Thermodynamics of Irreversible Processes*, Interscience, New York (1961)
7. De Groot S and Mazur P, *Non Equilibrium Thermodynamics*, North Holland, Amsterdam (1962)
8. Truesdell C and Toupin R A, *The Classical Field Theories*, Enc. of Physics, III/1, Springer, Berlin (1960)
9. Truesdell C and Noll W, *The Non-linear Field Theories of Mechanics*, Enc. of Physics, III/3, Springer, Berlin (1965)
10. Bird R, Armstrong R and Hassager O, *Dynamics of Polymeric Liquids*, vol.I, 2nd ed., Wiley, New York (1987)
11. Ericksen J L, *Tensor Fields*, Enc. of Physics, III/1, Springer, Berlin (1960)
12. Eringen A C, *Non-linear Theory of Continuous Media*, McGraw-Hill Book Company, New York (1962)

CONVECTION IN VISCOELASTIC FLUIDS

C. Pérez-García*, J. Martínez-Mardones† and J. Millán

Dpto. Física y Matemática Aplicada
Universidad de Navarra.
31080 Pamplona, Navarra, Spain.

1. INTRODUCTION

Convection takes place when a layer of fluid is heated from below. When a critical temperature difference is reached, the quiescent, purely conductive state, is replaced by some convective motions that organizes themselves to form a regular pattern that can be characterized by some wavelength.

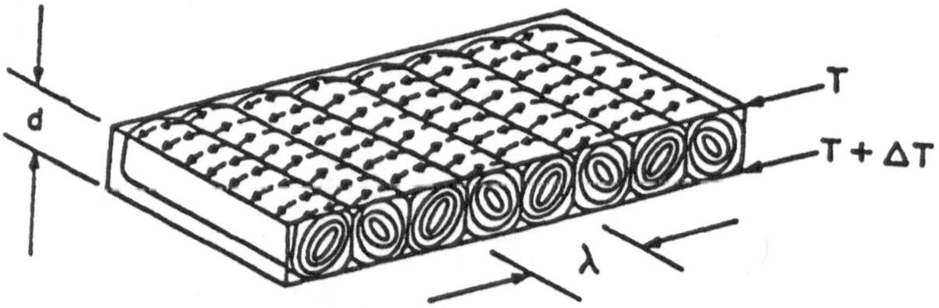

Figure 1. The cell pattern in convective motions.

This problem has been studied extensively [1][2] and can be considered as a prototype for studies of pattern forming systems, chaos, spatio-temporal intermittency, etc. [3][4]. The typical experimental configurations to study convection is not too complicated [5] and allow to measure with great precission the temperature and velocity fields inside the liquid layer, with or without movement.

When the convective cell is filled with a normal fluid, convection forms a pattern of rolls as illustrated in Fig. 1. However, when the fluid has some special properties, i.e., a binary mixture, temperature dependent transport coefficient, polymeric fluids,

*Also at Departament de Física, Universitat Autònoma de Barcelona, 08193 Bellaterra (Barcelona), Spain.

†On leave from Instituto de Física, Universidad Católica de Valparaíso, Casilla 4059, Valparaíso, Chile.

other symmetries and dynamical phenomena can appear as a consequence of convection [6]-[8].

The aim of the present work is to determine how certain rheological properties affect convection [9][10]. But it is also interesting to notice that the precision of measurement can give some idea of rheological properties [11] of polymeric fluids used in convection. As it has been discussed in this meeting, there is not a unique rheological model that cover the crowd of experimental data available in rheological studies of complex fluids. For the sake of simplicity we restrict the analysis to a model for viscoelastic fluid, the Oldroyd B model [12], that account for the mean experimental features and is well grounded on theoretical bases.

We analyse in detail the linear stability of convection in this fluids [13]-[17]. These results are completed with a weakly nonlinear perturbative analysis [18]-[20] of the convective motions in an Oldroyd B fluid. This scheme is coherent, because it is well known that this rheological model is mainly applicable when the shear stresses in the system is not too big, which is the case when one considers a weakly nonlinear analysis of convection.

In section 2 we recall briefly the main equations, boundary conditions and approximations that describe convection in a general fluid. The constitutive relation used along this paper is also discussed, as well as the main contributions to the convective equations. A complete linear analysis is made in section 3 in order to determine the different kinds of convective motions that can arise in the problem and the corresponding bifurcations. We emphasize the influence of the two main viscoelastic parameters in this bifurcation analysis. Using the usual techniques one can determine the normal forms (amplitude equations) that characterize the different bifurcations near threshold. This is made in section 4. The coefficients on these equations are calculated in the simplest cases and the solutions are discussed. Finally, section 5 is devoted to summarize the main conclusions and a discussion of the results and perspectives for future works.

2. EQUATIONS OF CONVECTION IN A VISCOELASTIC FLUID

We consider a horizontal fluid layer of depth d heated from below (Rayleigh-Bénard problem). This system may be described by the following equations [1]

$$\nabla \cdot \mathbf{v} = 0 \tag{1}$$

$$\rho_0[\partial_t \mathbf{v} + (\mathbf{v} \cdot \nabla)\mathbf{v}] = -\nabla p + \nabla \cdot \tau + \rho_0[1 - \alpha(T - T_0)]\mathbf{g} \tag{2}$$

$$[\partial_t T + (\mathbf{v} \cdot \nabla) T] = \kappa \nabla^2 T. \tag{3}$$

These equations account for incompressibility, momentum and energy balance under the so called Boussinesq approximation. That is a quite reasonable approximation that assume that thermal expansion only affects the external force term, that the transport thermal coefficients is constant and that the dissipation term in the energy equation is negligible compared to the thermal conduction term. In writing these equations we follow the usual notation: **v** velocity field, p the presure, τ the extra stress tensor, T the temperature, **g** is the acceleration due to gravity, α the thermal expansion coefficient, ρ_0 the density and κ the thermal diffusivity.

The system of Eqs. (1)-(3) is not complete. It must be supplemented by a relationship between the τ and the variable **v** , the so called constitutive equation. In the simplest case one assumes the linear Newtonian law $\tau = \eta\dot{\gamma}$, where η is the shear viscosity and $\dot{\gamma}$ is the rate of strain tensor $\dot{\gamma} = \nabla\mathbf{v} + (\nabla\mathbf{v})^T$. Maxwell [21] proposed the following generalization to account for the "elasticity", as well as for viscous effects in fluids

$$\tau + \lambda_1 \partial_t \tau = \eta(\dot{\gamma} + \lambda_2 \partial_t \dot{\gamma}). \tag{4}$$

where λ_1 is the relaxation time of the extra stress tensor, usually very small in normal fluids. But these two models are not sufficient for accounting the variety of rheological properties of more complex fluids. Oldroyd [12] made a detailed analysis of the characteristics for a constitutive equation, to be independent of local rotational properties. To account for this fact he formulated a serie of constitutive equations that have been a good guide to formulate and study viscoelastic properties. Here we do not intend to make an exhaustive analysis of this wide subject. We refer to the interested reader to good discussions in recent books [22]-[26]. For the sake of symplicity we take a model (the so called Oldroyd model B[12]) that, having the correct symmetry, account for viscoelastic properties for not too high strains. The constitutive equation for this model can be written in the form

$$\tau + \lambda_1 \mathrm{D}_t \tau = \eta(\dot{\gamma} + \lambda_2 \mathrm{D}_t \dot{\gamma}). \tag{5}$$

Here λ_2 is the retardation time, the finite time necessary for the elastic effects to become measurable. The derivative D_t is defined as

$$D_t \tau = \partial_t \tau + (\mathbf{v} \cdot \nabla)\tau - [(\nabla\mathbf{v})^T \cdot \tau + \tau \cdot (\nabla\mathbf{v})] \tag{6}$$

and leaves the constitutive equation (5) invariant under local rotations.

Then we study the system of equations (1)-(3),(5) for convection in a viscoelastic fluid. The quiescent stationary solution of these equations is symply $\mathbf{v}_s = \mathbf{0}$, $\tau_s = 0$ and $T_s = T_1 - (\Delta T/d)z$. To simplify the problem we will take the two-dimensional case. As the velocity field is solenoidal, a stream function ψ can replace the velocity field $(v_x, 0, v_z) = (\partial_z\psi, 0, -\partial_x\psi)$. This is a good approximation because the pattern that arises in an infinite system in Bénard convection is a system of convective rolls, as sketched in Fig.1. If one takes the x-direction perpendicular to the roll axis, the dynamics is purely two-dimensional, at least near the convective threshold. (Some secondary instabilities with y-variations are possible for sufficiently high heating [27]).

With this approximation the system of (1+3+1+6=11) eleven scalar variables reduces to (1+1+3=5) five scalar variables. In this 2D-approximation the extra stress pressure tensor has only three independent component $\tau_{xx}, \tau_{zz}, \tau_{xz} = \tau_{zx}$. But, in general, some combinations of these components are taken in rheological studies: $S(x,y,z) = \tau_{xx} - \tau_{zz}$ the primary normal stress difference and $U(x,y,z) = \tau_{xx} + \tau_{zz}$ the trace of the stress tensor.

As usual in a stability problem, the equations are rendered nondimensional by dividing the corresponding quantities by appropriated references: d, d^2/κ, κ/d, $\mu\kappa/d^2$ and βd for the lenght, time, velocity, extra stress tensor and temperature respectively. The second step is to consider perturbation around the stationary state $\psi = \psi_s + \psi' = \psi'$, $\tau = \tau_s + \tau' = \tau'$, $T = T_s + \theta$ where the prime ($'$) indicates the corresponding perturbation, and θ is the temperature perturbation. After introducing the form (6) into the balance and constitutive equations (1)-(3),(5) one arrives to the following system

$$\partial_t \nabla^2 \psi = -PR\partial_x \theta + P\Delta^2 \tau_{xz} + P\partial^2_{xz} S - J(\psi, \nabla^2 \psi) \tag{7}$$

$$\partial_t \theta = -\partial_x \psi + \nabla^2 \theta - J(\psi, \theta) \tag{8}$$

$$\begin{aligned}\Gamma\partial_t \tau_{xz} - \Gamma\Lambda\partial_t \Delta^2 \psi = \Delta^2 \psi - \tau_{xz} - \Gamma[J(\psi, \tau_{xz}) + (1/2)(S\nabla^2\psi) - \\ - (1/2)(U\partial^2_{xz}\psi)] + \Gamma\Lambda[J(\psi, \Delta^2\psi) + 2(\partial^2_{xz}\psi\nabla^2\psi)]\end{aligned} \tag{9}$$

$$\begin{aligned}\Gamma\partial_t S - 4\Gamma\Lambda\partial_t\partial^2_{xz}\psi = 4\partial^2_{xz}\psi - S - \Gamma[J(\psi, S) - 2(\tau_{xz}\nabla^2\psi) - \\ - 2(U\partial^2_{xz}\psi)] + \Gamma\Lambda[4J(\psi, \partial^2_{xz}\psi) - 2(\nabla^2\psi\Delta^2\psi)]\end{aligned} \tag{10}$$

$$\begin{aligned}\Gamma\partial_t U = -U - \Gamma[J(\psi, U) - 2(\tau_{xz}\Delta^2\psi - 2(S\partial^2_{xz}\psi)] \\ - 2\Gamma\Lambda[(2\partial^2_{xz}\psi)^2 + (\Delta^2\psi)^2]\end{aligned} \tag{11}$$

where, for convenience, the prime ($'$) is eliminated. Here $J(f,g) = \partial_z f \cdot \partial_x g - \partial_x f \cdot \partial_z g$ denotes the Jacobian operator, $\nabla^2 = \partial^2_x + \partial^2_z$ the Laplacian operator and $\Delta^2 = \partial^2_x - \partial^2_z$.

A group of four nondimensional parameters appear in these equations. These are

$R = \rho_0 g \alpha \Delta T d^3/\mu\kappa$ the Rayleigh number
$P = \mu/\rho_0\kappa$ the Prandtl number
$\Gamma = \lambda_1\kappa/d^2$ the relaxation parameter
$\Lambda = \lambda_2/\lambda_1$ the ratio between the retardation time and the relaxation time.

The Rayleigh number is the main parameter in convective studies and gives the ratio between buoyancy and dissipative effects. The Prandtl number measures the relative importance of viscous effects compared to thermal conduction effects. The relaxation parameter Γ is the ratio between the viscous relaxation time and the characteristic time of vertical convective motions. Finally, Λ gives an idea of the importance of elastic relaxation effects compared with the viscous relaxation effects.

These equations must be supplemented by boundary conditions (BC). As stressed from the beginning, we will take the simplest conditions, to have the insight on the main physical consequences of viscoelasticity and a mathematical analysis as simple as possible. Therefore we consider the following idealized BC: free and perfectly conducting upper and lower surfaces. (The system is laterally unbouded). The first mechanical condition lead to

$$\psi = \partial_z^2 \psi = 0 \quad \text{at} \quad z = 0,1 \tag{12}$$

and as a consequence

$$\tau_{xz} = \partial_z S = U = 0 \quad \text{at} \quad z = 0,1\,. \tag{13}$$

The second (thermal) BC is equivalent to

$$\theta = 0 \quad \text{at} \quad z = 0,1\,. \tag{14}$$

The advantage of these BC is that allow to obtain simple analytical solutions for Eqs.(7)-(11). By defining the vector $\Psi(x,z,t) = [\psi, \theta, \tau_{xz}, S, U]^T$ one can be write these equations in a more compact form

$$\partial_t \mathbf{L}\Psi = \mathbf{M}\Psi + \mathbf{N}(\Psi, \Psi) \tag{15}$$

where the matrices $\mathbf{L}$ and $\mathbf{M}$ are linear and the matrix $\mathbf{N}$ includes all the nonlinearities in Eqs.(7)-(11).

3. LINEAR STABILITY ANALYSIS

The first step in a stability analysis is to determine the influence of infinitesimal perturbation, i.e., those that allow to neglect the nonlinear terms $\mathbf{N}(\Psi, \Psi)$ on Eq.(15). Before starting the analysis it is interesting to notice that the variable U (Eq.(11)) does not play any role in the linear regime. This is so because we are dealing with an incompressible fluid, that requires $\nabla \cdot \mathbf{v} = 0$. The trace of the stress tensor U is linearly linked to $\nabla \cdot \mathbf{v}$ and then it is zero in Newtonian fluids. In viscoelastic models supplementary terms appear, both linear and nonlinear. However, the linear part is just a decay equation and, therefore, this variable cannot contribute to destabilize the system. As a consequence, the linear analysis can be restricted to the four variables

$(\psi, \theta, \tau_{xz}, S)$. The solution of the reduced system of four equations can be developed in a normal mode expansion

$$\Psi(x,z,t) = \begin{pmatrix} \psi \\ \theta \\ \tau_{xz} \\ S \end{pmatrix} = e^{st} \begin{pmatrix} \psi_0 \, sin(mkx) \, sin(n\pi z) \\ \theta_0 \, cos(mkx) \, sin(n\pi z) \\ \tau_{xz\,0} \, sin(mkx) \, sin(n\pi z) \\ S_0 \, cos(mkx) \, cos(n\pi z) \end{pmatrix} \tag{16}$$

where s is, in general, a complex parameter $s = \sigma + i\omega$. This development satisfy the BC (12)-(14). After introducing these solutions in the system of Eqs.(7)-(11), Eq.(15) leads to the general eigenvalue problem

$$det(M_{mn} - sL_{mn}) = s^j + \sum_{i=0}^{j-1} \mu_i s^i = 0, \qquad m,n \geq 1 \tag{17}$$

where the matrices L_{mn} and M_{mn} correspond to **L** and **M** after applying the expansion

$$L_{mn} = \begin{pmatrix} -q^2_{mn} & 0 & 0 & 0 \\ 0 & 1 & 0 & 0 \\ \Gamma\Lambda\delta^2_{mn} & 0 & \Gamma & 0 \\ -4mn\Gamma\Lambda\pi k & 0 & 0 & \Gamma \end{pmatrix} \tag{18}$$

$$M_{mn} = \begin{pmatrix} 0 & mkPR & -P\delta^2_{mn} & mn\pi kP \\ -mk & -q_{mn} & 0 & 0 \\ -\delta^2_{mn} & 0 & -1 & 0 \\ 4mn\pi k & 0 & 0 & -1 \end{pmatrix} \tag{19}$$

where $q^2_{mn} = m^2k^2 + n^2\pi^2$ and $\delta^2_{mn} = n^2\pi^2 - m^2k^2$.

For m and n different from 1, the characteristic polynomial (17) is a quartic function of s. But it can be proved that, as in R-B convection in Newtonian fluids, the most unstable modes are those with $m = n = 1$. In this case, one root of Eq.(17) is always negative ($s = -1/\Gamma$) and, therefore, it has a stable eigenvector. Then the study of the eigenvalue problem can be reduced to the cubic characteristic polynomial

$$P(s) = s^3 + as^2 + bs + c = 0\,. \tag{20}$$

The different roots of this polynomial, allow to distinguish a critical surface in the space of parameters, as sketched in Fig. 2 and explained in the following. The possible roots of $P(s)$ give to the bifurcations [28][30] explained in subsection 3.1.

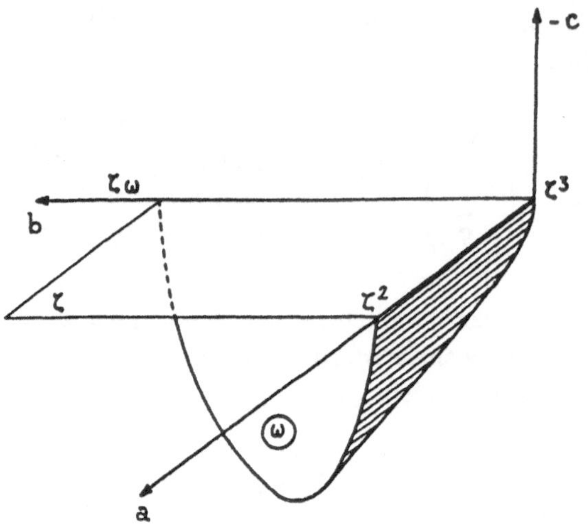

Figure 2. Critical surface in the space of parameters of the characteristic equation.

3.1. Bifurcation in a viscoelastic fluid

i) Codimension one stationary instability ζ [31]

This correspond to a simple zero eigenvalue (s=0), which is obtained when $c = 0$ and $b > 0$. From the first condition one can obtain the same that for a Newtonian fluid $R_s = q^6/k^2$ whose minimum is $R_{cs} = 27\pi^4/4 = 657.5$ for $k_{cs} = \pi/\sqrt{2} = 2.221$. (Here and in the following, the subscript $()_s$ indicate quantities for stationary instability). The second condition ($b > 0$) shows that the instability acting on the system is stationary provided that

$$\Gamma < \frac{(1+P)}{[q^2 P(1-\Lambda)]} \, . \tag{21}$$

ii) Codimension one oscillatory instability ω [32]

A different case appears for roots with $\sigma = 0$ and $\omega \neq 0$. These are pairs of complex conjugate roots that correspond to a Hopf bifurcation [11]. In this case an oscillatory instability with a frequency ω appears. This is possible when $c = a \cdot b$ with $b > 0$. From this first equality one derives the form of the oscillatory marginal curve

$$R_o = \frac{q^6 \Lambda}{k^2} + \frac{q^2}{k^2 \Gamma(1+\Lambda P)}(q^2(2\Lambda + \Lambda P + P^{-1}) + \frac{1+P}{\Gamma P}) \, . \tag{22}$$

(The subscript $()_o$ indicates quantities for the oscillatory convection). The condition $b > 0$ leads to the following equation

$$\omega^2 = \frac{q^2 P\Gamma(1-\Lambda) - (1+P)}{\Gamma^2(1+\Lambda P)} > 0\,. \tag{23}$$

It is obvious that this formula is valid if

$$\Gamma > (1+P)/(q^2 P(1-\Lambda))\,. \tag{24}$$

iii) Codimension two (CT) stationary bifurcation ζ^2 [33]

Another case can be found for a zero eigenvalue of multiplicity two ($s^2 = 0$). This is possible when $c = b = 0$ and from this one obtains

$$\omega_{CT} = 0, \quad \Gamma_{CT} = \frac{1+P}{q^2 P(1-\Lambda)}, \quad R_o = R_s = R_{CT} = \frac{q^6}{k^2}\,. \tag{25}$$

iv) Codimension two oscillatory bifurcation $\zeta\omega$ [34]

It could be possible to have a CT Hopf bifurcation when one has $s = 0$ and $s = \pm i\omega$. This could occurs when $c = a = 0$ and $b > 0$, but is not possible for a Jeffreys viscoelastic fluid where always $a > 0$.

3.2. Numerical results

The influence of the different parameters on bifurcation is analysed in more detail in this subsection. In particular we will pay some attention to the role of the parameter Λ that distinguish Jeffreys and Maxwell models.

In Fig. 3(a) some typical marginal curves $R(k)$ are plotted. The full curve concerns stationary stability. As explained above this is independent of P and Γ. Broken curves correspond to oscillatory stability for several values of Λ, with $P = 10.0$ and $\Gamma = 0.1$. For $\Lambda = 0$ (Maxwell model) the values obtained by Vest and Arpaci [14] and Sokolov and Tanner [15] are recovered.

Fig. 3(b)(c) show the dependence of the critical wavenumber and the critical Rayleigh number on Λ for $P = 10.0$ and $\Gamma = 0.1$. The critical wavenumber k_{co} decreases and the critical Rayleigh number R_{co} increases with increasing Λ. For $\Lambda^f = 0.3156$ the critical Rayleigh numbers coincide (the values that characterise these frontier points will be labelled with the superscript $()^f$ in the following), and the rest of the critical values are $k^f_{c,s} = 2.221$, $k^f_{c,o} = 2.810$ and $\omega^f_c = 5.279$. This is the point where the lowest threshold changes from oscillatory to stationary instabilities. For $\Lambda > \Lambda^f$ overstable motions cannot appear spontaneously in a system of infinite horizontal extent. However, overstable motions are still possible in a system with a fixed wavenumber. In this situation CT points can be reached. In particular, the quadratic minimun in the curve $R_o(k)$ disappear for $\Lambda(CT) = 0.3711$, $k_c(CT) = 2.761$, $R_c(CT) = 702.1$ and $\omega_c = 0$ (for $P = 10.0$ and $\Gamma = 0.1$) leading to a degenerate CT point.

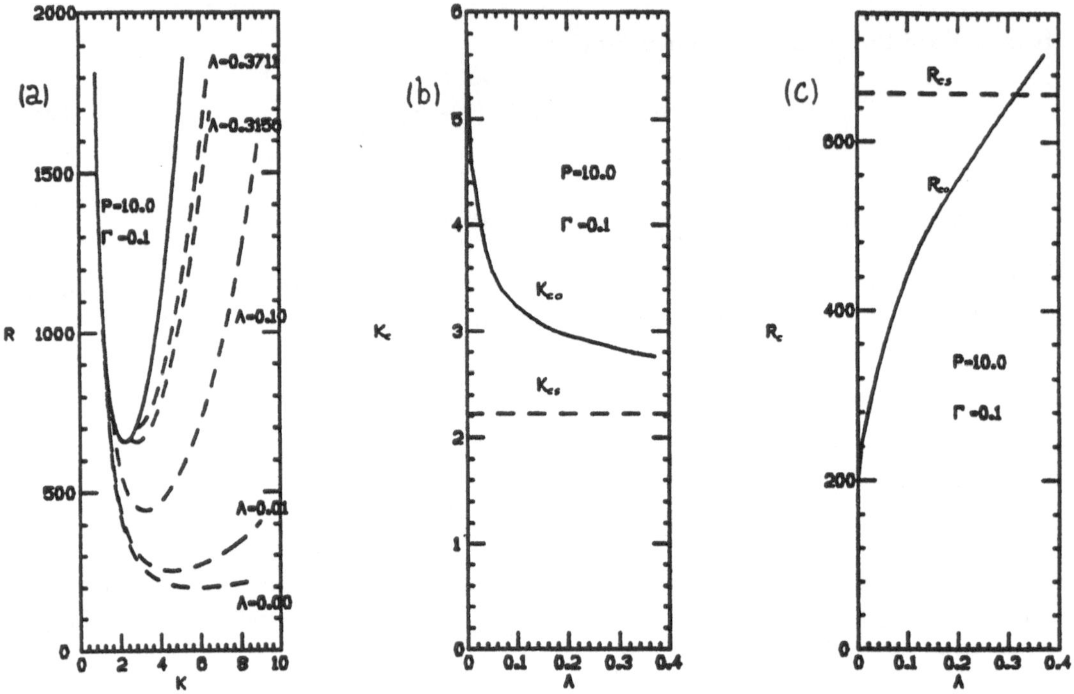

Figure 3. (a) Marginal curves for stationary and oscillatory instabilities. (b) Critical wavenumber for stationary and oscillatory instabilities, as a function of Λ. (c) Critical Rayleigh number for stationary and oscillatory instabilities as a function of Λ for $\Gamma = 0.1$ and $P = 10.0$.

We present now the results on the influence of the parameter Λ and P on that frontier between oscillatory and stationary convection. Fig. 4 gather the main results. In Fig. 4(a) we show the dependence of Γ^f as a function of the parameter Λ. In the region above these curves the system is unstable under oscillatory convection. (Stationary motions are the unstable modes below these curves.) For a fixed Λ, Γ^f decreases with increasing P. This means that the higher the viscous effects, the lower relaxation time necessary to start overstability for a fixed R. In the case of $\Lambda = 0$ (Maxwell fluid) and $P \to \infty$ the minimun value $\Gamma^f = 0.039$ is obtained [15]. The results of the Newton fluid is recovered in the limit of $\Lambda = 1$ independently of the value of Γ. This is the reason for the divergence of Γ^f in that limit.

The values of the critical wavenumber for oscillatory instability in the frontier $k^f_{c,o}$ as a function of Λ is quoted in Fig. 4(b). Notice that in the limit $\Lambda = 0$ (Maxwell model) the values of $k^f_{c,o}$ are greater than $k^f_{c,s}$ and the diference between them increases with P. For example for $P = 100$ and $\Lambda = 0$, $k^f_{c,o}/k^f_{c,s} = 5.6$ and $\omega^f_c = 579.8$, but that ratio decreases to 1.8 and the frecuency to ω^f_c decrease rapidly until in the limit $\Lambda \to 1$ (Newtonian fluid), one recovers $k_{co} \to k_{cs}$ independently of the value of P.

The oscillation frequency ω^f_c as a function of Λ is shown in Fig. 4(c). This frequency ω^f_c is very large in the limit $\Lambda \to 0$, while it tends to zero when $\Lambda \to 1$, because overstability disappears in this limit. From these results we remark that in the pure Maxwell model the critical wavenumber and the oscillation frecuency increase monotonically when P increases, reaching very unrealistic values. However, with a small retardation time or, equivalently, a small Λ, we recover more reasonable values for the critical parameters.

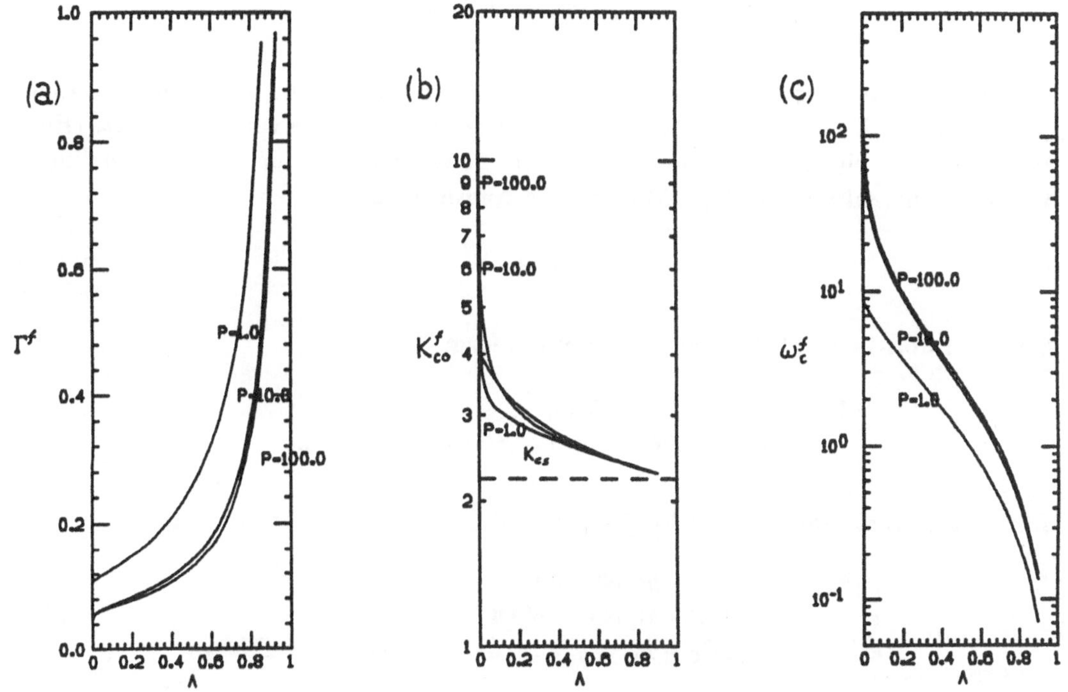

Figure 4. Dependence of (a) the relaxation time Γ^f, (b) the critical wavenumber k^f_{co}. (c) The critical frequency ω^f_c at the crossover between oscillatory and stationary convection, as a function of Λ, for various values of P.

4. NONLINEAR ANALYSIS

The linear analysis allows to determine the different dynamical situations when the system becomes unstable. However, it cannot provide information on the different patterns that arise beyond the instability point. Therefore, a nonlinear analysis must be used to determine the convective motions observed in experiments. One of the most useful weakly nonlinear method is the so-called Stuart-Landau method [35][36], which is based on the amplitude equations. It assumes basically that the constant amplitudes of the linear analysis $(\psi_0, \theta_0, \tau_0, S_0)$ in Eq.(16), are variable in space and in time above threshold. But these amplitudes depend on space and time on scales different from the usual ones [16]. Following the analysis of Stuart we can consider developments in series of a small parameter ϵ, defined as

$$\epsilon = \frac{R - R_c}{R_c} \tag{26}$$

which indicates the supercritical heating. The next step is to assume that the solutions of the nonlinear problem can be developed as

$$\Psi(x,z,t) = \epsilon^{1/2}\Psi_1 + \epsilon\Psi_2 + \epsilon^{3/2}\Psi_3 + ... \tag{27}$$

where Ψ_1 corresponds to the solution of the linear problem in which the constants $(\psi_0, \theta_0, \tau_0, S_0, U_0)$ are now assumed to depend on the slow time variable $\mathrm{T} = \epsilon t$. (Here for the sake of simplicity we do not consider the spatial variation). Correspondingly, the time differentiation in Eq.(15) should be transformed as

$$\partial_t \rightarrow \partial_t + \epsilon\partial_{\mathrm{T}} \; . \tag{28}$$

The operators in Eq.(15) are expanded in the form [18]

$$\begin{aligned} \mathbf{M} &= M_1 + \epsilon M_3 + ... \\ \mathbf{N} &= \epsilon N_2(\Psi_1.\Psi_1) + \epsilon^{3/2} N_3(\Psi_1, \Psi_2) + ... \end{aligned} \tag{29}$$

Introducing these equations in Eq.(15) one arrives to the hierarchy of equations

$$\begin{array}{ll} \epsilon^{1/2}: & (L\partial_t - M_1)\Psi_1 = 0 \\ \epsilon: & (L\partial_t - M_1)\Psi_2 = N_1(\Psi_1, \Psi_1) \\ \epsilon^{3/2}: & (L\partial_{\mathrm{T}} - M_1)\Psi_3 = M_3\Psi_1 + N_3(\Psi_1, \Psi_2) - L\partial_t\Psi_1 \; . \end{array} \tag{30}$$

The first equation is the linear problem, whose solutions are in the form

$$\Psi_1 = \begin{pmatrix} \psi(\mathrm{T})\, sink_c x \; sin\pi z \\ \theta(\mathrm{T})\, cosk_c x \; sin\pi z \\ \tau_{xz}(\mathrm{T})\, sink_c x \; sin\pi z \\ S(\mathrm{T})\, cosk_c x \; cos\pi z \end{pmatrix} e^{st} + c.c. \tag{31}$$

where *c.c.* indicates the complex conjugate.

Only one of the four amplitudes $(\psi(\mathrm{T}), \theta(\mathrm{T}), \tau_{xz}(\mathrm{T}), S(\mathrm{T}))$ is independent. These are linked by relationships that depend on matrices **L** and **M**. We apply this general scheme to the different bifurcations studied in section 3.

4.1 Stationary bifurcation

After introducing the general solution of the linear problem (31) with $s = 0$ into Eqs.(30) one arrives to relationships between the amplitudes in the form

$$\theta_s(\mathrm{T}) = -\frac{k_{cs}{}^2}{q_{cs}}\psi_s(\mathrm{T}), \quad \tau_s(\mathrm{T}) = -\delta_{cs}\psi_s(\mathrm{T}), \quad S_s(\mathrm{T}) = 4\pi k_{cs}\psi_s(\mathrm{T})\,. \tag{32}$$

By applying the multiple scale perturbative expansion (30), the following amplitude equation for $\psi(\mathrm{T})$ is obtained

$$\tau_s\dot{\psi}_s = \epsilon_s\psi_s - g_s\psi_s^3 \tag{33}$$

where now $\dot{()} = d()/d(\mathrm{T})$, $\epsilon_s = (R - R_{cs})/R_{cs}$ and the coefficients are

$$\tau_s = \frac{P+1}{q_{cs}^2 P} - \Gamma(1-\Lambda), \tag{34}$$

$$g_s = \frac{1}{8q_{cs}^2} + \frac{\Gamma^2(1-\Lambda)}{16k_{cs}^2 q_{cs}^4}(9\delta_{cs}^8 + 8\pi^2 k_{cs}^2 + 80\pi^4 k_{cs}^4 - 9q_{cs}^4\delta_{cs}^4 - 4\pi^2 k_{cs}^2 q_{cs}^4)\,. \tag{35}$$

Notice that τ_s is the relaxation time that is derived from the linear theory, using the relation [37]

$$\tau_s = \left(R_c\frac{\partial\sigma}{\partial R}\right)^{-1} \tag{36}$$

where σ is the growth rate above threshold. For $\Lambda = \Gamma = 0$ the result for a Newtonian fluid $\tau_s = (P+1)/q_{cs}^2 P$ is also recovered.

In general, the slowing varying amplitude $\psi(\mathrm{T})$ is complex, but it obeys Eq.(33) whose coefficients are real. The main coefficient is g_s in the cubic term when $\Lambda = 0$. When $\Gamma = \Lambda = 0$ or $\Lambda = 1$, Eq.(35) gives the value $g_s = 1/8q_{cs}^2$ of the Newtonian fluid. When it is positive Eq.(33) admits a stationary solution and the corresponding bifurcation is supercritical

$$\psi = \left(\frac{\epsilon_s}{g_s}\right)^{1/2}\,. \tag{37}$$

When $g_s/\tau_s \leq 0$ more terms in the developments (29) are needed. The case $g_s/\tau_s = 0$ leads to a tricritical point, and $g_s/\tau_s < 0$ to a subcritical bifurcation. After analysing Eq.(35) it is not too difficult to conclude that g_s/τ_s cannot be negative and, therefore, neither tricritical point nor subcritical bifurcation are possible in an Oldroyd fluid. The dependence of this coefficients on Γ for differents values of Λ is shown in Fig. 5.

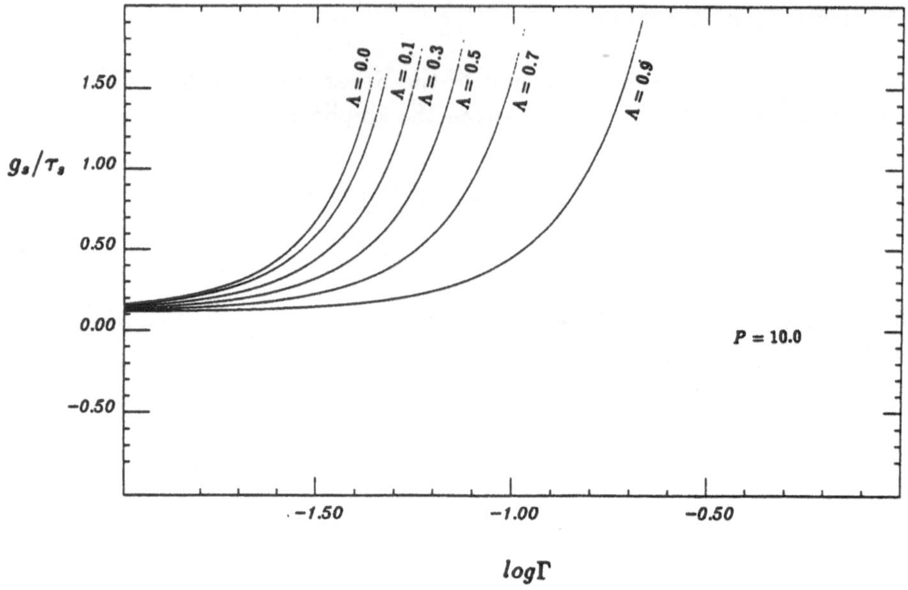

Figure 5. The coefficient g_s/τ_s as a function of Γ and Λ for $P = 10.0$.

This result is in contradiction with the claims in some recents works [19][20]. The discrepancy is not in the model used (Oldroyd B in our case, that includes as a particular case the Maxwell model used in those works) but on an unappropriated choice of the derivatives in the constitutive equation (5). The upper convected derivative introduces some extra non linear terms that eliminate the possibility of stationary bifurcations other than supercritical.

4.2. Hopf bifurcation

In the case of an oscillatory instability the solution Ψ_1 must be written in the form

$$\Psi_1 = \begin{pmatrix} \psi_o(\mathrm{T})\, sin k_{co}x \; sin \pi z \\ \theta_o(\mathrm{T})\, cos k_{co}x \; sin \pi z \\ \tau_{xz(o)}(\mathrm{T})\, sin k_{co}x \; sin \pi z \\ S_o(\mathrm{T})\, cos k_{co}x \; cos \pi z \end{pmatrix} e^{i\omega_c t} + c.c. \tag{38}$$

This solution of Eqs.(30) gives a relation between the amplitudes in the form

$$\theta_o(\mathrm{T}) = -\frac{k_{co}}{q_{co}^2 + i\omega_c\Gamma}\psi_o(\mathrm{T}), \tag{39}$$

$$\tau_{xz(o)}(\mathrm{T}) = \frac{4\pi k_{co}(1+i\omega_c\Gamma\Lambda)}{1+i\omega_c\Gamma}\psi_o(\mathrm{T}), \tag{40}$$

$$S_o(\mathrm{T}) = -\frac{\delta_{co}^2(1+i\omega_c\Gamma\Lambda)}{1+i\omega_c\Gamma}\psi_o(\mathrm{T})\,. \tag{41}$$

These expressions are taken to solve the perturbative scheme (30) and leads to the amplitude equation

$$\tau_o\dot{\psi}_o = \epsilon_o\psi_o - g_o \mid \psi_o \mid^2 \psi_o\,. \tag{42}$$

The main difference between Eq.(33) and Eq.(42) is that in the last the coefficients τ_o and g_o are complex, and now $\epsilon_o = (R - R_{co})/R_{co}$. Moreover, oscillatory convection is not possible for a Newtonian fluid under the usual conditions. In the particular case of a viscoelastic fluid this solution only exists when $\Gamma > (1+P)/[q^2P(1-\Lambda)]$. The coefficients τ_o and g_o take the form

$$\tau_o = \frac{1}{q_{co}^2 + i\omega_c} + \frac{q_{co}^2(q_{co}^2 + i\omega_c)}{k_{co}^2 P R_{co}} - \frac{\Gamma(1-\Lambda)q_{co}^4(q_{co}^2+i\omega_c)}{k_{co}^2 P R_{co}(1+i\omega_c\Gamma)^2}, \tag{43}$$

$$g_o = \frac{q_{co}^2}{4(q_{co}^4+\omega_c^2)} + \frac{\pi^2}{(8\pi^2+4i\omega_c)(q_{co}^2+i\omega_c)} + \frac{\Gamma^2(1-\Lambda)}{16(1+i\omega_c\Gamma)k_{co}^4 R_{co}} \tag{44}$$
$$(9\delta_{co}^8 + 8\pi^2 k_{co}^2 + 80\pi^4 k_{co}^4 - 9q_{co}^4\delta_{co}^4 - 4\pi^2 k_{co}^2 q_{co}^4)\left(\frac{2}{1+\omega_c^2\Gamma^2} + \frac{1}{(1+i\omega_c\Gamma)(1+2i\omega_c\Gamma)}\right).$$

The corresponding Hopf bifurcation admits solutions in the form

$$\psi_o = \mathcal{R}_o e^{i\phi}, \quad \mathcal{R}_o = \left(\frac{\epsilon_o}{Re\, g_o}\right)^{1/2}, \quad \phi = \left[\frac{\epsilon_o}{Im\,\tau_o} - Im\left(\frac{g_o}{\tau_o}\right)\mathcal{R}_o^2\right]\mathrm{T} + const. \tag{45}$$

Now the different kind of bifurcation is given by the sign of $Re\,(g_o/\tau_o)$. From Eqs.(15)-(16), one can prove that $Re\,(g_o/\tau_o) > 0$ and, therefore, the bifurcation is always supercritical in a Oldroyd B fluid. In Fig. 6 the dependence of this parameter on Γ for different values of Λ ($P = 10$) is shown. For a fixed Λ, a similar behaviour of $Re\,(g_o/\tau_o)$ as a function of Λ is obtained.

As in the case of stationary bifurcation these results do not agree with those obtained by Zielinska et al. [19] and Brand et al. [20]. The discrepancy here has the same origin that in the stationary case.

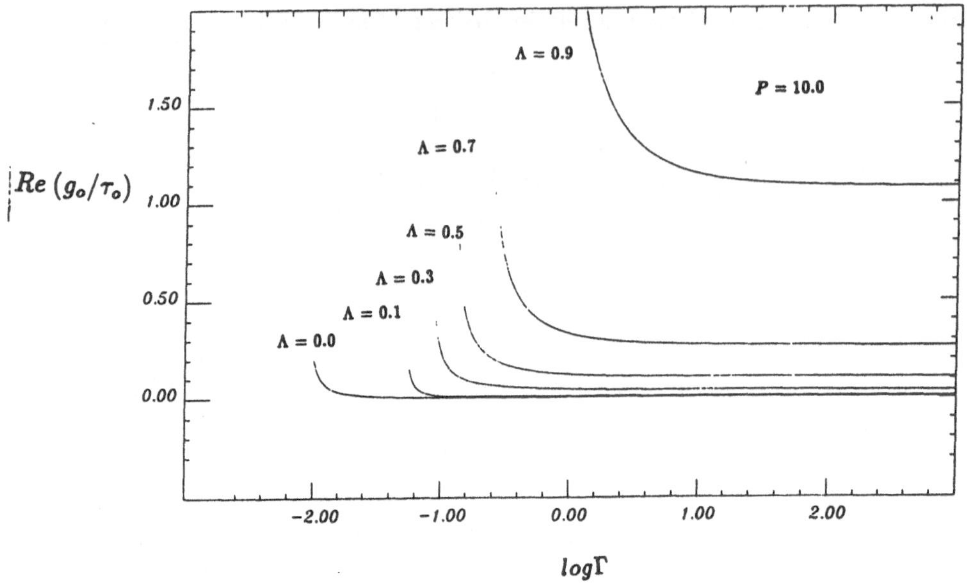

Figure 6. The coefficient $Re\,(g_o/\tau_o)$ as a function of Γ and Λ for $P = 10.0$.

5. CONCLUSIONS AND FINAL DISCUSSION

By means of a quite general model we have shown in the present work how convection can be affected by viscoelastic coefficients. The Oldroyd B viscoelastic model (with upper convected derivatives) have been taken because it allows to fit quite well viscoelastic properties of polymer solutions, at least for moderated shear stresses.

The retardation time Λ, introduced to generalize the Maxwell model, has an important influence on the linear analysis, both for stationary and oscillatory convection. A complete linear analysis, with the different kinds of bifurcations have been presented. The possibility to obtain degenerated CT point have been discussed, as well as the range of parameters for which stationary and oscillatory motions can coexist.

A weakly nonlinear analysis complete the present study. The main conclusion of this analysis is that the stationary bifurcation and the oscillatory bifurcation are always supercritical. Subcritical bifurcation and tricritical points are not allowed for Oldroyd B fluids, when the appropriated derivatives are taken.

The present work have been restricted to the unrealistic case of free-free boundary conditions. However, even in this simplified case many dynamical phenomena can be present (traveling or standing waves, coexistence of stationary and oscillatory motions). We hope that this analysis would stimulate new experimental works to determine the main features of convection in viscoelastic fluids. A complementary task, that could give some insight on rheological properties, is to take advantage of the precision of experiments in convection to determine some coefficients of polymeric fluids.

Acknowledgments

We have benefited from discussions with Prof. J. Casas-Vázquez (Universidad Autónoma de Barcelona), Prof. D. Joseph (University of Minnesota) and Prof. G. Lebon (Université de Liège). This work was supported by an EEC project SC1-0035-C and by a grant of the DGICyT (Spanish Ministry of Education) CE89-0002. One of us (J.M-M) acknowledges the support of the Instituto de Cooperación Iberoamericana (ICI)(Spain).

References

[1] S. Chandrasekhar, *Hydrodynamic and Hydromagnetic Stability* (Oxford Univ. Press, Oxford, 1961).

[2] J. C. Platten and J. Legros, *Convection in Liquids* (Springer, Berlin, 1984).

[3] P. Bergé, ed., *Le Chaos: Théorie et Experiences* (Eyrolles, Paris, 1989).

[4] J. E. Wesfreid and S. Zaleski, eds., *Cellular Structures in Instabilities, Lecture Notes in Physics*, vol. 210 (Springer, Berlin, 1984).

[5] H. Bénard, *Ann.Chim. Phys.* **23** 62 (1901).

[6] D. T. J. Hurle and E. Jakeman, *J. Fluid Mech.* **47** 667 (1971).

[7] S. Ciliberto, E. Pampaloni and C. Pérez-García, *Phys. Rev. Lett.* **61** 1198 (1988).

[8] I. A. Eltayeb, *Proc. R. Soc. Lond. A.* **356** 161 (1977).

[9] J. Martínez-Mardones and C. Pérez-García, *J. Phys.: Cond. Mat.* **2** 1281 (1990).

[10] R. W. Kolkka and G. R. Ierly, *J. Non-Newtonian Fluid Mech.* **25** 209 (1987).

[11] J. D. Ferry, *Viscoelastic Properties of Polymers* (Wiley, New York, 1980).

[12] J. G. Oldroyd, *Proc. Roy. Soc.* **A 200** 523 (1950).

[13] T. Green, *Phys. Fluids* **11** 1410 (1968).

[14] C. M. Vest and V. S. Arpaci, *J. Fluid Mech.* **36** 613 (1969).

[15] M. Sokolov and R. I. Tanner, *Phys. Fluids* **15** 534 (1972).

[16] D. Gözum and V. S. Arpaci, *J. Fluid Mech.* **64** 439 (1974).

[17] S. Carmi and M. Sokolov, *Phys. Fluids* **7** 544 (1974).

[18] S. Rosenblat, *J. Non-Newtonian Fluid Mech.* **21** 20 (1986).

[19] B. J. A. Zielinska, D. Mukamel and V. Steinberg, *Phys. Rev. A* **33** 1454 (1986).

[20] H. R. Brand and B. J. A. Zielinska, *Phys. Rev. Lett.* **57** 3167 (1986).

[21] J. C. Maxwell, *Proc. R. Soc. Lond.* **22** 46 (1873).

[22] R. B. Bird, R. C. Armstrong and O. Hassager, *Dynamics of Polymeric Fluids*, vol.1 (Wiley, New York, 1977).

[23] R. I. Tanner, *Engineering Rheology* (Oxford Univ. Press, Oxford, 1985).

[24] D. D. Joseph, *Fluid Dynamics of Viscoelastic Liquids* (Springer, New York, 1990).

[25] M. Doi and S. F. Edwards, *Theory of Polymer Dynamics* (Oxford Univ. Press, Oxford, 1986).

[26] R. G. Larson, *Constitutive Equations for Polymer Melts and Solutions* (Butterworths, Boston, 1988).

[27] F. H. Busse, *Rep. Prog. Phys.* **41** 1929 (1978).

[28] V. I. Arnold, *Chapitres Supplémentaires de la Théorie des Equations Différentielles Ordinaires* (Mir, Moscou, 1980).

[29] J. Guckenheimer and P. Holmes, *Nonlinear Oscillations, Dynamical Systems and Bifurcations of Vector Fields* (Springer, Berlin, 1983).

[30] G. Iooss and D. D. Joseph, *Elementary Stability and Bifurcation Theory* (Springer, New York, 1980).

[31] Lord Rayleigh, *Phil. Mag.* **32** 529 (1916).

[32] J. E. Marsden and M. MacCracken, *Hopf Bifurcation and its Applications* (Springer, Berlin, 1976).

[33] P. H. Coullet and E. A. Spiegel, *SIAM J. Appl. Math.* **43** 776 (1983).

[34] A. Arneodo, P. H. Coullet and E. A. Spiegel, *Geophys. Astrophys. Fluid Dynamics* **31** 1 (1985).

[35] J. T. Stuart, *J. Fluid Mech.* **9** 353 (1960).

[36] L. D. Landau, *C. R. Dokl. Acad. Sci. USSR* **44** 311 (1944).

[37] A. C. Newell and J. A. Whitehead, *J. Fluid Mech.* **38** 279 (1969).

FRACTIONAL RELAXATION EQUATIONS FOR VISCOELASTICITY AND RELATED PHENOMENA

Theo F. Nonnenmacher
Mathematische Physik, Universität Ulm
Albert–Einstein–Allee 11
7900 Ulm, Germany

Abstract

Standard relaxation equations contain integer-number differentials. Applying the fractional calculus we derive and solve fractional relaxation equations. Maxwell's rheological constitutive law will be discussed in detail, and the fractional form of the Maxwell equation will be investigated for different initial value conditions. A comparison with experimental relaxation data supports the fractional model.

1. Introduction, Formulation of the Problem

One of the first theoretical concepts to model viscoelasticity has been proposed (in 1867) by Maxwell [1], whose rheological constitutive law

$$\sigma + \tau_0 \frac{d\sigma}{dt} = \eta \frac{d\gamma}{dt} \tag{1}$$

relating stress (σ) to strain (γ) forms the basis of Maxwell's theory of elasticity (supplemented by the Kelvin-Voigt theory of retarded elasticity). Here, $\tau_0 = \eta / G_0$ with η= viscosity, $1/G_0$ equilibrium compliance. The Maxwell theory has first been applied to model viscoelasticity in relation to two fundamental processes:

(i) straining (tension or compression) of cylinders of various colloidal materials under constant stress, and
(ii) relaxation of stress of the same material held at constant strain.

For constant strain ($\gamma = \gamma_0 = const$), Eq. (1) integrates to

$$\sigma(t) = \sigma_0 e^{-t/\tau_0} \quad . \tag{2}$$

However, many experimental data sets could not be fitted by Maxwell's relaxation law. Instead, when Kohlrausch [2] carried out relaxation experiments on various materials in order to study mechanical creep he came to the result that many of his data could be fitted much better by the empirical law (stretched exponential)

$$\sigma(t) = \sigma_0 e^{-(t/\tau_0)^\beta} \quad , \quad (0 < \beta < 1) \tag{3}$$

indicating a slower decay than predicted by an exponential relaxation ($\beta = 1$). In 1970 Williams and Watts [3] postulated the same function for a successful description of dielectric relaxation in polymers. In recent years much attention has been focused on physical models prescribing the common feature that is responsible for generating the stretched exponential decay law.

During the first half of this century a quite different approach to viscoelasticity has been proposed:

- In 1921 Nutting [4] observed that stress-strain data sets of many complex materials closely obey the empirical law, the so-called Nutting equation

  $$\sigma = c_0 \gamma t^{-k} \tag{4}$$

  which (for constant strain $\gamma = const$) indicates inverse power-law relaxation.

- In 1936 Gemant [5] and in 1946 Bosworth [6] analyzed experimental results obtained from elasto-viscous bodies, and gave a definition of plasticity that incorporates the concept of "fractional differentiation". The idea behind all that is to find an appropriate approximation for viscoelastic bodies that are neither a Hookean solid nor a Newtonian fluid but something in between. This "principle of intermediacy" might be applied to interpolate between fluid behaviour

  $$(i) \qquad \sigma = \eta \frac{d\gamma}{dt} \quad , \eta = const$$

  and solid behaviour (Hooke's law)

  $$(ii) \qquad \sigma = a\gamma \quad , a = const.$$

  Consequently, for an intermediate body the guess was [5,7]

  $$(iii) \qquad \sigma = \chi_1 \frac{d^\mu \gamma}{dt^\mu} \quad , \quad \chi_1 = const \quad , \quad (0 < \mu < 1)$$

  Generally one may write [6]

  $$\sigma = f\left(\frac{d^\mu \gamma}{dt^\mu}\right)$$

  including all three cases (i) to (iii), i. e. $\mu = 1$ (case (i)), $\mu = 0$ (case (ii)) and for an intermediate body one expects $0 < \mu < 1$.

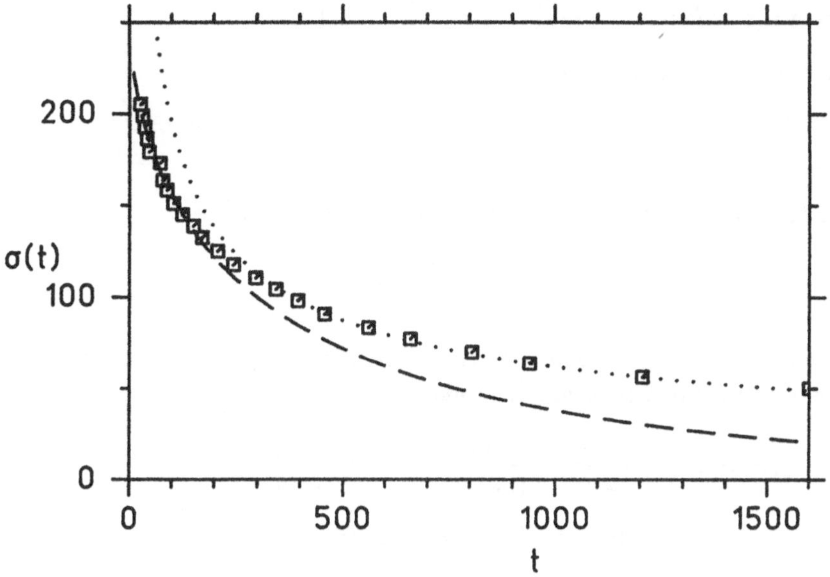

Fig. 1 Stress relaxation for constant strain: (...) Nutting Eq (4) for $c_0\gamma_0 = 1950$ and $k = 0.5$, (- - -) stretched exponential law (3) for $\sigma_0 = 250, \tau_0 = 350$ and $\beta = 0.6$, experimental data from Ref. [16]

In the following sections we extend the standard Maxwell theory and related relaxation equations by applying the fractional calculus to the standard relaxation equation $df/dt = -\lambda f(t)$ and to Maxwell's constitutive equation (1) in order to come up with the corresponding fractional equations. The solutions will be presented and will be compared with experimental data sets.

2. Fractional Calculus, Definitions, Historical Background

We first mention that the fractional calculus is old but little applied. Hence, it will be of some advantage to sketch briefly the main ideas. We are following closely Oldham and Spanier [8] and Ross [9].

When in the 17th century the (integer-number) differential calculus had been developed, Leibniz asked in a letter addressed to L'Hospital:

> Can the meaning of derivatives of integral order $d^n f(x)/dx^n$ be extended to have meaning when n is not an integer but any number (irrational, fractional or even complex-valued)?

L'Hospital responded:

What if n be 1/2? , $\frac{d^{1/2}f(x)}{dx^{1/2}}=?$ for $f(x)=x$

Leibniz, in a letter dated from Sept. 30, 1695, replied:

> It will lead to a paradox, from which one day useful consequences will be drawn.

124 years later (in 1819) Lacroix presented the result for $f(x)=x$:

$$\frac{d^{1/2}x}{dx^{1/2}}=\frac{2}{\sqrt{\pi}}\sqrt{x} \quad .$$

How can we understand this result? Integer derivatives of powers of x can be defined by making use of the gamma function, i.e. take

$$f(x)=x^m \to \frac{d^n x^m}{dx^n}=\frac{m!}{(m-n)!}x^{m-n}=\frac{\Gamma(m+1)}{\Gamma(m-n+1)}x^{m-n}$$

We generalize $n\to\nu$ and $\Gamma(n)\to\Gamma(\nu)$, and find (formally)

$$\frac{d^\nu x^m}{dx^\nu}=\frac{\Gamma(m+1)}{\Gamma(m-\nu+1)}x^{m-\nu} \overset{\nu=1/2}{\underset{m=1}{\longrightarrow}} \frac{\Gamma(2)}{\Gamma(3/2)}x^{1/2}=\frac{1}{\sqrt{\pi}/2}x^{1/2}.$$

Thus, generalizing the integer-number differential calculus to include fractional numbers $0<\nu<1$ one observes that the gamma function plays a crucial part.
In 1834 Liouville asked:

> If $d^n f(x)/dx^n=0$ has a complementary solution, why should not $d^\nu f(x)/dx^\nu=0$?

In 1847 Riemann generalized his theory of integration:

(i) classic, standard, integer-number integration theory
Consider a differential equation

$$y^{(m)}(t)=F(t) \qquad (m=1,2,3,\cdots \quad , \quad y^{(m)}=d^m y/dt^m)$$

and a given set of initial values

$$y^{(\ell)}(t=0)=y_\ell \qquad (0\le\ell\le m-1)\ .$$

If $F(t)$ is continuous over $[0,t]$, then one can integrate the initial value problem to obtain the unique solution

$$y(t)=\sum_{\ell=0}^{m-1} y_\ell t^\ell/\ell! + {}_0D_t^{-m}F(t)$$

where

$${}_0D_t^{-m}F(t):=\frac{1}{\Gamma(m)}\int_0^t d\tau(t-\tau)^{m-1}F(\tau)$$

defines *Riemann's classic* (integer-number) *integral operator.*

(ii) fractional generalization of Riemann's integral operator
If m is not an integer but any number $q \neq 0, -1, -2, \cdots$, then ${}_0D_t^{-q}$ may still be meaningful and defines for $q > 0$ the *Liouville-Riemann* (LR) *fractional integral operator*

$$ {}_0D_t^{-q} F(t) := \frac{1}{\Gamma(q)} \int_0^t d\tau (t-\tau)^{q-1} F(\tau) \qquad (q > 0) \tag{5}$$

(iii) fractional differential operator
The fractional differential operator ${}_0D_t^{\nu}$ for $\nu > 0$ is defined by

$$ {}_0D_t^{\nu} F(t) := \frac{d^n}{dx^n} ({}_0D_t^{\nu-n} F(t)) \qquad (\nu - n < 0) \tag{6}$$

i.e. within the context of the LR fractional calculus the operation "fractional differentiation" can be decomposed into a "fractional integration" ${}_0D_t^{-(n-\nu)}$ followed by an ordinary differentiation d^n/dx^n, where n is the least positive integer greater than ν.

We note that there are further definitions of fractional integral and differential operators like, for instance, a definition given by Weyl. But in what follows we are strictly dealing with the LR fractional calculus. In the next sections we will derive some fractional relaxation equations by starting out with the corresponding "classic" standard (integer-number) differential equations. Some caution, however, must be taken concerning the incorporation of initial values. Thus, for instance, a few investigators have represented fractional differential equations just by formally replacing the operator d/dt by d^{ν}/dt^{ν} in the original integer-number differential equation. Here, we present a consistent procedure of incorporating initial values into the fractional formulation by following the method already outlined for the derivation of fractional diffusion and wave equations [10] and for the fractional Boltzmann equation [11].

3. Fractional Relaxation Equations

3.1. Nutting Equation for Constant Stress

The empirical Nutting equation

$$\gamma(t) = \frac{\sigma}{c_0} t^k \tag{7}$$

relating stress (σ), strain (γ) and time(t) can be put in a form of a fractional differential equation. First we regard the case of constant stress condition $\sigma = const = \sigma_0$. If we fractionally differentiate (7) for $\sigma = \sigma_0$ with respect to time we come up with a fractional differential equation:

$$\begin{aligned}
\frac{d^\mu \gamma(t)}{dt^\mu} &= {}_0D_t^\mu \gamma(t) = \frac{d}{dt}\,{}_0D_t^{\mu-1}\gamma(t) = \frac{\sigma_0}{c_0}\frac{d}{dt}\,{}_0D_t^{\mu-1}t^k \\
&= \frac{\sigma_0}{c_0}\frac{d}{dt}\frac{1}{\Gamma(1-\mu)}\int_0^t (t-\tau)^{-\mu}\tau^k d\tau \\
&= \frac{\sigma_0}{c_0}\frac{d}{dt}\frac{1}{\Gamma(1-\mu)}\frac{\Gamma(1-\mu)\Gamma(k+1)}{\Gamma(k-\mu+2)}t^{k-\mu+1} \\
&= \frac{\sigma_0}{c_0}\frac{\Gamma(k+1)}{\Gamma(k-\mu+1)}t^{k-\mu} \qquad (0<\mu<1,\ k>-1)
\end{aligned}$$

Now, identifying the order μ of the fractional differential operator ${}_0D_t^\mu$ with the power-law exponent k, i.e. taking $k=\mu$, one obtains

$$\frac{d^\mu \gamma(t)}{dt^\mu} = \frac{\sigma_0}{c_0}\Gamma(\mu+1) = const.$$

Upon eliminating σ_0/c_0 via the Nutting equation (7) we find the following fractional differential equation for the strain $\gamma(t)$:

$$\frac{d^\mu \gamma(t)}{dt^\mu} = \Gamma(\mu+1)\gamma(t)t^{-\mu}$$

Of course, the solution of this equation is the Nutting equation for constant stress condition.

3.2. Nutting Equation for Constant Strain

For the constant strain condition $\gamma = \gamma_0 = const$ the Nutting equation reads

$$\sigma(t) = c_0\gamma_0 t^{-k} \tag{8}$$

delivering an inverse power-law decay for the stress function $\sigma(t)$. It is obvious that $\sigma(t)$ satisfies the homogenious (self-similar scaling) relation

$$\sigma(\lambda t) = \lambda^{-k}\sigma(t) \tag{9}$$

indicating that k might be interpreted as a similarity (fractal) dimension in the spirit of Mandelbrot [12]. Here we note that a general relation between Mandelbrot's definition of a fractal (or similarity) dimension and the fractional order μ of the fractional differential operator ${}_0D_t^\mu$ is not known. However, very recently [13] it has been shown that for a certain class of Lévy-type distribution functions such a relation does exist.

Taking fractional differentiation of (8) we find

$$\frac{d^\mu \sigma(t)}{dt^\mu} = c_0\gamma_0\frac{\Gamma(1-k)}{\Gamma(1-\mu-k)}t^{-\mu-k}, \quad 0<\mu<1,\ k<1$$

Again, identifying $k=\mu$ and eliminating $c_0\gamma_0$ via the Nutting equation (8) we obtain

$$\frac{d^\mu \sigma(t)}{dt^\mu} = \frac{\Gamma(1-\mu)}{\Gamma(1-2\mu)}\sigma(t)t^{-\mu}.$$

3.3. Exponential Relaxation Equation and its Fractional Generalization

We start with the standard exponential relaxation equation

$$\frac{df(t)}{dt} = -\lambda f(t) \tag{10}$$

and "transform" to the Riemann integral (as in Section 2, with $F(t) \to -\lambda f(t)$). The result is a fractional integral equation (FIE) ($0 < q \le 1$) with incorporated initial value $f_0 = f(t = 0)$:

$$f(t) = f_0 - \lambda \; {}_0D_t^{-q} f(t) \quad . \tag{11}$$

Inverting the FIE (11) one obtains the corresponding fractional differential equation (FDE) by applying the inverse operator ${}_0D_t^{\nu}$ with $0 < \nu < 1$ to Eq (11):

$${}_0D_t^{\nu} f(t) = {}_0D_t^{\nu} f_0 - \lambda \; {}_0D_t^{\nu} \; {}_0D_t^{-q} f(t), \quad {}_0D_t^{\nu} \; {}_0D_t^{-q} = {}_0D_t^{\nu-q}$$

Fractional derivative of a constant f_0 is (according to Eq (6))

$${}_0D_t^{\nu} f_0 = \frac{d}{dt} \; {}_0D_t^{\nu-1} f_0 = f_0 \frac{d}{dt} \frac{1}{\Gamma(1-\nu)} \int_0^t (t-\tau)^{-\nu} d\tau = \frac{f_0 t^{-\nu}}{\Gamma(1-\nu)} \tag{12}$$

Thus, for $\nu = q$ one finds

$${}_0D_t^{q} f(t) = \frac{f_0 t^{-q}}{\Gamma(1-q)} - \lambda f(t) \qquad (0 < q < 1) \quad , \tag{13}$$

which is the fractional differential equation (FDE) with incorporated initial value $f_0 = f(t = 0)$. In the limit $q \to 1$, Eq (13) is sent back to the original exponential relaxation equation (10), or more precisely to $df/dt = f_0 \delta(t) - \lambda f(t)$, if use will be made of $\lim_{q \to 1} t^{-q}/\Gamma(1-q) = \delta(t)$.

3.4. The Fractional Maxwell Equation

Following the procedure outlined in subsection 3.3. we now derive the fractional Maxwell equation by starting out with the standard equation (1)

$$\sigma(t) + \tau_0 \frac{d\sigma(t)}{dt} = \eta \frac{d\gamma(t)}{dt} \equiv \Phi(t) \tag{14}$$

which may be written as

$$\text{a)} \quad \eta \frac{d\gamma(t)}{dt} = \Phi(t) \quad , \qquad \text{b)} \quad \sigma(t) + \tau_0 \frac{d\sigma(t)}{dt} = \Phi(t) \; .$$

Applying Riemann's integration theory to a) and b), respectively, one gets

$$\begin{aligned} &\text{a')} \qquad \gamma(t) = \gamma_0 + (1/\eta) \; {}_0D_t^{-\mu} \Phi(t) \quad , \qquad (0 < \mu < 1) \\ &\text{b')} \qquad \sigma(t) = \sigma_0 - (1/\tau_0) \; {}_0D_t^{-q} \sigma(t) + (1/\tau_0) \; {}_0D_t^{-q} \Phi(t) \quad , \qquad (0 < q < 1) \end{aligned}$$

to be solved for the initial condition $\sigma(t=0)=\sigma_0=const$, $\gamma_0=const$. Now we apply the operator ${}_0D_t^{\mu}$ from the left to a') and obtain

$${}_0D_t^{\mu}\gamma(t) = {}_0D_t^{\mu}\gamma_0 + (1/\eta)\Phi(t) = \frac{\gamma_0 t^{-\mu}}{\Gamma(1-\mu)} + (1/\eta)\Phi(t) \; .$$

Here we eliminate $\Phi(t)$ and insert it into b'). The result is

$$\sigma(t) + \tau_0\, {}_0D_t^{q}\sigma(t) = \frac{\tau_0\sigma_0 t^{-q}}{\Gamma(1-q)} - \frac{\eta\gamma_0 t^{-\mu}}{\Gamma(1-\mu)} + \eta\, {}_0D_t^{\mu}\gamma(t) \tag{15}$$

representing the fractional version of Maxwell's rheological constitutive law with incorporated (finite) initial values $\sigma_0=\sigma(t=0)$ and $\gamma_0=const.$

Discussing stress relaxation experiments carried out under constant strain condition $\gamma=\gamma_0$ and for the initial condition $\sigma(t=0)=\sigma_0$ (finite constant), Eq (15) reduces (since for $\gamma(t)=\gamma_0$ the last two terms on the right hand side cancel) to

$$\sigma(t) + \tau_0\, {}_0D_t^{q}\sigma(t) = \frac{\tau_0\sigma_0 t^{-q}}{\Gamma(1-q)} \; . \tag{16}$$

This result looks quite similar to Eq (13) replacing there $f(t)\to\sigma(t)$ and $\lambda\to 1/\tau_0$. The solution of (16) can be found by going through Laplace and Mellin transform techniques [8, 10] and is given by

$$\sigma(t) = \sigma_0 \sum_{j=0}^{\infty} \frac{(-t^q/\tau_0)^j}{\Gamma(1+qj)} \quad , \qquad (0<q<1) \quad . \tag{17}$$

In the limit $q\to 1$ one just recovers the exponential solution. The solution (17) satisfies the initial value condition $\sigma(t=0)=\sigma_0$. For large t-values ($t\to\infty$) one finds from (17) asymptotically inverse power-law decay $\sigma(t)\sim t^{-q}$ (Nutting Eq (8)). In Figs. 2 and 3 we compare the result (17) with various experimental data sets and we find good agreement.

Solving the fractional Maxwell equation (15) for constant stress $\sigma=\sigma_0$ gives

$$\gamma(t) - \gamma_0 = \frac{\sigma_0 t^{\mu}}{\eta\Gamma(1+\mu)} \tag{18}$$

which is for the initial value $\gamma(t=0)=\gamma_0$ just the Nutting equation (7).

We remark that our method for deriving fractional relaxation equations, presented here, works well for all those initial value problems for which the initial value $\sigma(t=0)=\sigma_0$ is a finite constant (or zero). However, some investigators are interested in stress relaxation functions $\sigma(t)$ that diverge for $t\to 0$ ($\sigma(0)\to\infty$). Fractional differential equations for this sort of solutions have also been discussed in literature [8,14,15]. Concerning the Maxwell equation the starting point is again Eq (1) in which one replaces formally $d\sigma/dt \to {}_0D_t^{q}\sigma(t)$ and $d\gamma/dt \to {}_0D_t^{\mu}\gamma(t)$, respectively, leading to the fractional differential equation [14]

$$\sigma(t) + \tau_0\, {}_0D_t^{q}\sigma(t) = \eta\, {}_0D_t^{\mu}\gamma(t) \tag{19}$$

which can be transformed to a fractional integral equation by applying the fractional integral operator ${}_0D_t^{-q}$ from the left obtaining

$${}_0D_t^{-q}\sigma + \tau_0\, {}_0D_t^{-q}\, {}_0D_t^{q}\sigma = \eta\, {}_0D_t^{-q}\, {}_0D_t^{\mu}\gamma(t)$$

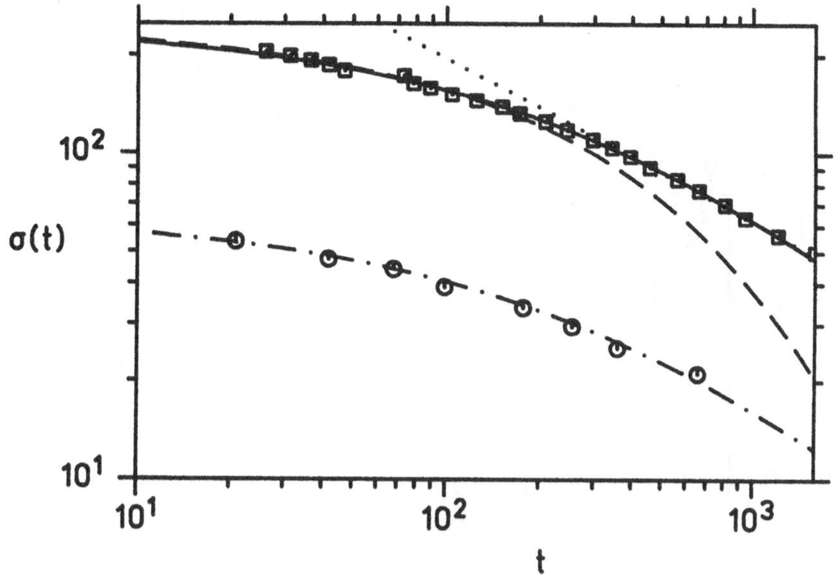

Fig. 2 Data points from Ref. [16] for two different initial conditions, (...) Nutting Eq (4) and (- - -) stretched exponential law (3) for the same parameters as in Fig. 1, (—) this theory, Eq (17), for $\sigma_0 = 250, \tau_0 = 30, q = 0.6$, (— · —·) this theory, Eq (17), for $\sigma_0 = 65, \tau_0 = 30, q = 0.6$.

where, ${}_0D_t^{-q}\, {}_0D_t^q \sigma(t) = \sigma(t) - c_1 t^{q-1}$. Considering again stress relaxation under constant strain condition $\gamma(t) = \gamma_0 = const$ we finally arrive at the fractional integral equation

$$\begin{aligned} {}_0D_t^{-q}\sigma(t) + \tau_0\sigma(t) - c_1\tau_0 t^{q-1} &= \frac{\eta\gamma_0}{\Gamma(1-\mu)}\, {}_0D_t^{-q} t^{-\mu} \\ &= \frac{\eta\gamma_0}{\Gamma(q-\mu+1)} t^{q-\mu} \quad . \end{aligned} \tag{20}$$

Here, c_1 plays the role of an integration constant which determines the singular behaviour of $\sigma(t)$ for $t \to 0$.

Taking $\gamma = \gamma_0$ (constant strain condition) in (19) we get the fractional differential equation (FDE)

$$\sigma(t) + \tau_0\, {}_0D_t^q \sigma(t) = \frac{\eta\gamma_0 t^{-\mu}}{\Gamma(1-\mu)} \quad . \tag{21}$$

The general solution of the homogeneous equation

$$\sigma_h(t) + \tau_0\, {}_0D_t^q \sigma_h(t) = 0$$

is given by [8]

$$\sigma_h(t) = c_0 t^{q-1} \sum_{j=0}^{\infty} \frac{(-t^q/\tau_0)^j}{\Gamma(qj+q)} \tag{22}$$

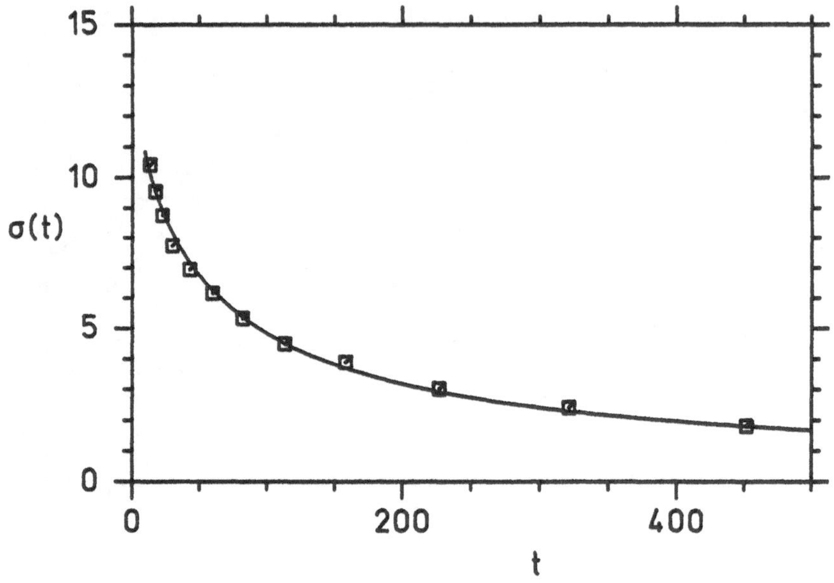

Fig. 3 Stress relaxation data for a plastic material from Ref. [7], (—) this theory, Eq (17), for the parameter choices $\sigma_0 = 14, \tau_0 = 18, q = 0.67$.

where c_0 is an arbitrary constant. Since $0 < q < 1$, this result diverges ($\sim t^{q-1}$) for $t \to 0$. The complete solution of the FDE (21) is directly obtained by Laplace and Mellin transform techniques and is given by

$$\begin{aligned}\sigma(t) = \; & \eta\gamma_0 \left(\frac{1}{\tau_0}\right)^{\mu/q} \sum_{j=0}^{\infty} \frac{(-1)^j}{\Gamma(qj+q-\mu+1)} \left(\frac{t}{\tau_0^{1/q}}\right)^{qj+q-\mu} \\ & + \frac{c_1}{\Gamma(q)} \left(\frac{1}{\tau_0}\right)^{1/q-1} \sum_{j=0}^{\infty} \frac{(-1)^j}{\Gamma(qj+q)} \left(\frac{t}{\tau_0^{1/q}}\right)^{qj+q-1} \end{aligned} \tag{23}$$

An appropriate choice of the integration constant $c_1 (= c_0\Gamma(q))$ shows that the second term on the right hand side (rhs) of (23) represents just the solution (22), $\sigma_h(t)$, of the homogeneous fractional differential equation, and the first part on the rhs of (23) is an inhomogeneous solution of (21) which is being discussed by Friedrich [15]. It follows that for $q = \mu$ the inhomogeneous part of the solution (23) becomes idential to (17) if we identify $\eta\gamma_0/\tau_0 = \sigma_0$.

4. Final Comments and Conclusions

We have discussed rheological behaviour of viscoelastic bodies in relation to two experimental situations: (i) the straining under constant stress, and (ii) the relaxation of stress under constant strain condition. The theoretical concept we developed is strictly based on the Liouville-Riemann fractional calculus. We have shown how the initial value problem for first-order differential equations can be generalized in a consistent way to the corresponding initial value problem for fractional differential equations. As a leading example we have studied Maxwell's constitutive law and its fractional generalization. The most general solutions of the fractional Maxwell equation have been presented for different initial value problems. We have analyzed experimental data sets which could not be interpreted by the standard Maxwell equation. However, these stress relaxation data could be fitted accurately (Fig. 2 and Fig. 3) over the measured data range by our solution of the fractional Maxwell equation, and thus giving support to the fractional theory.

One could object that the fractional calculus has a mathematical meaning at the best, but no physical interpretation whatever: so it seems to be unlikely to have a fractional differential in a basic physical law (differential equation). Indeed, Newtonian physics is so defined as to produce the integer-number differential concept which defines physical quantities like, for instance, the velocity to be the first differential of length with respect to time. Acceleration is defined as the second differential, and forces, energies etc., i.e. all physical properties defined within the context of Newtonian physics, have integral (and not fractional) indices of mass, length and time.

On the other hand, there is no doubt that the fractional calculus offers an extremely powerful mathematical technique for dealing with non-standard, unfamilar or even "pathological" functions which have been proposed in order to fit experimental observations. Power laws and Lévy (distribution) functions are prominent examples. Fox functions do probably represent the most general class of such functions. Wright functions, Mittag-Leffler functions, Lévy functions and other pathological functions are just subclasses of Fox functions. In a forthcoming paper [17] we are discussing Fox-function representations of the solutions of fractional relaxation equations.

Our conclusion is the following: even if the basic physical laws do not contain fractional differentials, it is still possible - in many cases - to transform the primary physical laws to "secondary" differential or integral equations of fractional order, which can be solved by applying Laplace and Mellin transform techniques. For instance, use has been made of the fractional calculus very recently [13] in order to formulate (and to solve) a fractional integral equation for a certain class of Lévy distribution functions being important for the physical interpretation of non-standard random walks (Lévy flights). Another example has been given by Douglas [18]. He transformed Feynman's path integration formulation of surface interacting polymers into an equivalent integral equation approach. An exact solution of the surface-interacting partition function has been obtained by using the fractional calculus.

References

[1] J.C. Maxwell, Phil. Trans. R. Soc. 157 (1867) 49

[2] R. Kohlrausch, Ann. Phys. 12 (1847) 393

[3] G. Williams and D.C. Watts, Trans. Faraday Soc. 66 (1970) 80

[4] P.G. Nutting, Proc. Amer. Soc. Test Mater 21 (1921) 1162

[5] A. Gemant, Physics 7 (1936) 311

[6] R.C.L. Bosworth, Nature 157 (1946) 447

[7] G.W. Scott Blair and J.E. Caffyn, Phil. Mag. 40 (1949) 80

[8] K.B. Oldham and J. Spanier, The Fractional Calculus, Academic Press 1974, New York

[9] B. Ross, Lecture Notes in Mathematics, Vol. 457, Springer Verlag (1975) 1 - 36

[10] W.R. Schneider and W. Wyss, J. Math. Phys. 30 (1989) 134

[11] T.F. Nonnenmacher and J.D.F. Nonnenmacher, Acta Physica Hungarica 66 (1989) 145

[12] B.B. Mandelbrot, The fractal geometry of nature, Freemann and Company 1983, New York

[13] T.F. Nonnenmacher, J. Phys. A: Math. Gen. 23 (1990) L 697

[14] R.L. Bagley and P.J. Torvik, J. Rheol. 30 (1986) 133

[15] Chr. Friedrich, this Proceedings

[16] R.K. Schofield and G.W. Scott Blair, Proc. Roy. Soc. A138 (1932) 707

[17] W.G. Glöckle and T.F. Nonnenmacher, to be published

[18] J.F. Douglas, Macromolecules 22 (1989) 1786

RELAXATION FUNCTIONS OF RHEOLOGICAL CONSTITUTIVE EQUATIONS WITH FRACTIONAL DERIVATIVES: THERMODYNAMICAL CONSTRAINTS

Christian Friedrich
Institut für Makromolekulare Chemie, Universität Freiburg i. Br.
Stefan - Meier - Straße 21, D-7800 Freiburg i. Br.

Summary: This paper deals with relaxation functions of rheological constitutive equations with fractional derivatives and with constraints on parameters contained in these constitutive equations. The constraints guarantee the consistency of the constitutive equations with thermodynamical principles like nonnegative rate of mechanical energy dissipation.

1. Intoduction

Gemant [1] and 10 years later Scott-Blair [2,3] were the first who described some dynamic phenomena like creep, stress relaxation or oscillatory shear of viscoelastic materials in terms of rheological constitutive equations with fractional derivatives. In his discussion the "Theory of Quasi - properties", Scott-Blair formulated the necessety to introduce such constitutive equations with fractional derivatives. His "Principle of Intermediacy" gave him the conviction that it is possible to express material behaviour laying between the Hookean solid and the Newtonian fluid in terms of derivatives laying between the zeroth order derivative of strain (the strain itself) and the first order derivative of strain (the strain rate). Scott-Blair determined the creep and relaxation functions of this easiest constitutive equation with fractional derivative.

The theory of fractional derivatives (better, derivatives of fractional order) is almost as old as the theory of the known calculus of integer order. A historical abstract ot this interesting development in the area of mathematics is given in [11].

Some time later, Slonimsky [4] generalized this typ of constitutive equation further and formulated models laying between the Hookean solid and the viscoelastic Kelvin-Voigt solid. A Generalized Function approach was used to describe fractional derivatives in a formal manner.

Smitt and de Vries [5] also calculated the relaxation function and other material functions arising in different rheological experiments. Like Scott-Blair, they found that the relaxation and retardation functions are of power law type for the very simple constitutive equations with fractional derivative.

Since the early 80's, US-rheologists started to generalize the fractional derivative approach in several directions. They successfully showed the connections between molecular theories and the empirical fractional calculus method to viscoelasticity [6], generalized this type of constitutive equation by introducing fractional derivative operators and looked for its behaviour in the frequency domain [7] and developed constitutive equations with fractional rates of deformation like fractional White-Metzner deformation rate tensors [8].

Bagley and Torvik [9] were the first who tried to calculate the complex modulus $G^*(\omega)$ and its components $G'(\omega)$ and $G''(\omega)$ as well as the relaxation function G(t) and the retardation function J(t) for a five-parameter model with two fractional derivatives of different orders. This model is called in this paper the fractional standard solid (FSS) includes the four-parameter fractional Maxwell model (FM). It's definition is given later. While it is easy to determine $G^*(\omega)$ (see also [5-7]), the determination of G(t) and J(t) was only possible in a numerical manner.These results are not instructive enough to get a feeling of its behaviour. Thermodynamic restrictions on parameters of the model were calculated on the basis of the functions $G'(\omega)$ and $G''(\omega)$ and, as will be shown, are restrictive.

Only after a way was found to solve the differential equations arising from fractional derivative models analytically [10], the questions of thrmodynamical admissibility can be answered with the help of restrictions concerning the relaxation or retardation functions.

The aim of this paper is to determine the thermodynamical constraints for the fractional Maxwll model, the fractional Standard Solid and other models on the basis of the relaxation function.

2. Fractional derivative models and its relaxation functions

In this part of the paper the relaxation function of the FM model will be developed with the help of the Laplace transform. The relaxation functions of the other models can be deduced in analogy.

First of all, the fractional Maxwell model will be introduced in an empirical manner, following Scott-Blair's "Principle of Intermediacy". Using a fractional derivative of order α instead of the first order stress derivative and a fractional derivative of order β instead of the first order strain derivative the following four-parameter Maxwell model with fractional derivatives is obtained:

$$\tau + \lambda^{\alpha} D^{\alpha}[\tau] = \lambda^{\beta} G_0 D^{\beta}[\gamma] \qquad 0<\alpha,\beta \leq 1 \qquad (1)$$

In Eq. (1), τ is the stress tensor, γ the deformation tensor, λ a time constant and G_0 the modulus. What concerns the definition of a fractional derivative D^{α} is refered to [11] and Appendix.

Eq. (1) can be transformed in an equation for the dimensionless relaxation function, g, if the following deformation history $\gamma_{12} = \gamma_0 h(t)$ is introduced, generating a shear stress component τ_{12}. γ_{12} is the shear component of the strain tensor, γ_0 is the jump height and h(t) the unit jump function. These conditions lead to the following equation:

$$g + D^{\alpha}[g] = D^{\beta}[1] = \frac{x^{-\beta}}{\Gamma(1-\beta)} \qquad x>0 \qquad (2)$$

where x = t/λ is the dimensionless time and g is defined in the following way: g = G(t)/G_0 and G(t) = τ_{12}/γ_0. If Eq. (2) is transformed in the Laplace domain (see Eq. (4) of Appendix), then Eq. (3) is obtained for the Laplace transform of the relaxation function L [g] = $\bar{g}$.

$$\bar{g} + y^{\alpha}\bar{g} - c = y^{\beta-1} \qquad (3)$$

In this equation 'y' is the dimensionless Laplace variable. The inverse Laplace transform [10] yields Eq. (4) under the assumption that the function g at the abscissa x = 0 can have a finite value (this means c = 0).

$$g(t) = (t/\lambda)^{\alpha-\beta} \sum_{k=0}^{\infty} \frac{(-1)^k}{\Gamma(\alpha k+\nu)} (t/\lambda)^{\alpha k} \qquad x = t/\lambda, \quad \nu = \alpha-\beta+1 \qquad (4)$$

The assumption c = 0 will be used throughout this paper without loss of generality.

It can be seen that the relaxation function consists of two parts: the power law part and a infinite series known under the name Generalized Mittag-Leffler Function (GMLF). The behaviour of this function is explained in [10,14] as are some details in the Appendix. Eq. (4) can be presented as follows if the behaviour (A5) is taken into account and a normalizing factor $\Gamma(\nu)$ is introduced (this normalization is equivalent to a decomposition of the modulus G_0 in following way: $G_0 = G_{00}/\Gamma(\nu)$).

$$g(x) = \quad \Gamma(\nu) \quad x^{\alpha-\beta} \quad E_{\alpha,\nu}(-x^{\alpha}) \qquad \begin{array}{l} x = t/\lambda \\ \nu=\alpha-\beta+1 \end{array} \tag{5}$$

Special cases of this relaxation function are possible for fixed parameters.

$$\begin{aligned}
&\alpha = \beta = 1 \\
&\qquad g = \exp(-x) \\
&\alpha = 1/2, \beta = 1 \\
&\qquad g = x^{-1/2}[1 - \Gamma(1/2)\, x^{1/2} \exp(x)\ \mathrm{erfc}(x^{1/2})\ ] \\
&\alpha = \beta = 1/2 \\
&\qquad g = \exp(x)\ \mathrm{erfc}(x^{1/2})
\end{aligned} \tag{6}$$

These solutions are also given in [9] or [12].
Before considering the compatibility of Eq. (6) with thermodynamics, the behaviour of g(x) at large times is explained. It is seen from the asympthodic expansion for GMLF according to Eqs. (A6) and (A7), that in the cases α and/or ν smaller than 1, the GMLF is no longer of exponential type. It displays power law behaviour. These asympthodic expansions yield in connection with Eq. (5) the Eqs. (7) and (8).

$$g(x) \quad \propto \quad \frac{\Gamma(1+\alpha)}{\Gamma(1-\alpha)}\, x^{-(1+\alpha)} \qquad x \to \infty \qquad \text{for } \beta = 1 \tag{7}$$

$$g(x) \quad \propto \quad \frac{\Gamma(\nu)}{\Gamma(1-\beta)}\, x^{-\beta} \qquad x \to \infty \qquad \text{for } 0 < \beta < 1 \tag{8}$$

Now the fractional Standard Solid model will be considered. This is a FM model with an additional term of zeroth order on the right side (it is the deformation only) modelling the solid-like behaviour with the help of the dimensionless equilibrium modulus $g_{\infty} = G_{\infty}/G_0$. If is used the same procedure as in the case of FMM the following equation is obtained.

$$g_{SS} = g_M + g_{\infty}[1 - E_{\alpha,1}(-x^{\alpha})] \tag{9}$$

g_{SS} is the dimensionless relaxation function of this model which can be presented as a sum of the relaxation function of the FMM, g_M, according to Eq. (5) and an additional term with g_{∞} as the prefactor. For this function analogous results for their behaviour at long times can beoutlined.

3. Thermodynamic compatibility

If a rheological constitutive equation is formulated, the thermodynamic behaviour of this "model material" is also prescribed. Because thermodynamics impose restrictions on physically realizable processes by the second law, it is necessary to consider the thermodynamic compatibility of a given rheological constitutive equation. That means, we have to look after conditions under which the constitutive equation in general or the parameter functions or parameters in special in it guarantee nonnegative rate of mechanical energy dissipation δ_m. This can be formulated [13] by the following equation

$$\delta_m = \Phi - \rho \dot{f} \geq 0, \qquad \text{with } \Phi = \tau_{12} \dot{\gamma}_{12}{}^2 \tag{10}$$

where Φ is the stress power which is in the case of relaxation after a step - strain experiment equal to zero, f the rate of change of the free energy and ρ the density. It can be seen that in the case of relaxation, the demand for nonnegative mechanical dissipation rate is equivalent to the nonpositive rate of free energy. This means that relaxation is associated with the release of free energy stored durind the jump.

Investigations concerning the energy storage during harmonic excitations [9] or arbitrary deformation histories [15] demand that the relaxation function should be a positive definite function. This means in detail [15] that

- $G(t) \geq 0$ for all t and
- $G(t)$ is monotonic nonincreasing.

If Eq. (5) is to be analyzed for thermodynamic compatibility, we must demand that both terms of the relaxation function, the power law term and the GMLF term, respond to these restrictions. For the power law term follows: $\lambda > 0$ and $\alpha - \beta \leq 0$. For the GMLF can be deduced (without proof) from the results of monotonicity of MLF [14,16] and on the basis of Eqs. (A6) and (A7) that this function is nonincreasing if $G_0 \geq 0, \lambda > 0$ and $\beta \geq \alpha > 0$.
For the relaxation function follows:

$$G_0 \geq 0, \ \lambda > 0, \ \beta \geq \alpha > 0 \tag{11}$$

The analysis of Eq. (9) is more complicated. For the Maxwellian part of this equation the conditions (11) are valid. The second part of Eq. (9) is a nondecreasing function. So.it is obvious that Eq. (9) violates the second law of thermodynamics for the parameter combination (11). This fact is also displayed in Fig. 3. In the long time range we have the following asympthodic behaviour confirming the violation.

$$g_{SS} - g_\infty = \frac{\Gamma(\nu)}{\Gamma(1-\beta)} x^{-\beta} - g_\infty \frac{\Gamma(1)}{\Gamma(1-\alpha)} x^{-\alpha} \tag{12}$$

From this equation it is seen that only the parameter combination $\alpha = \beta$ leads to a behaviour not violating the thermodynamic constraints. Under these more restrictive conditions, the relaxation function of the fractional standard solid model obeys Eq. (13).

$$g_{SS} - g_\infty = (1 - g_\infty) E_{\alpha,1}(-x^\alpha), \quad \alpha = \beta \tag{13}$$

Does a fractional model with a g_∞- term exist which is compatible to thermodynamics under the same conditions like given by correlations (11)? The answer is yes if the term D^α [1] is added on the right side in the FSS model. This equation reads then as follows:

$$g + D^\alpha[g] = g_\infty(1 + D^\alpha[1]) + D^\beta[1] \tag{14}$$

That means that a fractional derivative of strain must be added with an order of differentiation which corresponds to that of the stress derivative. The relaxation function belonging to Eq. (14), g_{SSm} (modified FSS model), and compatible with thermodynamics for the parameter combination (11) is the following Eq. (15):

$$g_{FSSm} - g_\infty = x^{\alpha-\beta} E_{\alpha,\nu}(-x^\alpha) \tag{15}$$

By this, the thermodynamic constraints concerning the simplest rheological constitutive equations with more than one fractional derivative are given. While the determination of the constraints of the parameters G_0, G_∞ and λ is trivial the determination of constraints concerning the parameters α and β is only possible on the basis of an analysis of GMLF.

4. Simulation of material properties

In this part of the paper the material behaviour given by different relaxation functions will be simulated and presented graphically.

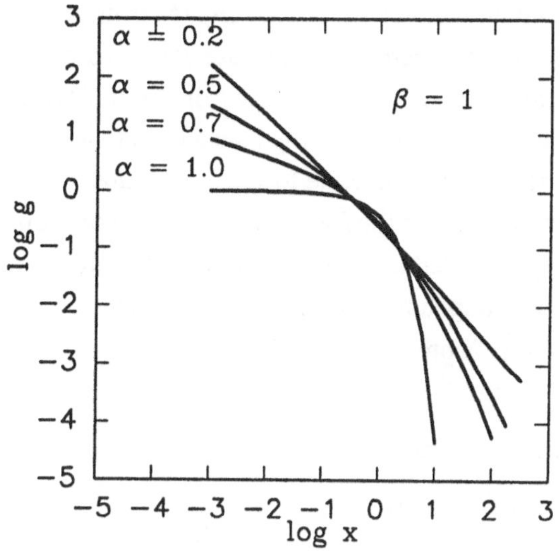

Fig. 1: Dimensionless relaxation function, g, of the fractional Maxwell model vs. dimension less time for different orders, α, of the stress derivative.The strain derivative is of order $\beta = 1$

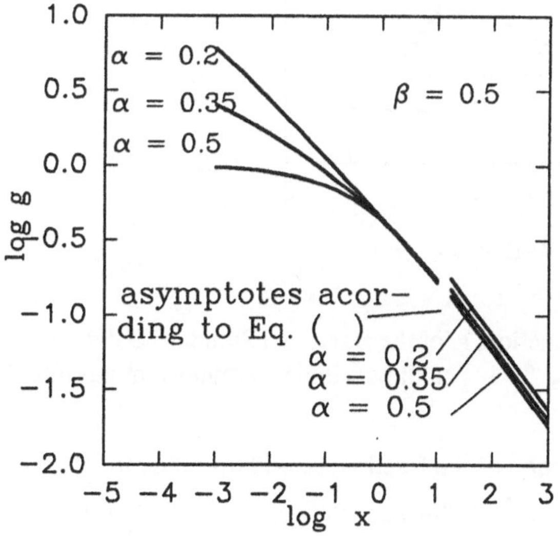

Fig.2: Dimensionless relaxation function, g, of the FMM vs. dimensionless time for different orders, α, of the stress derivative. The strain derivative is of order $\beta = 1/2$.

The Fig.1 gives a clear impression about the type change of Eq. (5) from exponential (α=β= 1) to power law ($0 < \alpha,\beta < 1$). It is seen that in the case $\beta = 1$ (ordinary first order derivative of strain = strain rate) the variation of α influences on the short time behaviour as well as on the long time tail. The asympthodes obey the laws given in the figure . In this relation it should be mentioned that it is more accurate to speak of double power law.
The Figure 2 corresponds to the case $\beta < 1$ for the fractional Maxwell model and shows a different behaviour. It is immediatly seen that in this case α influences only on the short time taile. At long times, the order of decay depends only on β as indicated by the formula in the figure. The double power law character is preserved.

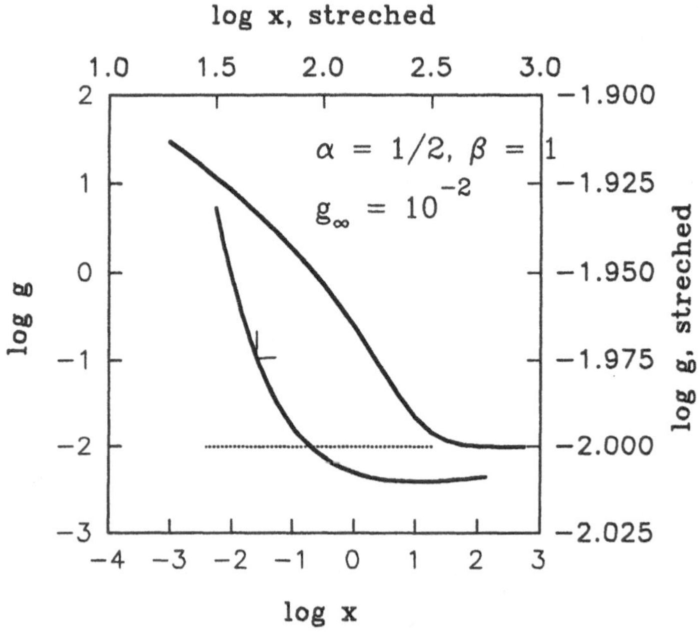

Fig.3: Dimensionless relaxation function, g, of the fractional Standard Solid model vs. dimensionless time for α=1/2, β=1. The dimensionless equilibrium modulus is g_∞=0.01.

Figure 3 shows the relaxation function of the fractional standard solid model according to Eq. (9) for the parameters $\alpha = 1/2$, $\beta = 1$ and $g_\infty = 0.01$. If this function is considered in the same scale as the other relaxation function in Figs. 1 and 2 no violation of monotonicity is seen. Only in a finer scale can be seen that approaching a distinct time the strong monotonic character is lost. This is the consequence of the not right choosen parameters α and β and therefore the rate of mechanical energy dissipation is no longer positive in the merked time range.

5. Appendix

5.1 Definition of fractional integration and differentiation

The fractional integration can be introduced as follows

$$I^{\alpha}[f] = \frac{1}{\Gamma(\alpha)} \int_a^x \frac{f(y)}{(x-y)^{1-\alpha}} dy \qquad 0 < \alpha \leq 1 \qquad (A1)$$

The definition of the operation of fractional differentiation, i. e., the inverse operation to the fractional integration, can be arrived at in several ways. Here we will follow Ref. [11] and expand the range of α to negative values and therefore introduce the operator of fractional differentiation D^{β}.

$$D^{\beta} = I^{-\beta} = D^{n}\left[I^{n-\beta}[f] \right] \qquad n > \beta > 0 \qquad (A2)$$

D^n is the usual n-th order derivative. An equivalent presentation of this formula is the following:

$$D^{\beta}[f] = \sum_{k=0}^{n-1} \frac{(x-a)^{k-\beta}}{\Gamma(k-\beta+1)} f^{(k)}(a) + \frac{1}{\Gamma(n-\beta)} \int_a^x \frac{f^{(n)}(y)}{(x-y)^{1-(n-\beta)}} dy \qquad (A3)$$

In this equation $f^{(n)}$ designates $D^n[f]$. Throughout this paper a = 0 is used as the lower limit of integration

In addition, it is necessary to introduce the Laplace transform $L[f] = \bar{f}$ of a fractional derivative . This is given by Eq. (A4)

$$L\{D^{\beta}[f]\} = s^{\beta} L\{f\} - \sum_{k=0}^{n-1} s^{k} D^{-(1-\beta)-k}[f]\Big|_{t=0} \qquad (A4)$$

where s is the Laplace variable.

5.2 Definition of the Mittag-Leffler Function and Generalized Mittag-Leffler Function

The Mittag-Leffler Function $E_\alpha(x)$ is a special case of Generalized Mittag-Leffler Function $E_{\alpha,\beta}(x)$ for $\beta = 1$.
From the following definition [14]

$$E_{\alpha\beta}(x) = \sum_{k=0}^{\infty} \frac{(x)^k}{\Gamma(\alpha k + \beta)} \qquad \alpha, \beta > 0 \tag{A5}$$

can be seen that only in the case $x \leq 0$ this function decreases monotonically. In this case the following asympthodic expansions for large times can be given [10,14,16]:

$$E_{\alpha,\alpha}(-x) \propto \frac{\alpha}{\Gamma(1-\alpha)} x^{-2} \qquad \alpha \neq 1 \tag{A6}$$

$$E_{\alpha,\nu}(-x) \propto \frac{1}{\Gamma(\nu-\alpha)} x^{-1} \qquad \nu \neq \alpha \tag{A7}$$

References

1. Gemant A (1936) Physics 7: 311 - 317
2. Scott-Blair GW, Veinoglou BC, Caffyn JE (1947) Proc Roy Soc Ser A 189: 69 - 87
3. Scott-Blair GW, Caffyn JE (1949) Phil Mag 40: 80 - 94
4. Slonimsky GL (1967) J Polym Sci C 16: 1667 - 1672
5. Smit W, de Fries H (1970) Rheol Acta 9: 525 - 534
6. Bagley RL, Torvik PJ (1983) J Rheol 27: 201 - 210
7. Rogers L (1983) J Rheol 27: 351 - 372
8. Vanarsdale WE (1985) J Rheol 29: 851 - 857
9. Bagley RL, Torvik PJ (1986) J Rheol 30: 133 - 155
10. Friedrich C to appear in Rheol Acta
11. Oldham KB, Spanier J (1974) The Fractional Calculus. Academic Press, New York and London
12. Tschoegl NW (1989) The Phenomenological Theory of Linear Viscoelastic Behaviour - An Introduction. Springer - Verlag, Berlin, Heidelberg, New York
13. Astarita G, Marrucci G (1974) Principles of Non - Newtoanian Fluid Mechanics. McGraw-Hill, London and New York
14. Bateman H (1955) Higher Transcendental Functions, Volume III. McGraw Hill Book Company, New York
15. Rabotnov JN (1977) Elements of the Mechanics of Memorysolids. Nauka, Moscow (in Russian)
16. Pollard H (1948) Bull Am Math Soc 54: 1115 - 1121

A SIMPLE ONE DIMENSIONAL MODEL SHOWING GLASS LIKE DYNAMICAL BEHAVIOR

J.J. Brey and M.J. Ruiz-Montero
Física Teórica. Facultad de Física. Universidad de Sevilla
41080 Sevilla. Spain

1. Introduction

In this paper we discuss a simple model showing many of the dynamical properties of supercooled fluids and glasses near the glass transition. Such systems display a rich variety of phenomena [1,2], including strongly nonexponential decay in response to infinitesimal perturbations, nonlinear response to finite perturbations, non-Arrhenius temperature dependence of relaxation times, and hysteresis effects. It is also characteristic the cooling rate dependence of low temperature properties.

In spite of the great deal of work carried out during the last years, the above properties are far from being well understood, and little is known about the physical processes governing the behavior of glasses and supercooled fluids. Lacking a solid general formalism, it seems worth while to study simple specific models that mimic the dynamics of real fluids near the glass transition. An interesting goal is to identify what is required for a model to describe glassy dynamical properties.

Our model is based in the one introduced by Bell [3] to describe some of the equilibrium properties of liquid water. It is a lattice model for which only the statics was specified. Very recently [4], we have defined a dynamics for it, in terms of a master equation with transition probabilities obeying detailed balance. The equilibrium properties of the model can be computed analytically quite easily, while the kinetic ones are provided with limited effort by Monte Carlo simulation. In spite of its simplicity, we will see that the model presents many of the characteristic properties of supercooled liquids and glasses.

2. Description of the model.

2.1 Statics

We consider a one dimensional lattice of N equally spaced sites. Each site can be either occupied by a particle or it can be empty. The number of particles in the lattice is M. There are two kinds of interactions among particles. Two nearest neighbor particles interact with an energy $-\varepsilon$. Two particles separated by a single empty lattice site may be bonded with energy $-(\varepsilon+\omega)$. Both, ε and ω are taken to be positive.

In order to visualize the states of the system it is useful to introduce the following representation [3]. A site between two bonded particles will be considered as occupied by a bond (b). The other empty sites are considered as holes (h). Representing the particles by m, a possible configuration of the system is

...h m m h h m b m h h m b m h...

$-\varepsilon$ $\quad -(\varepsilon+\omega)$ $\quad -(\varepsilon+\omega)$

We have also indicated the interaction energies. Assuming periodic boundary conditions, the energy of a given state can be expressed as

$$E(N,M,Z,B) = -(M-Z)\,\varepsilon - B\,\omega \tag{1}$$

where B is the number of bonds and 2Z is the number of particle-holes contacts.

The equilibrium thermodynamical properties of the system can be obtained in different ways, for instance by constructing the isothermal-isobaric partition function [3,4], or by using the transfer matrix method [3]. In the limit of M going to infinity, one gets the following expression for the Gibbs free energy G:

$$G = Mk_BT \ln \frac{J}{ADJ^{-1}+A+(J-1)^{-1}} \tag{2}$$

with $A=e^{\beta\varepsilon}$, $D=e^{\beta\omega}$, $J=e^{\beta pl}$ and $\beta=(k_BT)^{-1}$. Here k_B is the Boltzmann constant, p is the pressure, l is the distance between neighboring sites, and T is the temperature.

All the equilibrium properties can now be derived from Eq. (2). Their expressions will not be given here. It must be noticed that, since we are dealing with a one-dimensional model with short range interactions the system does not present any equilibrium phase transition.

The transfer matrix method also allows the calculation of equilibrium correlations. For instance, we can define a site variable α_j by:

$\alpha_j=0$ if site j is occupied by a bond or a hole
$\alpha_j=1$ if site j is occupied by a particle

Then, the following expression is obtained for the correlation function of α [3,5]:

$$c_l \equiv <(\alpha_j-<\alpha_j>)(\alpha_{j+l}-<\alpha_{j+l}>)>$$
$$= \frac{1}{J^l}\;\frac{M}{N}\,(K^l)_{11} - \left(\frac{M}{N}\right)^2 \qquad (3)$$

where the angular brackets denote equilibrium average, and K is the transfer matrix

$$K=\begin{pmatrix} XA & X^{1/2} & (XAD)^{1/2} \\ X^{1/2} & 1 & 0 \\ (XAD)^{1/2} & 0 & 0 \end{pmatrix} \qquad (4)$$

with

$$X=\frac{J^2(J-1)}{A(J-1)(J+D)-J} \qquad (5)$$

As an example, we have plotted in Figures 1 and 2 the correlation functions for $k_BT=400$ and $k_BT=200$, respectively. In both cases we have taken $\varepsilon=200$, $\omega=500$ and N/M=2. These are also the values used in the Monte Carlo simulation to be presented later on. As expected, the correlation length is seen to increase as the temperature decreases. Of course, it tends to infinity as T goes to zero, reflecting the fact that the system is reaching a perfect crystalline structure. For all the conditions considered in our simulation the correlation length is small as compared with the length of the system.

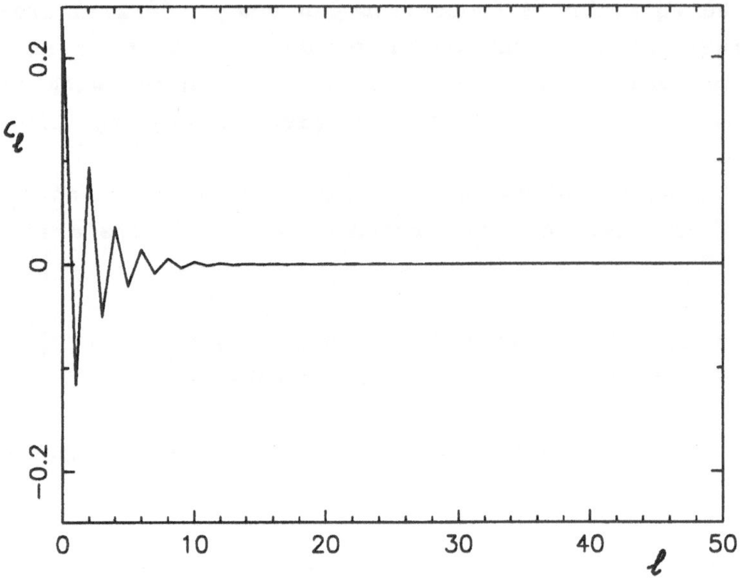

Fig. 1. *Correlation function for $k_B T$=400, ε=200, ω=500, and N/M=2. All the energies are measured in arbitrary units.*

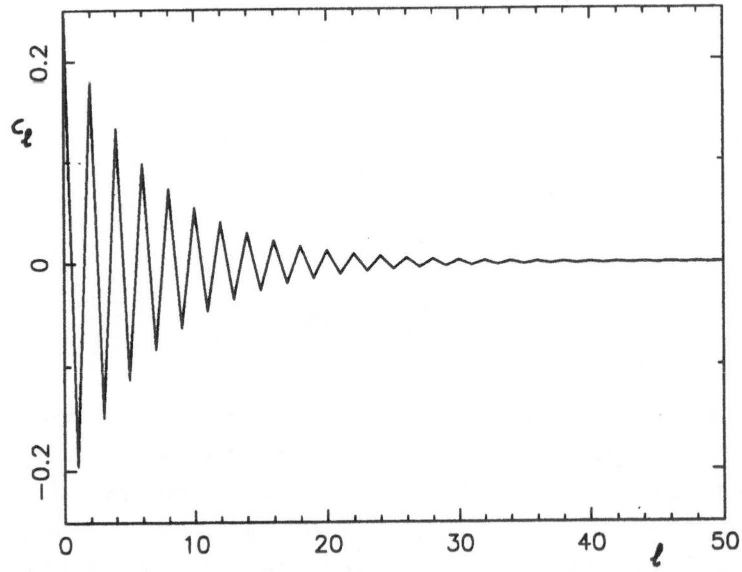

Fig. 2. *The same as Fig. 1, but $k_B T$=200.*

2.2 Dynamics

The dynamics of the system is assumed to be governed by a master equation for the probability $P(\underline{a},t)$ of finding the system in a given configuration $\underline{a}$ at time t:

$$\frac{\partial}{\partial t} P(\underline{a},t) = \sum_{\underline{a}'} \left[W(\underline{a}' \to \underline{a})\ P(\underline{a}',t) - W(\underline{a} \to \underline{a}') P(\underline{a},t) \right] \tag{6}$$

where $W(\underline{a} \to \underline{a}')$ is the transition rate from the state $\underline{a}$ to the state $\underline{a}'$, and the sum extends over all the possible states of the lattice.

We consider a fixed number of both particles and sites. The allowed transitions consist in the creation or destruction of a bond, and in the motion of a particle to a next empty site. A particle can not jump to a site that is occupied by a bond. Before jumping, the bond must be destroyed. Finally, two particles can form a bond only if there is a hole between them. For the allowed transitions the specific form chosen for the transition rate is

$$W(\underline{a} \to \underline{a}') = \frac{1}{2\tau_o} \left[1 - \tanh \frac{\Delta E}{2k_B T} \right] . \tag{7}$$

Here $\Delta E = E(\underline{a}') - E(\underline{a})$, and τ_o is a constant fixing the natural time unit of our system. The transition probabilities defined by Eq. (7) verify the detailed balance condition [6]. Besides, the associated Markov process is irreducible, in the sense that every state can be reached from every other state. Therefore [7], any arbitrary initial distribution of configurations will approach in time the canonical equilibrium distribution

$$P(\underline{a}) \propto e^{-\beta E(\underline{a})} \tag{8}$$

with $E(\underline{a})$ given by Eq. (1).

The mechanism for relaxation of the model at low temperatures is easily understood. At such temperatures there are small populations of holes and particle-particle contacts. In order to create a new bond, two particles must be previously separated by a hole. For most of the configurations, this implies the migration of holes and particles through the system. In this process, bonds must be destroyed, and because bonds are energetically favored, the relaxation of the system is hindered. We want to point out that we will see that the above mechanism does not reduce to simple diffusion, leading to an Arrhenius temperature dependence of the average relaxation time, as it is the case for the one-spin facilitated model [8] introduced by Fredrickson and Andersen [9,10].

3. Dynamical properties

We have investigated some dynamical properties of the model through Monte Carlo simulations in a lattice of 200 sites and 100 particles. For all the temperatures and times relevant in this study we have checked that the results do not show significant finite size effects. This has been done by comparing with the exact analytical expressions for the equilibrium properties (valid in the limit of an infinite system), and also by changing the size of the system in the simulation. The details of the algorithm have been given elsewhere [4], and they will not be reproduced here.

3.1 Linear response

The linear response in energy to an infinitesimal long wavelength perturbation can be characterized by the function

$$\phi(t)= \frac{\langle E(t)E(0)\rangle-\langle E\rangle^2}{\langle E^2\rangle-\langle E\rangle^2} , \tag{9}$$

that is defined such that $\phi(0)=1$. The behavior of $\phi(t)$ has been investigated in the range $200<k_BT<700$, and in all the cases a monotonous decay towards zero has been obtained. Fig. 3 shows the results for $k_BT=280$.

The simulation data where fitted to a Kohlrausch-Williams-Watts (KWW) expression

$$\phi(t) = \exp\left[- \left(\frac{t}{\tau_w} \right)^{\gamma} \right] ,$$

(10)

where τ_w and γ are adjustable parameters. The fit was carried out by considering only the data in the interval $0.03 \le \phi(t) \le 0.3$. The solid line in Fig. 3 is the best KWW fit for that case, and it corresponds to τ_w=43.3 and γ=0.37. This value of γ indicates a broad spectrum of relaxation times, far from a pure exponential behavior. Besides, γ is seen to decrease as the temperature is lowered.

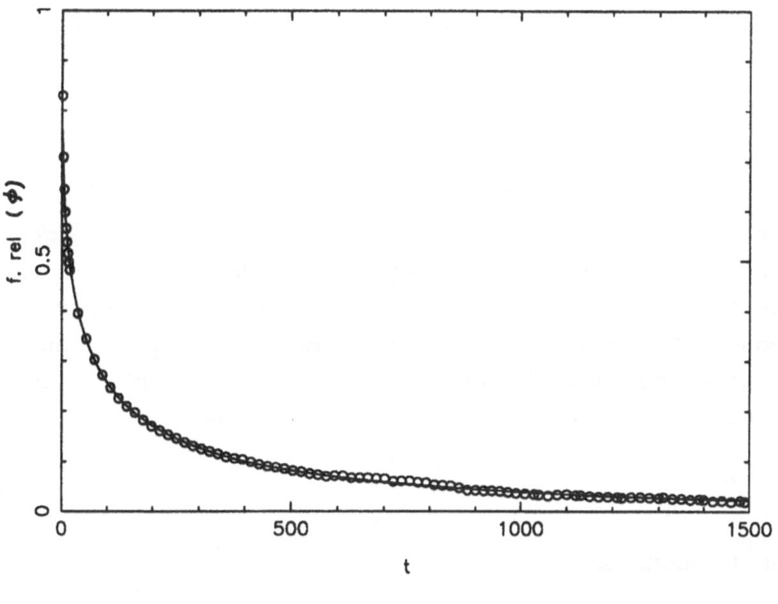

Fig.3. *Linear response function for k_BT=280. The parameters of the model are the same as in Fig. 1. The solid line is the best KWW fit.*

As said in the Introduction, nonexponential linear relaxation functions are characteristic of viscous fluids and glasses. Also , a peculiarity of these systems is the temperature dependence of γ, that is usually referred to as the lack of "thermorheological simplicity", to indicate that the temperature dependence of ϕ can not be reduced to a single time scaling.

The time τ_w also depends on the temperature, and so does the average relaxation time defined by

$$\overline{\tau} = \int_0^\infty dt\, \phi(t) \quad . \tag{11}$$

The results for $\overline{\tau}$ can be accurately fitted [11] with the Vogel-Fulcher equation

$$\overline{\tau} = \exp\left(\frac{E_o}{T-T_o} \right) \tag{12}$$

in all the temperature range considered. The values obtained for the adjustable parameters are E_o=1073 and $k_B T_o$=78. It follows that the apparent activation energy is clearly a decreasing function of the temperature. This behavior also agrees qualitatively with the one observed in viscous fluids. On the other hand, the Adam-Gibbs expression

$$\overline{\tau} = \exp\left(\frac{E_1}{T\, S_c} \right) \quad , \tag{13}$$

where S_c is the equilibrium configurational entropy, is not verified by our model. This is not surprising, since the entropy is determined by the statics of the model, while the average relaxation time depends on the chosen dynamics. When using a master equation description, there are many possible dynamics that are compatible with a given statics, even though detailed balance is required [6].

3.2 Nonlinear response

The time-dependent response of the model following an instantaneous quench has also been investigated. We have performed such experiments starting from the configuration with a hole between every two particles, i.e.,

...m h m h m h m h m h m h...

It formally corresponds to the equilibrium state of the system at $\beta=-\infty$.

After quenching to a finite temperature, we have monitored the energy as a function of time, and obtained the (nonlinear) relaxation function

$$\psi(t) = 1 - \frac{\overline{E}(t)}{\langle E \rangle_T} \;, \tag{14}$$

where $\langle E \rangle_T$ is the equilibrium average energy at the final temperature T, and $\overline{E}$ is the (nonequilibrium) average energy at time t. It is $\psi(0)=1$, and $\psi(\infty)=0$.

At all the quenching temperatures investigated ($50 \le k_B T \le 1000$), $\psi(t)$ decays to zero for sufficiently long times. Nevertheless, for high values of T ($k_B T>150$) the decay is not monotonous. An example, corresponding to $k_B T=1000$, is shown in figure 4. There, it is seen that the system overshoots the equilibrium value, and $\psi(t)$ becomes negative. The final relaxation takes place through values of the average energy that are smaller than the equilibrium value corresponding to the final temperature.

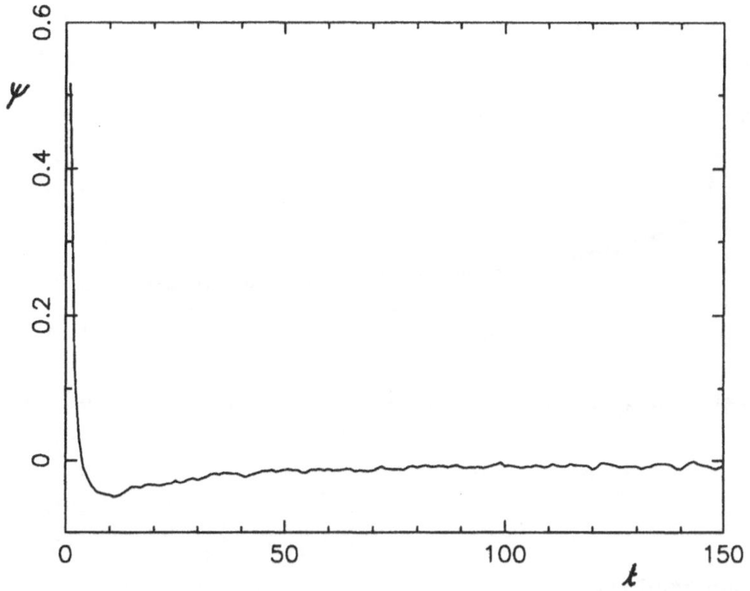

Fig. 4. *Nonlinear response function for $k_B T=1000$, corresponding to an instantaneous quench from $\beta=-\infty$.*

For low temperature quenches ($k_B T<150$), the decay is monotonous, although highly nonexponential. We have found [11] that the data can be again well fitted with a KWW expression

$$\psi(t)= \exp\left(-\frac{t}{\tau_q}\right)^{\gamma'} .$$

(15)

Figure 5 shows the simulation data for $k_BT=120$ and the fit with $\tau_q=5.3$ and $\gamma'=0.17$. A similar result was found by Fredrickson and Brawer [13] for the facilitated Ising model. In reference [11] the relation between the linear and nonlinear relaxations in our model is analyzed in terms of the Narayanaswamy theory [14,1].

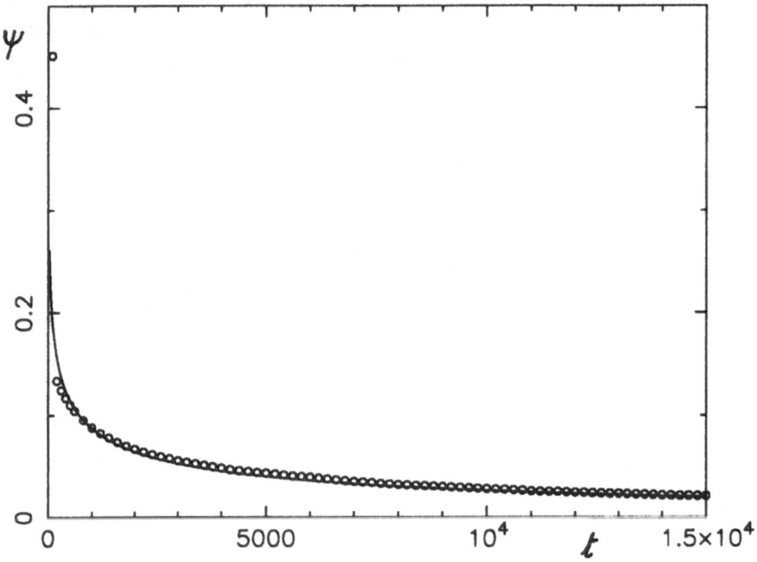

Fig.5. *The same as Fig. 4, but for $k_BT=120$. The solid line is the best KWW fit.*

3.3 Continuous cooling

In another set of experiments, we have cooled the system with a constant cooling rate

$$r= \frac{d}{dt} (k_BT) ,$$

(16)

starting from an equilibrium distribution of states. We assumed that the master equation (6) also holds when dealing with a time dependent temperature.

Fig. 6 shows the evolution of the average energy per particle for r=1, 0.1 and 0.01. Also plotted is the equilibrium value of the average energy as a function of the temperature. At high temperatures, the evolution of the energy follows the equilibrium curve, but the energy departs from its equilibrium value when sufficiently low temperatures are reached. The temperature at which the departure takes place decreases as the cooling rate decreases. For very low temperatures the energy reaches a value that is practically constant, and it is not affected by later cooling. The difference between this value of the energy and the energy of the ground state is the so called residual energy, and it is a function of the cooling rate. More precisely, the residual energy becomes smaller as the cooling rate r is decreased.

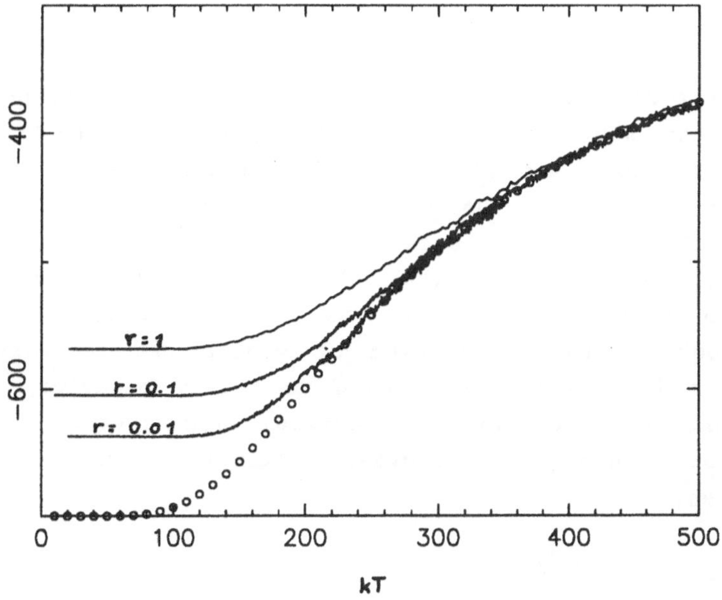

Fig. 6. *Evolution of the average energy per particle for constant cooling rates r=1, 0.1, 0.01. The circles are the equilibrium values of the energy.*

From the above discussion it follows that our model presents a phenomenon that is similar to the laboratory glass transition observed in real glasses [2]. Since our model has no thermodynamical or kinetic phase transition, it follows that the existence of an underlying transition is not a necessary condition for a system to show glass like behavior. This is in agreement with the general ideas expressed by Fredrickson [2], following the discussion of several models proposed for the liquid-glass transition.

An interesting problem that has raised some discussion in the last years is the functional dependence of the residual properties on the cooling rate [15,16], in the limit of very small cooling rates. The residual energy of our model can be fitted quite well to the two most used laws, namely a logarithmic and a power law dependence [4], in certain ranges of the cooling rates. No clear conclusion can be reached about which of the two laws describes best the dependence of the residual energy on the cooling rate.

4. Discussion

The one dimensional bonded fluid discussed in this paper exhibits many of the kinetic properties that are characteristic of real supercooled fluids and glasses. In particular, we have investigated the linear response, the nonlinear response to an instantaneous quench, and the behavior under continuous cooling. In spite of its simplicity, that allows rather efficient computer simulations, the relaxation of the model turns out to be quite complex.

We believe that this kind of models can be very useful to investigate the mechanisms leading to glassy behavior, and also to shed light on the way of deriving macroscopic equations for supercooled fluids and glasses, starting from a microscopic description.

Acknowledgements

We acknowledge financial support from DGICYT Project PB89-0618 (Spain).

References

1.- G.W. Scherer, *Relaxation in Glasses and Composites*, Wiley, New York, 1986.

2.- G.H. Fredrickson, Ann. Rev. Phys. Chem. 39, 181 (1988)

3.- G. H. Bell, J. Math. Phys. 10, 1753 (1968)

4.- J.J. Brey and M.J. Ruiz-Montero, Phys. Rev. B (1990)

5.- J.J. Brey and M.J. Ruiz-Montero, unpublished.

6.- K. Kawasaki, in *Phase Transitions and Critical Phenomena*, vol. 2, C. Domb and M.S. Green eds., Academic Press, London, 1972, p. 443.

7.- N.G. Van Kampen, *Stochastic Processes in Physics and Chemistry*, North Holland, Amsterdam, 1981.

8.- G.H. Fredrickson, in *Dynamical Aspects of Structural Change in Liquids and Glasses*, C.A. Angell and M. Goldstein eds., Ann. N.Y. Acad. Sci., New York, 1986, p. 185.

9.- G.H. Fredrickson and H.C. Andersen, Phys. Rev. Lett. 53, 1244 (1984).

10.- G.H. Fredrickson and H.C. Andersen, J. Chem. Phys. 83, 5822 (1985)

11.- J.J. Brey and M.J. Ruiz-Montero, to be published.

12.- G. Adam and J.H. Gibbs, J. Chem. Phys. 43, 139 (1965)

13.- G.H. Fredrickson and S.A. Brawer, J. Chem. Phys. 84, 3351 (1986)

14.- O.S. Narayanaswamy, J. Am. Ceram. Soc. 54, 491 (1971)

15.- D.A. Huse and D.S. Fisher, Phys. Rev. Lett. 57, 2203 (1986)

16.- J.J. Brey and A. Prados, Phys. Rev. A 42, 765 (1990), and to be published.

STATISTICAL CONFORMATION OF A POLYMER IN A NEMATIC MEDIUM UNDER A SHEAR FLOW USING THE ROUSE MODEL

Y. Thiriet*, R. Hocquart*, F. Lequeux* and J.F. Palierne⁺
*Laboratoire de Spectrométrie et d'Imagerie Ultrasonores,
Unité de Recherche Associée au CNRS n°851, Université Louis Pasteur,
4, rue Blaise Pascal, 67070 Strasbourg Cedex, France.
⁺Institut Charles Sadron, C.R.M.-E.A.H.P.,
6, rue Boussingault, 67083 Strasbourg Cedex.

I. INTRODUCTION

Several authors have investigated the conformation and the behaviour of flexible polymers in dilute solution in a nematic matrix [1-4] or in the isotropic phase of a nematic liquid [5,6]. These systems exhibit some peculiar rheological properties, particularly the effective viscosity of the nematics which is modified by the presence of polymer chains. Such a modification due to the geometrical anisotropy of the macromolecules at rest or under the action of an hydrodynamic field has been considered theoretically by Brochard [7].

In the following approach, we have assumed that the internal elasticity of the polymers is anisotropic owing to the coupling of the order parameter which is a feature of the nematic state. The model proposed is a generalisation of Brochard's model.

The statistical behaviour of a chain under the influence of a shear flow is studied in the frame of a Rouse model. We have established a stochastic LANGEVIN equation in order to calculate the mean square distances along the chain as well as the maximal stretching of the polymer as a function of the orientation of the director with respect to the flow field.

The case of an isotropic matrix is also considered as a particular case of the former one.

II. DESCRIPTION OF THE MODEL AND THE LANGEVIN EQUATION

We assume that the polymer chains are dilute in a nematic matrix. This solution is submitted to a simple shear flow whose velocity is $\underline{v} = (gy,0,0)$ where g is the shear rate and the nematic director $\underline{n} = (\cos\alpha, \sin\alpha, 0)$.(See fig. 1)

Fig. 1

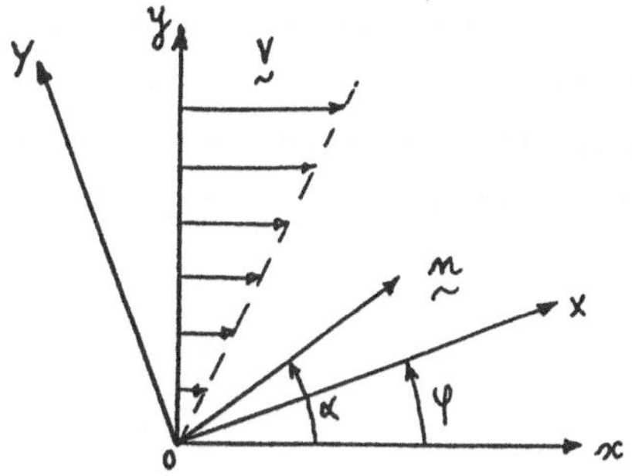

Let $r(s,t)$ be the position of the chain at time t, s being the mean curvilinear coordinate on the chain at rest ($0 < s < L$).

The chain is considered as an elastic spring and the force per s unit is given by the relation :

$$T = k_1 \frac{\partial^2 r}{\partial s^2} + k_2\, n \cdot \left(n \cdot \frac{\partial^2 r}{\partial s^2} \right) \qquad (1)$$

where k_1 and k_2 are two elastic constants : the first one is the classical constant while the second represents the anisotropic coupling constant with the order parameter.

Note that k_1 must be negative for a nematic polymer and $k_1 + k_2 > 0$ in order to insure the stability of the chain.

The position r obeys the general Langevin equation :

$$-\xi \left(\frac{\partial r}{\partial t} - D \cdot r \right) + T + A = 0 \qquad (2)$$

where D is the flow rate tensor which in our case writes $D_{xy} = g$, ξ is the friction coefficient, and A a random force acting on the rest unit of length which satisfies the autocorrelation function :

$$\langle A_i(s,t)\ A_j(s',t') \rangle = A^2 \delta_{ij}\ \delta(s-s')\ \delta(t-t') \qquad (3)$$

The Langevin equation projected on the orthogonal axes $\overrightarrow{OX}$, $\overrightarrow{OY}$, $\overrightarrow{OZ}$ gives the following system of equations :

$$-\xi \left[\frac{\partial X}{\partial t} - \tfrac{1}{2}\, gX\sin 2\varphi - gY\cos^2\varphi \right] + k_1 \frac{\partial^2 X}{\partial s^2} + k_2 \left[\frac{\partial^2 X}{\partial s^2}\cos^2(\alpha-\varphi) + \tfrac{1}{2}\, \frac{\partial^2 Y}{\partial s^2}\sin 2(\alpha-\varphi) \right] + A_X = 0$$

$$-\xi \left[\frac{\partial Y}{\partial t} + gX\sin^2\varphi + \tfrac{1}{2}\, gY\sin 2\varphi \right] + k_1 \frac{\partial^2 Y}{\partial s^2} + k_2 \left[\tfrac{1}{2}\, \frac{\partial^2 X}{\partial s^2}\sin 2(\alpha-\varphi) + \frac{\partial^2 Y}{\partial s^2}\sin^2(\alpha-\varphi) \right] + A_Y = 0$$

$$-\xi \left[\frac{\partial Z}{\partial t} \right] + k_1\, \frac{\partial^2 Z}{\partial s^2} + A_Z = 0$$

In the following development, we are especially interested by the stationary movement (steady state).

A standard way of solving such a system of equations is to apply the double Fourier transform $\mathfrak{r}(p,v)$ of the vector $\mathfrak{r}(s,t)$. We have :

$$\mathfrak{r}(p,v) = \int_{-\infty}^{+\infty}\int_{0}^{L} \mathfrak{r}(s,t)\, e^{j2\pi vt} \cos p\frac{\pi s}{L}\, dt\, ds$$

or inversely :

$$\mathfrak{r}(s,t) = \frac{\mathfrak{r}(0,t)}{L} + \frac{2}{L}\sum_{p=1}^{\infty}\int_{-\infty}^{+\infty} \mathfrak{r}(p,v) e^{-j2\pi vt} \cos p\frac{\pi s}{L}\, dv$$

One can see that in such a transformation the condition at both ends of the chain, i.e.$(\partial\mathfrak{r}/\partial s)_{s=0,L}=0$, are automatically satisfied at any time.

In the same way, the relation (3) becomes :

$$\langle A_i(p,v)\ A_j(q,v')\rangle = A^2\ \frac{L}{2}\ \delta_{ij}\ \delta_{pq}\ \delta(v+v')$$

and the movement equations yield :

$$[j2\pi v - \tfrac{1}{2} g \sin2\varphi + \omega_1 + \omega_2\cos^2(\alpha-\varphi)]X + [-g\cos^2\varphi + \tfrac{1}{2}\omega_2\sin2(\alpha-\varphi)]Y = A_X/\xi$$

$$[j2\pi v + \tfrac{1}{2} g \sin2\varphi + \omega_1 + \omega_2\sin^2(\alpha-\varphi)]Y + [g\sin^2\varphi + \tfrac{1}{2}\omega_2\sin2(\alpha-\varphi)]X = A_Y/\xi$$

$$[j2\pi v + \omega_1]Z = A_Z/\xi$$

$$\omega_1 = \frac{k_1}{\xi}\frac{p^2\pi^2}{L^2} > 0 \qquad \omega_2 = \frac{k_2}{\xi}\frac{p^2\pi^2}{L^2} < 0$$

III. EXPRESSIONS OF SOME MEAN SQUARE DISTANCES

In order to calculate the mean square distances along the chain, one has to determine at first the expression of the autocorrelation function of $\mathfrak{r}(s,t)$. For instance, after Fourier transforms we get :

$$\langle X(p,t)X\ (q,t)\rangle = A^2\ \frac{L}{4}\ \delta_{pq}\ I(p)$$

or reciprocally :

$$\langle X(s,t)\ X(s',t)\rangle = \sum_{p} \frac{A^2}{\xi^2 L}\cos p\frac{\pi s}{L}\ \cos p\frac{\pi s'}{L}\ I(p) + \frac{X^2(0,t)}{L^2}$$

where $I(p) = \dfrac{1}{2\omega_1+\omega_2} + \dfrac{\omega_2\sin^2(\alpha-\varphi)+\frac{1}{2}g\sin2\alpha}{\omega_1(\omega_1+\omega_2)+\frac{1}{2}\omega_2 g\sin2\alpha} + \dfrac{g^2\cos^2\alpha+\omega_1^2-\frac{1}{2}\omega_1 g[\sin2\alpha+\sin2(\alpha-\varphi)]}{(2\omega_1+\omega_2)[\omega_1(\omega_1+\omega_2)+\frac{1}{2}\omega_2 g\sin2\alpha]}$

and after straightforward calculation we have :

$$\langle (X(s,t) - X(s',t))^2 \rangle = \frac{A^2}{\xi^2 L}\sum_{p=1}^{\infty}\left(\cos p\frac{\pi s}{L} - \cos p\frac{\pi s'}{L}\right)^2 I(p)$$

Let us call : $C_x = \sum_{p=1}^{\infty}\left(\cos p\frac{\pi s}{L} - \cos p\frac{\pi s'}{L}\right)^2 I(p)$

This expression can be split up in a sum of series as follows :

$$C_x = \sum_{1}^{\infty}\left(\frac{A_1}{p^2} + \frac{A_2}{p^2-\beta^2} + \frac{A_3}{p^2(p^2-\beta^2)} + \frac{A_4}{p^4(p^2-\beta^2)}\right)\left(\cos p\frac{\pi s}{L} - \cos p\frac{\pi s'}{L}\right)^2$$

where

$A_1 = 1/\lambda_4$

$A_2 = -\lambda_3\sin^2(\alpha-\varphi)/\lambda_1\lambda_2 + \lambda_1/\lambda_2\lambda_4$

$A_3 = g\sin2\varphi/2\lambda_1\lambda_2 + \lambda_3 g[\sin2\alpha + \sin2(\alpha-\varphi)]/2\lambda_1\lambda_2\lambda_4$

$A_4 = g^2\cos^2\varphi / \lambda_1\lambda_2\lambda_4$

$\beta^2 = \lambda_3 g \sin2\alpha / 2\lambda_1\lambda_2$

$\lambda_1 = k_1\pi^2/\xi L^2 \quad \lambda_2 = (k_1+k_2)\pi^2/\xi L^2 \quad \lambda_3 = -k_2\pi^2/\xi L^2 \quad \lambda_4 = (2k_1+k_2)\pi^2/\xi L^2.$

Another expansion of C_x can be useful especially if $\beta \neq 0$ and $\alpha \in]0,\pi/2[$. This expansion has the following form :

$$C_x = \sum\left(\frac{A_5}{p^2} + \frac{A_6}{p^4} + \frac{A_7}{p^2-\beta^2}\right)\left(\cos p\frac{\pi s}{L} - \cos p\frac{\pi s'}{L}\right)^2$$

where

$$A_5 = 1/\lambda_4 - \frac{\sin2\varphi}{\lambda_3\sin2\alpha} - \frac{\sin2\alpha + \sin2(\alpha-\varphi)}{\lambda_4\sin2\alpha} - \frac{4\lambda_1\lambda_2\cos^2\varphi}{\lambda_3^2\lambda_4\sin^2 2\alpha}$$

$$A_6 = -\frac{2g\cos^2\varphi}{\lambda_3\lambda_4\sin2\alpha}$$

$$A_7 = -\frac{\lambda_3}{\lambda_1\lambda_2}\sin^2(\alpha-\varphi) + \frac{\lambda_1}{\lambda_2\lambda_4} + \frac{\sin2\varphi}{\lambda_3\sin2\alpha} + \frac{\sin2\alpha + \sin2(\alpha-\varphi)}{\lambda_4\sin2\alpha} + \frac{4\lambda_1\lambda_2\cos^2\varphi}{\lambda_3^2\lambda_4\sin^2 2\alpha}$$

and after the calculation of the series, we obtain :

$$C_x = A_5 F + A_6 G + A_7 H$$

where :

$F = (\pi/2)|x-x'|$

$G = (\pi/4)(x+x')(x-x')^2 - (1/8)(x^2-x'^2)^2 - (\pi/12)|x-x'|^3$

$H = (\pi/2\beta)[(\cos x-\cos x')^2 \cot g\,\beta\pi + (\sin x+\sin x')(\cos x-\cos x') + \sin|x-x'|]$

$x = \pi s/L \qquad x' = \pi s'/L$

In the case $\beta \neq o$ and $\alpha \in]0,-\pi/2[$, we get the corresponding expansion by changing β in $j\beta$.

To determine the expression of the other coordinate mean square distances, we follow the same procedure and we get :

$$\langle (Y(s,t)- Y(s',t))^2\rangle = \frac{A^2}{\xi^2 L}\sum_1^{\infty}\left(\cos p\frac{\pi s}{L} - \cos p\frac{\pi s'}{L}\right)^2 J(p) = \frac{A^2}{\xi^2 L} C_Y$$

C_y is obtained from C_x by changing φ in $\varphi + \pi/2$

$$\langle (Z(s,t) - Z(s',t))^2\rangle = \frac{A^2}{\xi^2 L}\cdot\frac{F}{\lambda_1} = \frac{A^2}{2\xi k_1}|s-s'| = \frac{a}{3}|s-s'|$$

where a is the monomer length, using the wellknown relation for the Gaussian chain. In the direction Oz, the statistic is not perturbed.

IV. STUDY OF THE CHAIN STATISTIC

The behaviour of a chain depending on the medium in which it is dilute, we consider now the static as well as the dynamical aspect of the statistics of the polymer in an isotropic and nematic phase.

For this purpose, we have investigated the following cases :

- isotropic matrix at rest
- nematic matrix at rest
- isotropic matrix in a shear flow
- nematic matrix in a shear flow whose polarisation is such as $\alpha = 0$ and $\alpha = \pi/2$.

We first calculate the general expressions of the mean square distances for $\beta = 0$.

Under these conditions we have :

$$C_X = \sum\left(\frac{A_3}{p^4} + \frac{A_4}{p^6} + \frac{A_1+A_2}{p^2}\right)\left(\cos p\frac{\pi s}{L} - \cos p\frac{\pi s'}{L}\right)^2$$

$$C_X = (A_1+A_2)F + A_3 G + A_4 E$$

where

$E = \pi|x-x'|^5/240 + \pi^2(x^2-x'^2)^2/24 - \pi(x-x')^2(x+x')[(x+x')^2 + 2(x^2+x'^2)]/48$
$\quad + (x^2-x'^2)^2(x^2+x'^2)/48$

and $$\langle (X(s,t) - X(s',t))^2 \rangle = \frac{A^2}{L\xi^2} C_X = \frac{2k_1 a}{3L\xi} C_X$$

Finally the two interesting mean square distances may be written :

$$\langle (X(s,t)-X(s',t))^2 \rangle = \frac{2aL}{3\pi^2} \left[\frac{\lambda_1}{\lambda_2} - \frac{\lambda_3}{\lambda_2} \sin^2(\alpha-\varphi) \right] F$$

$$+ \frac{2aL}{3\pi^2} \left[\frac{g\sin 2\varphi}{2\lambda_2} + \frac{\lambda_3 g}{2\lambda_2 \lambda_4} (\sin 2\alpha + \sin 2(\alpha-\varphi)) \right] G$$

$$+ \frac{2aL}{3\pi^2} \frac{g^2 \cos^2\varphi}{\lambda_2 \lambda_4} E$$

$$\langle (Y(s,t)-Y(s',t))^2 \rangle = \frac{2aL}{3\pi^2} \left[\frac{\lambda_1}{\lambda_2} - \frac{\lambda_3}{\lambda_2} \cos^2(\alpha-\varphi) \right] F$$

$$- \frac{2aL}{3\pi^2} \left[\frac{g\sin 2\varphi}{2\lambda_2} - \frac{\lambda_3 g}{2\lambda_2 \lambda_4} (\sin 2\alpha - \sin 2(\alpha-\varphi)) \right] G$$

$$+ \frac{2aL}{3\pi^2} \frac{g^2 \sin^2\varphi}{\lambda_2 \lambda_4} E$$

The calculation of the extremum of the mean square distances for $\alpha = 0$ and $\alpha = \pi/2$ is easy. As a matter of fact, the general forms of the mean square distances are :

$$\langle (X(s,t) - X(s',t))^2 \rangle = P + Q \sin 2\varphi + R \cos^2\varphi$$

$$\langle (Y(s,t) - Y(s',t))^2 \rangle = P - Q \sin 2\varphi + R \sin^2\varphi.$$

The extremum of these functions are such as :

$$\mathrm{tg}2\varphi = \frac{2Q}{R}$$

and in this case we obtain the reduced expressions $\bar{X}$ and $\bar{Y}$ such as :

$$\bar{X} = \langle (X(s,t) - X(s',t))^2 \rangle = P + R/2 \left(1 + \sqrt{1+(2Q/R)^2} \right)$$

$$\bar{Y} = \langle (Y(s,t) - Y(s',t))^2 \rangle = P + R/2 \left(1 - \sqrt{1+(2Q/R)^2} \right)$$

Note that P and R being always positive, the maximum of

$$\langle (X(s,t) - X(s',t))^2 \rangle$$

corresponds to the minimum of

$$\langle (Y(s,t) - Y(s',t))^2 \rangle.$$

a) Case $\alpha = 0$

The coefficients of the sinusoidal functions are :

$$P_1 = \frac{2aL}{3\pi^2} F \quad ; \quad Q_1 = \frac{2aL}{3\pi^2} \frac{Gg}{\lambda_4} \quad ; \quad R_1 = \frac{2aL}{3\pi^2} \frac{g^2 E}{\lambda_2 \lambda_4} \left[1 + \frac{\lambda_3 \lambda_4}{g^2} \frac{F}{E}\right]$$

and the angular position of the maximum of $\langle (X(s,t) - X(s',t))^2 \rangle$ is defined by :

$$\mathrm{tg}\ 2\varphi_1 = \frac{2Q_1}{R_1} = \frac{2\lambda_2 G}{Eg\left(1+\frac{\lambda_3 \lambda_4}{g^2} \frac{F}{E}\right)}$$

b) Case $\alpha = \pi/2$

The coefficients of the sinusoidal functions are :

$$P_2 = \frac{2aL}{3\pi^2} \frac{\lambda_1}{\lambda_2} F \quad ; \quad Q_2 = \frac{2aL}{3\pi^2} \frac{\lambda_1}{\lambda_2 \lambda_4} Gg \quad ; \quad R_2 = \frac{2aL}{3\pi^2} \frac{g^2 E}{\lambda_2 \lambda_4} \left[1 - \frac{\lambda_3 \lambda_4}{g^2} \frac{F}{E}\right]$$

and the angular position of the minimum of $\langle (X(s,t) - X(s',t))^2 \rangle$ is defined by :

$$\mathrm{tg}\ 2\varphi_2 = \frac{2Q_2}{R_2} = \frac{2\lambda_1 G}{Eg\left(1- \frac{\lambda_3 \lambda_4 F}{g^2\ E}\right)}$$

In order to determine the stretching along the chain, we have chosen the following partition :

$$s = 0 \ , \ s' = L/3 \quad ; \quad s = L/3 \ , \ s' = \frac{2L}{3} \quad ; \ s = 0 \ , \ s' = L$$

Under these conditions we may calculate the corresponding functions G, F and E

$$s = 0 \ , \ s' = \frac{L}{3} \ \rightarrow \ G = \frac{\pi^4}{3^3.8} \quad F = \frac{\pi^2}{6} \quad E = \frac{53\pi^6}{3^7.80}$$

$$s = L/3 \ , \ s' = 2\frac{L}{3} \ \rightarrow \ G = \frac{7\pi^4}{3^4.8} \quad F = \frac{\pi^2}{6} \quad E = \frac{61\pi^6}{3^6.80}$$

$$s = 0 \ , \ s' = L \ \rightarrow \ G = \frac{\pi^4}{24} \quad F = \frac{\pi^2}{2} \quad E = \frac{\pi^6}{240}$$

Moreover the functions G, F, and E are symmetric with respect to the middle of the chain

$$G(s,s') = G(L-s,\ L-s') \ ; \ F(s,s') = F(L-s,\ L-s') \ ; \ E(s,s') = E(L-s,\ L-s')$$

1) Isotropic matrix at rest

According to Brochard, k_2 is a function of the order parameter and therefore in the isotropic phase $k_2 = 0$.

We put now $k_2 = g = \varphi = 0$ in the general expressions of the mean square distances and note that they are all equal to aL/9 in the cases (s = 0, s' = L/3) and (s = L/3, s' = 2L/3), while the end to end mean square distance is three times the preceding value. These results show that the stretching is homogeneous along the chain.

2) Nematic matrix at rest

We put now $g = \alpha = \varphi = 0$ and $k_2 \neq 0$.

The expression of the mean square distances are then :

$$\overline{X} = \frac{2aL}{3\pi^2} F \frac{\lambda_1}{\lambda_2} \quad ; \quad \overline{Y} = \frac{2aL}{3\pi^2} F$$

- For s = 0, s' = L/3 :

$$\overline{X} = \frac{aL}{9} \frac{\lambda_1}{\lambda_2} \quad ; \quad \overline{Y} = \overline{Z} = \frac{aL}{9}$$

- For s = L/3, s' = 2L/3 :

$$\overline{X} = \frac{aL}{9} \frac{\lambda_1}{\lambda_2} \quad ; \quad \overline{Y} = \overline{Z} = \frac{aL}{9}$$

- For s = 0, s' = L

$$\overline{X} = \frac{aL}{3} \frac{\lambda_1}{\lambda_2} \quad ; \quad \overline{Y} = \overline{Z} = \frac{aL}{3}$$

The stretching along the chain is homogeneous and the central part of the chain in as stretched as the other part of the chain.

3) Isotropic matrix in a shear flow

In the dynamic regime $g \neq 0$, but the order parameter is null and $k_2 = 0$.

This condition implies $\lambda_1 = \lambda_2 = \lambda_4/2$ and $\lambda_3 = 0$ and we get :

$$P = \frac{2aL}{3\pi^2} F \quad ; \quad Q = \frac{aL}{3\pi^2} \frac{Gg}{\lambda_1} \quad ; \quad R = \frac{aL}{3\pi^2} \frac{Eg^2}{\lambda_1^2} \quad ; \quad tg2\varphi = \frac{2G\lambda_1}{Eg}$$

- For s = 0, s' = L/3

$$\left.\begin{matrix}\overline{X} \\ \overline{Y}\end{matrix}\right] = \frac{aL}{9} + \frac{aL}{9} \cdot \frac{53\pi^4}{160.3^6} \frac{g^2}{\lambda_1^2} \left(1 \pm \sqrt{1 + tg^2 2\varphi}\right)$$

$$tg2\varphi = \frac{20.3^4}{53.\pi^2} \cdot \frac{\lambda_1}{g}$$

- For s = L/3, s' = 2L/3

$$\left.\begin{matrix}\overline{X}\\ \overline{Y}\end{matrix}\right] = \frac{aL}{9} + \frac{aL}{9} \cdot \frac{61\pi^4}{160.3^5} \frac{g^2}{\lambda_1^2} \left(1 \pm \sqrt{1 + tg^2 2\varphi} \right)$$

$$tg2\varphi = \frac{140.3^2}{61.\pi^2} \cdot \frac{\lambda_1}{g}$$

- For s = 0, s' = L

$$\left.\begin{matrix}\overline{X}\\ \overline{Y}\end{matrix}\right] = \frac{aL}{3} + \frac{aL}{3} \cdot \frac{\pi^4}{480} \frac{g^2}{\lambda_1^2} \left(1 \pm \sqrt{1 + tg^2 2\varphi} \right)$$

$$tg2\varphi = \frac{20\lambda_1}{\pi^2 g}$$

Contrary to the previous case, the stretching on the chain is not homogeneous, the center of the molecule being less compact than near the ends. Furthermore, we observe a variation of the angular position along the chain.

4) Nematic matrix in a shear flow [8] :

The velocity field can be parallel ($\alpha = 0$) or orthogonal ($\alpha = \pi/2$) to the director $\underline{n}$.

a) Case $\alpha = 0$

The expressions of the mean square distances yield :

$$P_1 = \frac{2aL}{3\pi^2} F \quad ; \quad Q_1 = \frac{2aL}{3\pi^2} \frac{Gg}{\lambda_4} \quad ; \quad R_1 = \frac{2aL}{3\pi^2} \frac{Eg^2}{\lambda_2 \lambda_4} \left(1 + \frac{\lambda_3 \lambda_4}{g^2} \frac{F}{E} \right)$$

$$tg2\varphi_1 = \frac{2G\lambda_2}{gE\left(1 + \frac{\lambda_3 \lambda_4}{g^2} \frac{F}{E} \right)}$$

- For (s = 0, s' = L/3) :

$$\left.\begin{matrix}\overline{X}\\ \overline{Y}\end{matrix}\right] = \frac{aL}{9} + \frac{aL}{9} \cdot \frac{\lambda_3}{2\lambda_2} \left(1 + \frac{53\pi^4}{40.3^6} \frac{g^2}{\lambda_3 \lambda_4} \right) \left(1 \pm \sqrt{1 + tg^2 2\varphi_1} \right)$$

$$tg2\varphi_1 = \frac{20.3^4}{53.\pi^2} \cdot \frac{\lambda_2}{g} \left(1 + \frac{40.3^6 \lambda_3 \lambda_4}{53.\pi^4 g^2} \right)^{-1}$$

- For (s = L/3, s' = 2L/3) :

$$\left.\begin{matrix}\overline{X}\\ \overline{Y}\end{matrix}\right] = \frac{aL}{9} + \frac{aL}{9} \cdot \frac{\lambda_3}{2\lambda_2} \left(1 + \frac{61\pi^4}{40.3^5} \frac{g^2}{\lambda_3 \lambda_4} \right) \left(1 \pm \sqrt{1 + tg^2 2\varphi_1} \right)$$

$$tg2\varphi_1 = \frac{7.20.3^2\lambda_2}{61.\pi^2\ g}\left(1 + \frac{40.3^5\ \lambda_3\lambda_4}{61.\pi^4\ g^2}\right)^{-1}$$

- For s = 0, s' = L :

$$\left.\begin{matrix}\bar{X}\\ \bar{Y}\end{matrix}\right] = \frac{aL}{3} + \frac{aL}{3}\cdot\frac{\lambda_3}{2\lambda_2}\left(1 + \frac{\pi^4\ g^2}{120\ \lambda_3\lambda_4}\right)\left(1 \pm \sqrt{1 + tg^2 2\varphi_1}\right)$$

$$tg2\varphi_1 = \frac{20\ \lambda_2}{\pi^2 g}\left(1 + \frac{120\ \lambda_3\lambda_4}{\pi^4\ g^2}\right)^{-1}$$

Once more, the stretching along the chain is not homogeneous and the angular position from one part of the molecule to the other is not constant.

b) Case $\alpha = \pi/2$

The expressions of the mean square distances give :

$$P_2 = \frac{2aL}{3\pi^2}\frac{\lambda_1}{\lambda_2}\ ; \quad Q_2 = \frac{2aL}{3\pi^2}\frac{\lambda_1\ Gg}{\lambda_2\lambda_4}\ ; \qquad R_2 = \frac{2aL}{3\pi^2}\frac{Eg^2}{\lambda_2\lambda_4}\left(1 - \frac{\lambda_3\lambda_4}{g^2}\frac{F}{E}\right)$$

$$tg2\varphi_2 = \frac{2G\lambda_1}{Eg\left(1 - \frac{\lambda_3\lambda_4}{g^2}\frac{F}{E}\right)}$$

- For s = 0, s' = L/3 :

$$\left.\begin{matrix}\bar{X}\\ \bar{Y}\end{matrix}\right] = \frac{aL\lambda_1}{9\lambda_2} - \frac{aL}{9}\cdot\frac{\lambda_3}{2\lambda_2}\left(1 - \frac{53.\pi^4\ g^2}{40.3^6\ \lambda_3\lambda_4}\right)\left(1 \pm \sqrt{1 + tg^2 2\varphi_2}\right)$$

$$tg2\varphi_2 = \frac{20.3^4}{53.\pi^2}\cdot\frac{\lambda_1}{g}\left(1 - \frac{40.3^6\lambda_3\lambda_4}{53.\pi^4\ g^2}\right)^{-1}$$

- For s = L/3, s' = 2L/3 :

$$\left.\begin{matrix}\bar{X}\\ \bar{Y}\end{matrix}\right] = \frac{aL\lambda_1}{9\lambda_2} - \frac{aL}{9}\cdot\frac{\lambda_3}{2\lambda_2}\left(1 - \frac{61.\pi^4\ g^2}{40.3^5\ \lambda_3\lambda_4}\right)\left(1 \pm \sqrt{1 + tg^2 2\varphi_2}\right)$$

$$tg2\varphi_2 = \frac{2.70.3^2}{61.\pi^2}\cdot\frac{\lambda_1}{g}\left(1 - \frac{40.3^5\lambda_3\lambda_4}{61.\pi^4\ g^2}\right)^{-1}$$

- For s = 0, s' = L :

$$\left.\begin{matrix}\bar{X}\\ \bar{Y}\end{matrix}\right] = \frac{aL\lambda_1}{3\lambda_2} - \frac{aL}{3}\cdot\frac{\lambda_3}{2\lambda_2}\left(1 - \frac{\pi^4\ g^2}{120\ \lambda_3\lambda_4}\right)\left(1 \pm \sqrt{1 + tg^2 2\varphi_2}\right)$$

$$tg2\varphi_2 = \frac{20\lambda_1}{\pi^2 g}\left(1 - \frac{120\ \lambda_3\lambda_4}{\pi^4\ g^2}\right)^{-1}$$

Same conclusion as for the case $\alpha = 0$.

CONCLUSION

In this study, we focused our attention on the internal statistic of a chain in a nematic phase.

The stretching of different parts of similar chains submitted to a shear flow is not homogeneous and depends on the position of the part along the chain.

In fact, the stretching of the macromolecule is more important in the middle than near the ends of the molecule.

We note the same behaviour in an isotropic phase as was already mentioned by Rabin [9] for dilute solutions of polymers in an isotropic matrix, but in an elongational field.

At constant shear rate, we observe that the stretching direction of similar portions of a chain depends of their localization on the chain and differs from the orientation of the end to end stretching.

At last, the stretching direction as a function of the shear rate intensity is very different in an isotropic dilute solution compared to the one obtained in a nematic dilute solution.

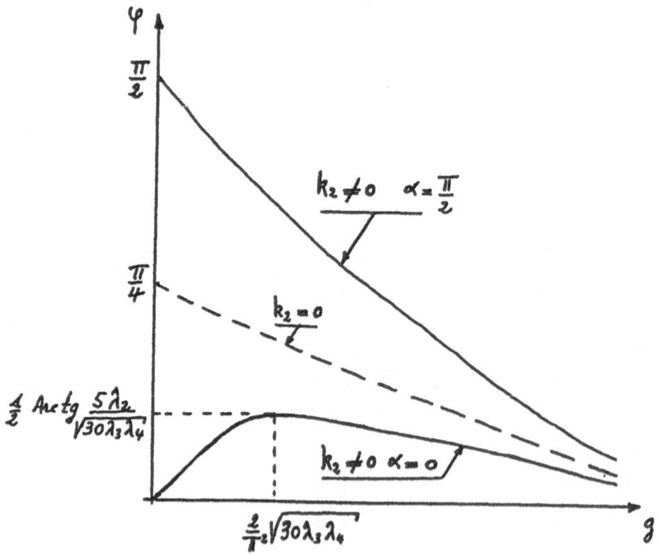

Figure. Principal results of the stretching orientation as a function of the shear rate for the end to end chain.

REFERENCES

1. A. Dubault, C. Casagrande and M. Veyssie, Molec. Crystals Liq. Crystals Lett., 41, 239 (1977).
2. A. Dubault, C. Casagrande, M. Veyssie and B. Deloche, Phys. Rev. Lett., 41, 935 (1980).
3. A. Dubault, R. Ober, M. Veyssie and B. Cabane, J. Phys. Paris, 46, 1227 (1985).
4. P. Martinoty, A. Dubault, C. Casagrande and M. Veyssie, J. Phys. Lett., 44, 935 (1983).
5. Y. Dormoy, J.L. Gallani and P. Martinoty, Molec. Crystals Liq. Crystals., 170, 135 (1989).
6. Y. Dormoy, V. Reys and P. Martinoty, Liq. Crystals, 4, 409 (1989).
7. F. Brochard, J. Polym. Sci., 17, 1367 (1979).
8. F. Lequeux and R. Hocquart, J. Phys. Paris, to be published.
9. Y. Rabin, J. Chem. Phys., 88, 4014 (1987).

On the modelling of stationary heat transfer by the use of dissipative networks

by Gerd BRUNK

Rheological networks and electrical equivalent circuits are often applied to render a graphic access to the macroscopic constitutive description of physical systems. Thus we obtain some insight in the essential physical properties of basic conceptions used in constitutive theory and at the same time consistency with thermodynamical restrictions as passivity is provided. Here we shall look at the idea of temperature in non-equilibrium systems as a generalisation of equilibrium temperature. For this purpose we connect the concept of contact temperature developped by MUSCHIK and coworkers /1/ with the idea of noise thermometry. By this way we can study the one-dimensional heat transfer which is involved in mechanical or electrical random processes as well between discrete systems as in continuous ones. We find – by the introduction of the fluctuation–dissipation theorem – as a consequence the (linear) FOURIER law of heat conduction and moreover an interpretation of contact temperature in the case of onedimensional modelling of heat transfer by radiation. We realize how to model a thermal contact on mesoscopic level, we clear up the meaning of temperature in non–equilibrium and we show the non–equivalence of different contacting with respect to temperature definition. Our starting point shall be the fluctuation–dissipation–theorem for onedimensional systems as detected by NYQUIST /2/ in 1928. This states the spectra of applied current j of a resistor or of force F of a dashpot to be

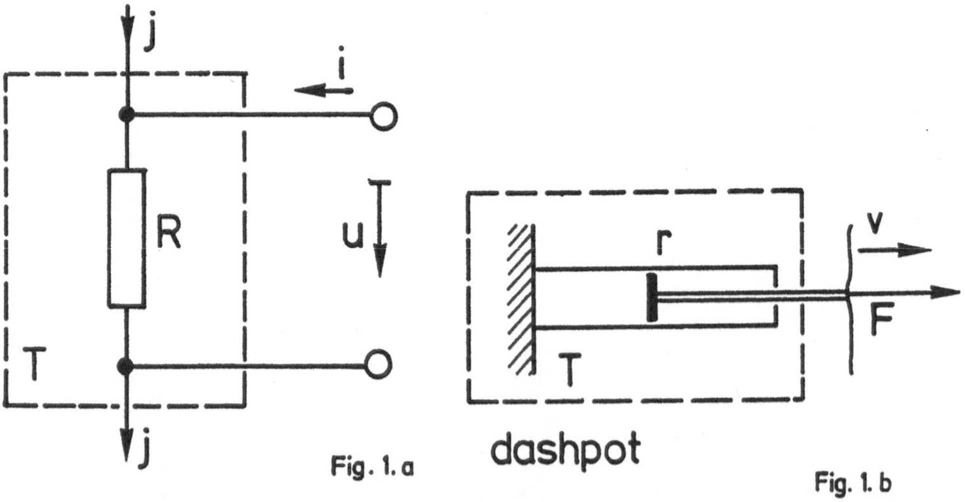

Fig 1.a: Notations and equivalent circuit corresponding to NYQUIST's fluctuation–dissipation–theorem, eq. (1.1), and Fig. 1.b: Notations and rheological model concerning thermal force fluctuations, eq. (1.2).

$$S_{jj}(\omega) = 2\,kT\,G \tag{1.1}$$

or

$$S_{FF}(\omega) = 2\,kT\,r \tag{1.2}$$

resp. where $G = 1/R$ means the conductance and r the damping coefficient (cf. figs. 1.a and 1.b). If we must take into account quantum mechanical effects the spectra must be corrected by the factor

$$g(\eta) = |\eta|\,(\exp|\eta|-1)\,; \qquad \eta := \frac{\hbar\omega}{kT} \tag{1.3}$$

Generally we have in multidimensional systems the matrixvalued spectrum of equivalent applied current vector **j** /3/, /4/

$$\mathbf{S}_{jj}(\omega) = 2\,kT\,\mathbf{Y}_h(\omega)\,. \tag{2}$$

Here is

$$\mathbf{Y}_h := \tfrac{1}{2}\,(\mathbf{Y}+\mathbf{Y}^{+}) \tag{3}$$

the hermitean part of the conductivity matrix **Y** – see fig. 2 for the electrical case.

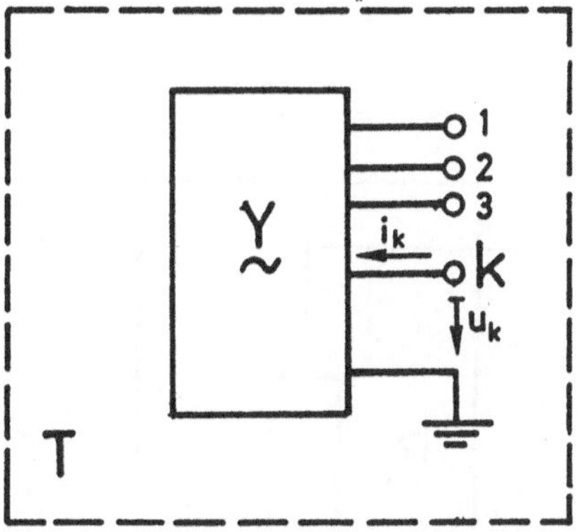

Fig. 2

Fig. 2: Multipole black–box configuration corresponding to thermal current fluctuations, eq. (2).

NYQUIST has deduced the equation (1.1) using an imaginary experiment sketched in fig. 3 and its generalisation

$$S_{jj}(\omega) = 2\,kT\,Y_r(\omega)\,, \qquad Y_r := \mathrm{Re}\,Y\,, \tag{4}$$

– being a special case of eq. (2) – by the study of the system traced by fig. 4.

In both imaginary experiments two systems in thermal equilibrium $T_2=T_1$ are connected by a well defined (electrical) contact.

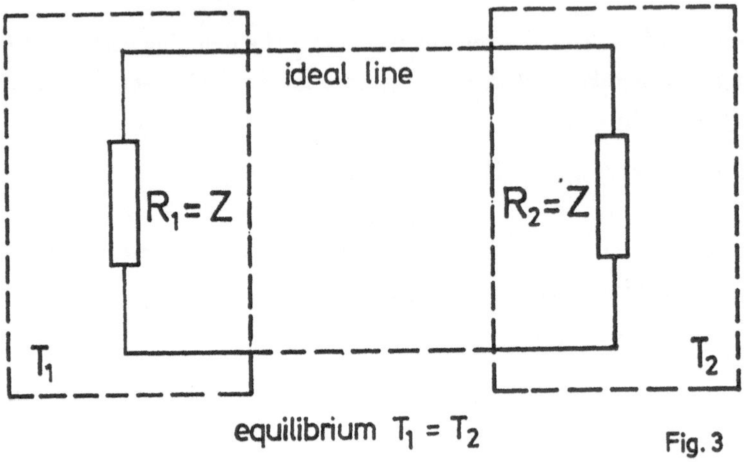

Fig. 3

Fig. 3: NYQUIST's first imaginary experiment for deducing the termal fluctuation spectrum of a pure ohmic resistor Z.

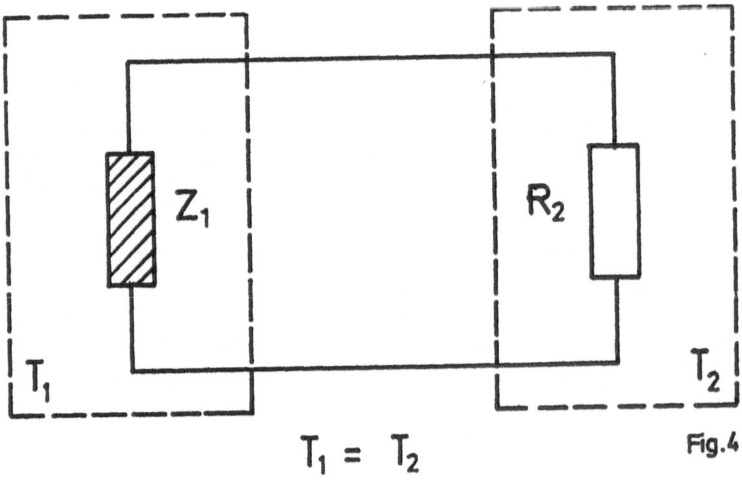

Fig.4

Fig 4: NYQUIST's second imaginary experiment for deducing the thermal noise of an arbitrary impedance Z_1 or admittance $Y_1 := 1/Z_1$.

Regarding these configurations we are suggested to ask what will happen in non-equilibrium. As the mathematical description of mechanical and electrical linear systems is largely analogous we confine us to the electrical case. The simplest configuration of two systems at different temperatures is presented in fig. 5.

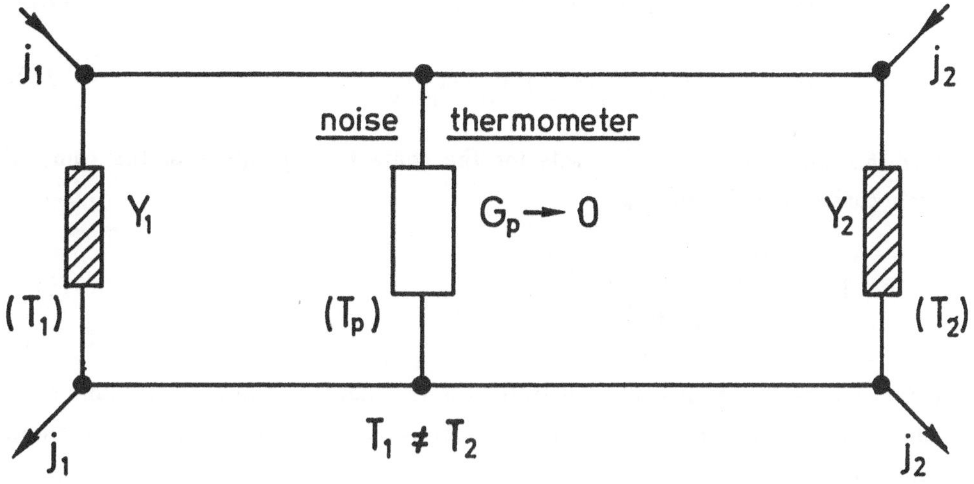

Fig 5: Compound system of two admittances at different temperatures connected in parallel to define the contact temperature measured by a noise thermometer.

Using eq. (4) we get from the description of stationary stochastic processes in linear circuits the power density spectrum

$$\tilde{P}_{1 \rightarrow 2}(\omega) = |Z_t|^2 (S^1_{jj} Y_{2r} - S^2_{jj} Y_{1r})$$

$$= 2k\, |Z_t|^2\, Y_{1r}\, Y_{2r}\, (T_1 - T_2)\ . \qquad (5)$$

Here means

$$Z_t := (Y_1 + Y_2 + G_p)^{-1} \qquad (6)$$

the total impedance of the coupled system, Y_1 and Y_2 are the admittances of the parts, G_p is the conductivity of a thermometer system, T_1 and T_2 are the equilibrium temperatures of heating baths of the parts.

The heat transfer described by eq. (5) shows the characteristic form of NEWTON's cooling laaw by the proportionality to $T_2 - T_1$. The contact temperature of a non—equilibrium system is defined by the condition that the exchange between the thermometer being in equilibrium and the system in non-equilibrium over the contact vanishes if the equilibrium temperature of the thermometer equals the contact temperature; thus the heating power $\tilde{P}_-$ absorbed by the thermometer must balance that one $\tilde{P}_+$ emitted. The corresponding spectra are

$$\tilde{P}_-(\omega) = 2k\,|Z_t|^2\,G_p(Y_{1r}\,T_1 + Y_{2r}\,T_2) \tag{7.1}$$

and

$$\tilde{P}_+(\omega) = 2k\,|Z_t|^2\,G_p(Y_{1r} + Y_{2r})\,G_p\,T_p\,. \tag{7.2}$$

From this we deduce a mixing rule for the contact temperature of the compound system consisting of Y_1 and Y_2:

$$T_p = \frac{Y_{1r}\,T_1 + Y_{2r}\,T_2}{Y_{1r} + Y_{2r}} \tag{8}$$

Another realisation of the contact between the thermometer and the compound system is the connecting in series of the thermometer resistor. From this we obtain another mixing rule

$$T_s = \frac{Z_{1r}\,T_1 + Z_{2r}\,T_2}{Z_{1r} + Z_{2r}} \tag{9}$$

Generally the contact temperatures T_p and T_s do not correspond. The equality $T_p = T_s$ holds if and only if $|Z_1| = |Z_2|$ where Z_1 and Z_2 are the impedances, $Z_k \equiv Y_k^{-1}$.

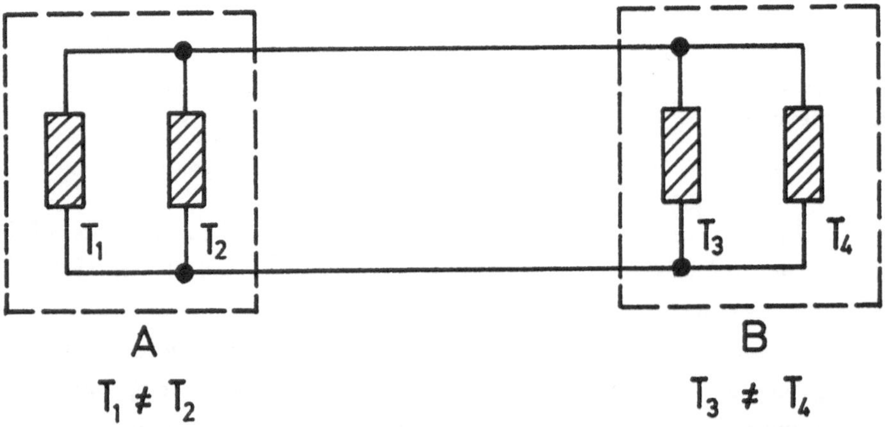

Fig. 6

Fig 6: Two simple compound systems interacting electrically and thus effecting heat exchange by thermal fluctuations.

The next step to be executed is the contact between two compound systems, A and B of them being characterized by its contact temperature T_{pA} and T_{pB} – cf. fig. 6 for the simplest case. The corresponding spectrum of the heat exchange power is

$$\tilde{P}_{A \to B} = 2k\,|Z_t|^2\,Y_{Ar}\,Y_{Br}\,(T_{pA} - T_{pB}) \,, \tag{10}$$

reproducing NEWTON's cooling law with respect to the difference of the contact temperatures.

From the treatment of discrete systems we shall pass to continuous one-dimensional ones in the usual way studying chainlike mechanical or electrical networks as damped

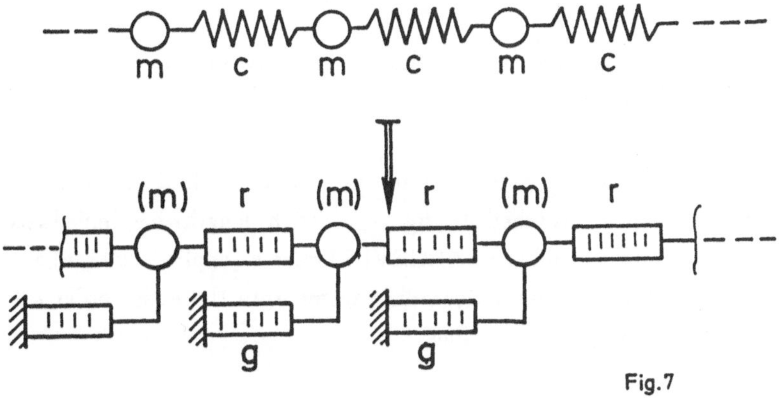

Fig. 7: Undamped and damped oscillator chain as models for fluctuating continuous mechanical systems.

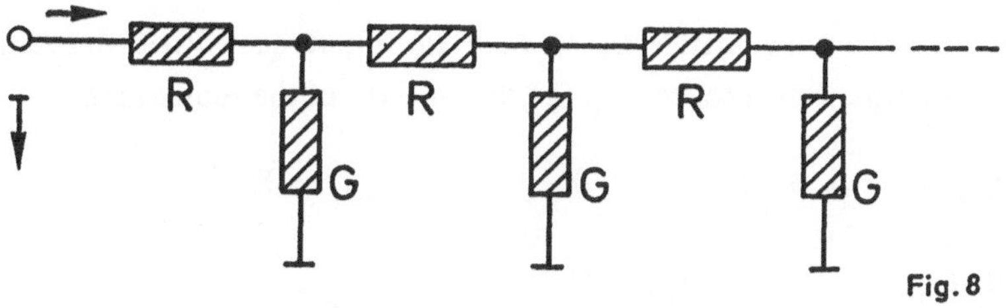

Fig. 8: Network modelling a line as example for a fluctuating continuous electrical system.

oscillator chains – fig. 7 – or lines – fig. 8. As we analyze spectra we characterize the systems by its lengthwise impedance and its admittance across per unit length. Further we shall execute deductions only in terms of the electrical case because of the analogy of the mechanical one which exists here as well as for discrete systems. We shall compare current with force, voltage with velocity, capacity with mass, inductivity with

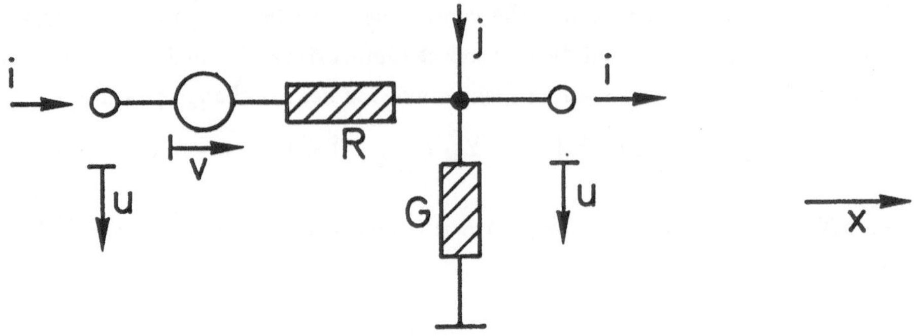

Fig.9

Fig 9: Circuit with applied noise voltage v and noise current j modelling a fluctuating element of line with length $\epsilon \equiv \Delta x \rightarrow 0$.

compliance and conductivity with damping. At first we look at an element of a line in local equilibrium – as sketched in fig. 9 – with lengthwise impedance $R = r\epsilon$, admittance across $G = g\epsilon$ and the LANGEVIN forces v (applied voltage) and j (current source) describing thermal noise. Regarding terms upto the linear order $\epsilon \equiv \Delta x$ of the length element we obtain the equations

$$u_2 = u_1 - v - r\,\epsilon\, i_1 \tag{11.1}$$

$$i_2 = i_1 + j - g\,\epsilon\,(u_1 - v) \tag{11.2}$$

connecting the output quantities u_2, i_2 with the input quantities. According to the fluctuation – dissipation–theorem, eq. (4), the spectra of the sources are

$$S_{vv}(x,\omega) = \epsilon\, 2\,kT(x)\, r_r(\omega) \qquad r_r := \mathrm{Re}\, r \tag{12.1}$$

and

$$S_{jj}(x,\omega) = \epsilon\, 2\,kT(x)\, g_r(\omega) \qquad g_r := \mathrm{Re}\, g \tag{12.2}$$

For the determination of the correlation of the input and output quantities $i_{1,2}$ and $u_{1,2}$ with the noise sources j and v we restrict us to the case of adaption, that means the line shall be closed at both ends by the wave impedance Z. Because of the linearity of the system we obtain $i_{1,2}$ and $u_{1,2}$ by superposition of portions excited by the sources j and v and the remainder uncorrelated portions. From this consideration follow the cross spectra of the input quantities to be

$$S^1_{iv}(x,\omega) = -S^1_{uv}(\alpha,\omega)/Z_1 = \epsilon \cdot 2\,kT \frac{r_r}{Z_1+Z_2} \tag{13.1}$$

and

$$S^1_{ij}(x,\omega) = -S^1_{uj}(x,\omega)/Z_1 = -\epsilon \cdot 2\,kT \frac{g_r Z_2}{Z_1+Z_2} \ . \tag{13.2}$$

Here Z_1 and Z_2 are the impedances of the input side and the output side resp. of the line. As we have claimed adaption we have

$$Z_1 = Z_2 = Z := \sqrt{r/g} \ . \tag{14}$$

By the limit $\epsilon \longrightarrow 0$ we obtain the coupled differential equations of the spectra:

$$\partial_x S_{uu} = -(r^* S_{ui} + r\, S_{iu}) \tag{15.1}$$

$$\partial_x S_{ii} = -(g^* S_{iu} + g\, S_{ui}) \tag{15.2}$$

$$\partial_x S_{ui} = -(g^* S_{uu} + r\, S_{ii}) + kT\,(Y^* r_r + Z\, g_r) \tag{15.3}$$

$$\partial_x S_{iu} = -(g\, S_{uu} + r^* S_{ii}) + kT\,(Z^* g_r + Y\, r_r) \tag{15.4}$$

The symbol $(\cdot)^*$ means the complex conjugate. $Y := 1/Z = \sqrt{g/r}$

From the last two eqs. follows the differential equation for the transmitted power per frequency interval

$$\partial_x \tilde{P}(x,\omega) \equiv \frac{1}{2}\,\partial_x(S_{iu} + S_{ui}) = g_r(kTZ_r - S_{uu}) + r_r\,(kTY_r - S_{ii}) \tag{16}$$

The condition eq. (14) seems at the first glance fairly restrictive, but it is correct not only for the infinite long line, but approcimately also if the length of the line exceeds sufficiently the decay length. Therefore the eqs. (15) and (16) retain their significance in this case. Now we may ask for the relation between the system temperature T(x) – which equals the temperature of a local real or imaginary heating bath in equilibrium – and the contact temperature as defined above and further for the relation between power transmitted and the temperature field. For noise thermometers we use as before pure ohmic resistors connected in parallel (conductivity $G_p \longrightarrow 0$) or in series (resistance $R_s \longrightarrow 0$) proportioned in such a way that they do not influence the transmission. From the balance of absorbed and emitted power we obtain the temperature

$$2\,kT_p = S_{uu} \frac{|Y_1 + Y_2|^2}{Y_{1r} + Y_{2r}} \tag{17.1}$$

measured at the resistor connected in parallel and the other one

$$2\,kT_s = S_{ii}\,\frac{|Z_1 + Z_2|}{Z_{1r} + Z_{2r}} \tag{17.2}$$

As we have adaption, eq. (14), it follows

$$kT_p = S_{un}/Z_r \quad \text{and} \quad kT_s = S_{ii}/Y_r\,. \tag{18}$$

If we impose the condition $\partial_x q = 0$ of stationary "heat conduction" in each frequency intervall we find the relation

$$\partial_x\,(S_{iu} + S_{ui}) = 0 \tag{19}$$

and from eq. (16) we get

$$g_r\,(kT\,Z_r - S_{uu}) + r_r\,(kT\,Y_r - S_{ii}) = 0\,. \tag{20}$$

Because $g_r > 0$ and $r_r > 0$ we deduce from eq. (19)

$$kT = S_{uu}/Z_r = S_{ii}/Y_r \tag{21}$$

If we compare this result with eq. (18) we see that the contact temperatures T_p and T_s equal the local system temperature. Inserting this in eqs. (15.1) and (15.2) yields for the transmitted power per frequency interval the expression

$$\tilde{P}(x,\omega) := \frac{1}{2}\,(S_{ui} + S_{iu}) = -\,\tilde{\lambda}(\omega)\,\partial_x T \tag{22}$$

with the heat conductivity per frequency interval being

$$2\,\tilde{\lambda}(\omega) := k\,\frac{g_i\,|r| + r_i\,|g|}{g_i r_r + r_i g_r}\,\mathrm{Re}\,[rg^*]^{-\frac{1}{2}}$$

$$= \frac{k}{\sqrt{2}}\,\frac{g_i\,|r| + r_i\,|g|}{g_i r_r + r_i g_r}\,\frac{[\,r_r g_r + r_i g_i + |r|\,|g|\,]^{\frac{1}{2}}}{|r|\,|g|} \tag{23}$$

Thus we have indeed a FOURIER law for the heat transfer wherein the heat conductivity is defined by line properties.

There is quite different situation if we impose by means of heating baths an arbitrary temperature distribution T(x) on the line and also arbitrary tempertures T_1 and T_2 on the load impedances at the boundaries $x=0$ and $x=\ell$. Then we have generally no stationarity, eq. (19) does not hold. The current and the voltage can be represented by superposition of portions due to the currents applied at the boundary loads (j_1 and j_2) and to that ones within the line itself by thermal fluctuations. As these excitations ar uncorrelated the spectra can be superposed, too. By example we have

$$S_{uu}(x,\omega) = S^1_{uu}(x,\omega) + S^2_{uu}(x,\omega) \tag{24}$$

where the first and the second term are due to the excitations at $\xi < x$ and $\xi > x$ resp. As we use adapted loads at the boundaries eq. (14) holds and it follows from

$$\tilde{u}^1(x,\omega) = Z\,\tau^1(x,\omega)\,, \qquad \tilde{u}^2(x,\omega) = -Z\,\tau^2(x,\omega)$$

that the relations

$$S^1_u(x,\omega) = |Z|^2\, S^1_{ii}(x,\omega)$$

and (25)

$$S^2_u(x,\omega) = |Z|^2\, S^1_{ii}(x,\omega)$$

are valid. (If we keep by an imaginary experiment the line at T=0 the spectra, eq. (25), are due to only the influence of the boundaries.) Therefore we get the equation

$$S_{uu}(x,\omega) = |Z|^2\, S_{ii}(x,\omega)\,. \tag{26}$$

By insertion of eq. (26) in eq. (18.1) we deduce looking at eq. (18.2)

$$kT_p = S_{uu}/Z_r = S_{ii}|Z|^2/Z_r = S_{ii}/Y_r = kT_s =: k\Theta \tag{27}$$

The both contact temperatures conincide and are different from the system temperature, naturally also in the special case $T(x) \equiv 0$. Because the statement of eq. (27) is formally the same as that one of eq. (21) we have the same consequence, that is FOURIER's law

$$\tilde{P}(x,\omega) = -\tilde{\lambda}(\omega)\,\partial_x\,\Theta \tag{28}$$

with $\tilde{\lambda}(\omega)$ according to eq. (23), but with the contact temperature Θ unequal to the temperature T imposed by the heating bath. As all the calculations hold by analogy as well for the mechanical as for the electrical case we have found at the same time a mechanical model of heat conduction and an one–dimensional model of heat transfer by radiation in a dissipative medium if we take account the quantum correction, eq. (1.3). In the non–dissipative case the definition of "heat conductivity", eq. (23), is impossible and we have no temperature field $\Theta(x)$; at the same time the so-called stationarity condition, eq. (19), is identically fulfilled.

References:

/1/ MUSCHIK, W./BRUNK, G.: A Concept of Non-Equilibrium Temperature, Int. J. Engng. Sci., Vol 15 (1977), 377 - 389

MUSCHIK, W.: Empirical Foundation and Axiomatic Treatment of Non-Equilibrium Temperature, Arch. Rat. Mech. Anal. 66 (1977), 379 - 401

/2/ NYQUIST, H.: Thermal agitation of elctric charge in conductors, Phys. Rev. 32 (1928), 10 ff

/3/ CALLEN, H.B./GREENE, R.F.: On a Theorem of Irreversible Thermodynamics,
I: Phys. Rev. 86 (1952), 702 ff.
II: Phys. Rev. 88 (1952), 1387 ff.

KUBO, R.: Statistical Mechanical Theory of Irreversible Processes, J. Phys. Soc. of Japan 12 (1957), 570 ff.

/4/ BRUNK, G.: On Energy Exchange between Multidimensional Fluctuating Systems in Equilibrium, in: Recent Developments in Nonequilibrium Thermodynamics (Eds. Casas-Vazquez/Jou/Rubi): Fluids and Related Topics, p. 383 - 386. Lecture Notes in Physics 253, Springer, Berlin-New York 1986

Prof. Dr.-Ing. Gerd Brunk
1. Institut für Mechanik, C 8
Technische Universität Berlin
Straße des 17. Juni 135
D-1000 BERLIN 12

THERMOMECHANICS OF POROUS MEDIA FILLED WITH A FLUID

Włodzimierz Derski
Institute of Fundamental Technological Research
Polish Academy of Sciences
ul. Mielżyńskiego 27, 61-725 Poznań

Porous materials make their appearance in a wide variety of settings, natural and artificial, and in diverse technological applications. As a consequence a number of problems arise dealing, among others, with statics and strength, fluid flow and heat conduction, and the dynamics of such materials [10], and in the area of exploration geophysics [8], the steadily growing literature bearing witness to the importance of the subject.

For propagation of stress-waves in a fluid-saturated porous medium the basic formulation is due to Biot [1], whose model consisted of an elastic matrix permeated by a network of interconnected spaces filled with liquid. An account of further researches based on Biot's theory is given in [5]. Biot subsequently extended his formulation to include more general response of the solid phase [2].

In the author's opinion, the equations of motion formulated by Biot give some results contradicting each other as far as their physical interpretation is concerned. A new system of equations being different from that proposed by Biot was obtained by the author in 1977 on the basis of general principles of continuum mechanics [4]. The purpose of this study is to set the complete system of equations basing our considerations on the non-equilibrium thermodynamics. Earlier attempts along these lines were made by Zolotarev [9] as well by Deresiewicz and Pecker [3]. Zolotarev confined himself to media forwhich the compressibility of the interstitial phase is much smaller than that of the skeleton, in addition to which dynamic coupling between the phases and heat diffusion within each phase were neglected.

The thermo-mechanical coupling in the poroelastic medium turns out to be of much greater complexity than that in the classical case of an impermeable elastic solid, since besides thermal and mechanical interaction within each phase, thermal and mechanical coupling occurs between the phases. Thus, a mechanical or thermal change in one phase results in mechanical and thermal changes throughout the aggregate.

Deresiewicz and Pecker followed the equations of motion proposed by Biot [1], took the two phases to be at different temperatures at every point of the medium and next postulated local costitutive relations without justifying them. In order to interpret the coefficients in their relations they referred to Onsager's theory. In this paper we confine ourselves to the same temperatures throughout two phases at each point of the medium, but we shall construct the functions of state on the basis of the thermodynamics of irreversible processes. In this way we shall be able to define every coefficient in our theory, which is not clear enough in that proposed by Deresiewicz and Pecker.

Following Biot [1], we take our physical model to consist of a homogeneous, isotropic, elastic matrix whose interstices are filled with a compressible viscous liquid. Both, solid and liquid form continuous (and interacting) regions and, while viscous stresses in the liquid are neglected, the liquid is assumed capable of exerting a velocity--dependent friction force on the skeleton. The mathematical model consists of two superposed, continuous phases, each filling separatly the entire space occupied by the aggregate. Thus, at each point of space there are two distinct elements. Each of them is characterized by its own displacement and stress, but by the same temperature; during a thermo-mechanical process they may interact with each other.

It should be noted that the present treatment falls within the general theory of multiphase materials which, under the name "Theory of Mixtures", has been the subject of investigations from the standpoint of continuum mechanics for over two decades 6 . However, whereas the purpose there has been to set down the most general formulation encompassing the various phenomena, it is our aim, within the framework of the linear thermodynamics, to identify and demonstrate analytically specific aspects of the material behaviours.

1. BIOT'S EQUATIONS OF MOTION

In 1956 Biot [1] proposed a theory of a porous solid containing fluid. In order to obtain the equations of motion he made use of the Lagrange equations:

$$(1.1) \quad \frac{\partial}{\partial t}\left(\frac{\partial W_k}{\partial v_i}\right) + \frac{\partial W_d}{\partial v_i} = q_i , \qquad \frac{\partial}{\partial t}\left(\frac{\partial W_k}{\partial w_i}\right) + \frac{\partial W_d}{\partial w_i} = Q_i ,$$

where $\underset{\sim}{v}(v_i)$ and $\underset{\sim}{w}(w_i)$ denote the velocities of the solid skeleton and of the fluid, respectively; $\underset{\sim}{q}(q_i)$ and $\underset{\sim}{Q}(Q_i)$ are the generalized forces acting on the solid skeleton and the fluid; finally, W_k and W_d denote the kinetic energy and the dissipation function given by the formulae:

$$(1.2) \quad 2W_k = \rho_{11} v_l v_l + 2\rho_{12} v_l w_l + \rho_{22} w_l w_l ,$$

$$(1.3) \quad 2W_d = b(v_l - w_l)(v_l - w_l), \qquad b > 0,$$

where

$$\rho_{11} = \rho_s + \rho_a , \qquad \rho_{22} = \rho_f + \rho_a , \qquad \rho_{12} = -\rho_a < 0.$$

Here ρ_s and ρ_f are the masses of the skeleton and the fluid, respectively, referred to the unit volume of aggregate; ρ_{12} represents a mass coupling between fluid and solid. Moreover, W_d is the well-known Rayleigh dissipation function.

The generalized forces $\underset{\sim}{q}$ and $\underset{\sim}{Q}$ are expressed as the stress gradients

$$(1.4) \quad q_i = \sigma_{ij,j} , \qquad Q_i = \sigma_{,i} , \qquad \sigma = -pf,$$

where $\underset{\sim}{\sigma}(\sigma_{ij})$ is the stress tensor in the solid skeleton, p is the fluid pressure and f stands for the porosity of the medium, the same for the volume as well as for the surface in the theory under discussion.

By introducing (1.4) into (1.1) and using (1.2), (1.3), one obtains the following equations of motion:

$$(1.5) \quad \begin{aligned} \sigma_{ij,j} &= \rho_{11}\dot{v}_i + b(v_i - w_i) + \rho_{12}\dot{w}_i ; \qquad (\dot{\ }) = \frac{\partial}{\partial t} , \\ \sigma_{,i} &= \rho_{22}\dot{w}_i + b(w_i - v_i) + \rho_{12}\dot{v}_i . \end{aligned}$$

In the present paper we propose another theory of a porous elastic solid containing fluid in which the mass coupling (negative in Biot's theory) does not exist, and the mass coefficients entering into the kinetic energy have a quite new meaning.

2. KINEMATIC AND PHYSICAL COMPONENTS OF THE MEDIUM

We assume that the medium can be identified with a continuum in which the mathematical expectancy of finding a solid or a fluid particle at a given point is constant, i.e. we assume that the medium is statistically homogeneous. We also postulate that our medium is characterized by two porosities: the surface porosity f_A and the volume porosity f_Ω defined as follows:

$$(2.1)\quad f_A = \frac{A_p}{A}, \qquad f_\Omega = \frac{\Omega_p}{\Omega},$$

where A_p and Ω_p are the surface and the volume of pores contained, respectively, in a smooth surface A and a volume Ω.

Configurations in our medium at every instant t are defined by two real velocity fields: $\underset{\sim}{v}(x_i,t)$ and $\underset{\sim}{w}(x_i,t)$. The first velocity describes the motion of the skeleton with a certain amount of liquid imprisoned in it. The second velocity describes the motion of a free fluid. This situation can be explained by means of a simple model shown in Fig. 1, in which the shaded area represents the skeleton, and the dotted area the flui imprisoned in the skeleton, while the hatched area corresponds to the free fluid.

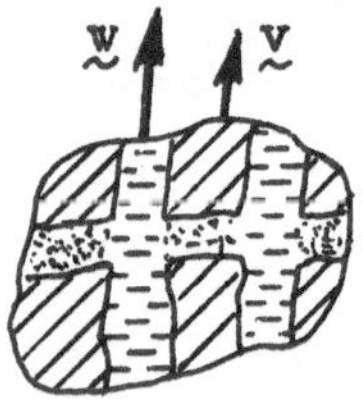

Fig. 1

The total mass density of the aggregate is equal to the sum

$$(2.2)\quad \rho = \rho_s + \rho_f .$$

The free fluid moving with its own velocity w has the mass density $\bar{}$ defined by the relation

$$(2.3)\quad \bar{\rho} = \rho_f f_A .$$

To this end the density of the mass moving with the velocity of the skeleton is equal to the difference: $(\rho - \bar{\rho})$.

The densities of mass introduced above correspond to the real velocities $\underset{\sim}{v}$ and $\underset{\sim}{w}$; we call them the kinematic components of the medium.

Let Ω denotes the volume of the aggregate bounded by a smooth surface A with the unit outward normal vector $\underset{\sim}{n}(n_i)$ (see Fig.2). For our volume the momenta of kinematic components may be written as follows:

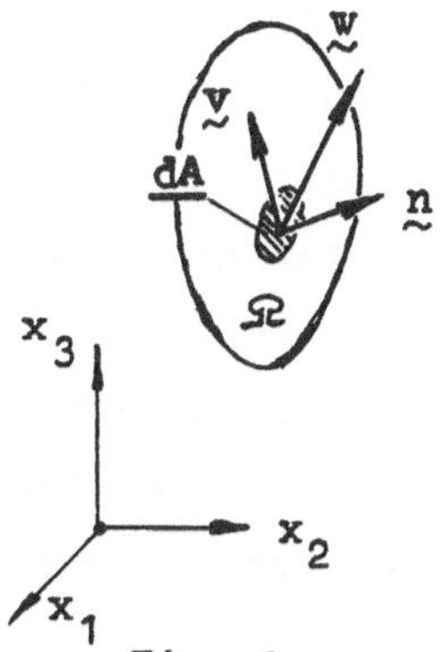

Fig. 2

(2.4) $\underset{\sim}{P}^s = \int_{(\Omega)} (\varrho - \bar{\varrho})\, \underset{\sim}{v}\, d\Omega\,, \qquad \underset{\sim}{P}^f = \int_{(\Omega)} \bar{\varrho}\, \underset{\sim}{w}\, d\Omega,$

where $P^s(P_i^s)$ and $P^f(P_i^f)$ are the momenta of the skeleton with non-free fluid and of the free fluid, respectively. The kinetic energy has the form:

(2.5) $2K = \int_{(\Omega)} [(\varrho - \bar{\varrho})\, v_1 v_1 + \bar{\varrho}\, w_1 w_1] d\Omega\,.$

In the case of physical components the appropiate momenta are:

(2.6) $\hat{\underset{\sim}{P}}^s = \int_{(\Omega)} \varrho_s \underset{\sim}{v}\, d\Omega, \qquad \hat{\underset{\sim}{P}}^f = \int_{(\Omega)} [\bar{\varrho}\underset{\sim}{w} + (\varrho_f - \bar{\varrho})\underset{\sim}{v}] d\Omega = \int_{(\Omega)} \varrho_f \hat{\underset{\sim}{w}}\, d\Omega\,,$

where $\hat{\underset{\sim}{w}}$ is the barocentric velocity of the fluid defined by the formula

(2.7) $\varrho_f \hat{\underset{\sim}{w}} = \bar{\varrho}\underset{\sim}{w} + (\varrho_f - \bar{\varrho})\underset{\sim}{v}.$

It is easily seen that the barocentric velocity of the fluid deals with the Biot's velocity of the fluid. In our considerations we make use of the kinematic components of the medium.

3. FIRST LAW OF THERMODYNAMICS. PRINCIPLE OF OBJECTIVITY.

As it was already mentioned we assme the temperature of the skeleton and the fluid to be the same at every point of the medium under consideration. we assmume that the medium is isotropic and statistically homogeneous. In such a medium we single out a certain volume Ω, large in the comparison with the dimenaions of the pores, bounded by a smooth surface A oriented outwards by a normal unit vector $\underset{\sim}{n}$. The action of the environment on the chosen volume in marked in the Fig.3, where $\underset{\sim}{T}^s$ and $\underset{\sim}{T}^f$ are the vectors of internal forces applied to the skeleton and the fluid, respectively, $\underset{\sim}{q}$ is the vector of the heat flux, $\underset{\sim}{X}$ is the vector of the body force. The energy balance in the chosen volume, which expresses the first law of thermodynamics, can be written as follows:

$$(3.1)\quad \dot{W}_k + \dot{U} = P + \dot{Q}\ ; \qquad (\dot{\ }) = \frac{d}{dt} = \frac{\partial}{\partial t} + v_1 \frac{\partial}{\partial x_1},$$

where W_k is the kinetic energy known from the previous section (2.5), U stands for the internal energy, P denotes the mechanical power, and Q is the heat.

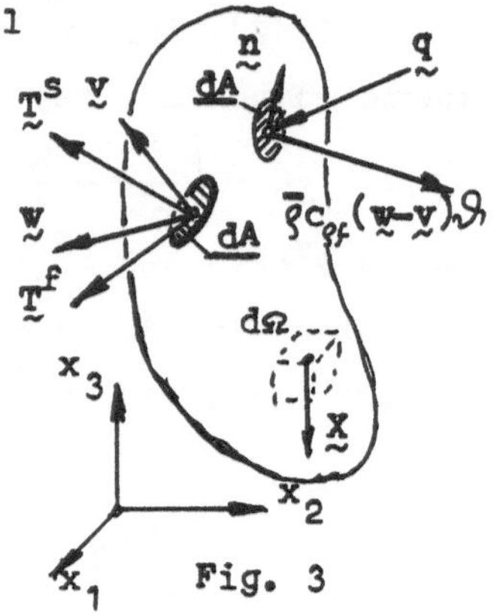

Fig. 3

The mechanical power is equal to the sum of the body forces power and that of the external forces, i.e.

$$(3.2)\quad P = \int_{(\Omega)} (\varrho - \bar{\varrho}) X_1 v_1 d\Omega + \int_{(\Omega)} \bar{\varrho} X_1 w_1 d\Omega + \int_{(A)} T_1^s v_1 dA + \int_{(A)} T_1^f w_1 dA,$$

where the surface forces define the Cauchy's formulae:

$$(3.3)\quad T_i^s = \sigma_{ij} n_j\ , \qquad T_i^f = \sigma n_i\ ; \qquad (\sigma_{ij}) = (\sigma_{ij}), \qquad \sigma = -f_A p\ ,$$

with symmetric stress tensor in the skeleton.

The internal energy, which is the attribute of the mass, express the relation

$$(3.4)\quad U = \int_{(\Omega)} (\varrho - \bar{\varrho}) U^{\varrho} d\Omega + \int_{(\Omega)} \bar{\varrho} U^{\varrho} d\Omega\ .$$

The non-mechanical power $\dot{Q}$ is a function of the heat flux $\underset{\sim}{q}$ per surface A and of the heat flux transported by the mass of free fluid flowing through the surface A. The mechanical power is calculated at constant entropy, whereas the non-mechanical power must be evaluated at the constant mass (constant volume) contained in the volume Ω . In accordance with our assumptions we find

$$(3.5)\quad \dot{Q} = -\int_{(A)} q_1 n_1 dA + \int_{(A)} \bar{\varrho} c_{\varrho f} (w_1 - v_1) \vartheta n_1 dA = -\int_{(\Omega)} \left[q_{1,1} - \bar{\varrho} c_{\varrho f} (w_1 - v_1)\, \vartheta_{,1} \right] d\Omega\ ,$$

where $c_{\varrho f}$ is the specific heat of the fluid per unit mass at constant volume and

$$(3.6)\quad \vartheta = T - T_o$$

is the temperature difference with T_o as the reference temperature or, in other words, the temperature of the natural state of the medium.

After a few regular transformation the equation of the first law shall take a form of an integral

$$(3.7)\quad \int_{(\Omega)} I(\varrho, \bar{\varrho}, v_i, w_i, \sigma_{ij}, \sigma, \dot{\varepsilon}_{ij}, \dot{\Theta})\,d\Omega = 0,$$

where

$$(3.8)\quad \dot{\varepsilon}_{ij} = \tfrac{1}{2}(v_{i,j} + v_{j,i}), \qquad \dot{\Theta} = w_{l,l}$$

is the strain-rate tensor for small deformations of the skeleton and the small volumetric change of the fluid, respectively.

The medium is assumed to be isotropic, the volume has been chosen in an arbitrary manner, and for this reason the integrant in the equality (3.7) must be equal to zero at every point. Thus obtained equation is the local form of the first law of thermodynamics, e.i.

$$(3.9)\quad \dot{U}_1 = -\left\{\frac{\partial(\varrho - \bar{\varrho})}{\partial t} + [(\varrho - \bar{\varrho})v_k]_{,k} + \frac{\partial\bar{\varrho}}{\partial t} + (\bar{\varrho}w_k)_{,k}\right\}U^{\varrho} +$$
$$- \frac{1}{2}\left\{\frac{\partial(\varrho - \bar{\varrho})}{\partial t} + [(\varrho - \bar{\varrho})v_k]_{,k}\right\}v_l v_l - \frac{1}{2}\left\{\frac{\partial\bar{\varrho}}{\partial t} + (\bar{\varrho}w_k)_{,k}\right\}w_l w_l +$$
$$+ [\sigma_{lk,k} + (\varrho - \bar{\varrho})X_l - (\varrho - \bar{\varrho})\dot{v}_l]v_l + [\sigma_{,l} + \bar{\varrho}X_l - \bar{\varrho}\dot{w}_l]w_l +$$
$$+ (\sigma_{lk}\dot{\varepsilon}_{lk} + \sigma\dot{\Theta}) - [q_{l,l} - \bar{\varrho}c_{\varrho f}(w_l - v_l)\vartheta_{,l}],$$

where U_1 is the internal energy per unit volume defined by the formula

$$(3.10)\quad \dot{U}_1 = \varrho\frac{\partial U^{\varrho}}{\partial t} + \varrho v_l U^{\varrho}_{,l} + \bar{\varrho}(w_l - v_l)U^{\varrho}_{,l}\ .$$

Let us now require, following the principle of objectivity, that the expression (3.9) be invariant with respect to the rigid motion of the medium. To this effect we consider a rectangular uniform motion

$$(3.11)\quad \underset{\sim}{v} \longrightarrow \underset{\sim}{v} + \underset{\sim}{a},$$

where $\underset{\sim}{a}$ is an arbitrary constant velocity vector. We assume that this motion does not lead to any changes of the quantities entering in in the first law of thermodynamics (3.9). In consequence we arrive at the equations:

$$(3.12)\quad \frac{1}{2}\left\{\frac{\partial(\varrho - \bar{\varrho})}{\partial t} + [(\varrho - \bar{\varrho})v_k]_{,k}\right\} + \frac{1}{2}\left[\frac{\partial\bar{\varrho}}{\partial t} + (\bar{\varrho}w_k)_{,k}\right] = 0,$$

$$(3.13)\quad \left\{\frac{\partial(\varrho - \bar{\varrho})}{\partial t} + [(\varrho - \bar{\varrho})v_k]_{,k}\right\}v_l + \left[\frac{\partial\bar{\varrho}}{\partial t} + (\bar{\varrho}w_k)_{,k}\right]w_l +$$
$$- [\sigma_{lk,k} + (\varrho - \bar{\varrho})X_l - (\varrho - \bar{\varrho})\dot{v}_l] - (\sigma_{,l} + \bar{\varrho}X_l - \bar{\varrho}\dot{w}_l) = 0,$$

which should be satisfied for every value of $\underset{\sim}{a}$ at every ponit of the medium. Thus, we obtain two local equations of the mass continuity:

$$(3.14)\quad \frac{\partial(\varrho - \bar{\varrho})}{\partial t} + [(\varrho - \bar{\varrho})v_k]_{,k} = \varrho^*, \qquad \frac{\partial\bar{\varrho}}{\partial t} + (\bar{\varrho}w_k)_{,k} = -\varrho^*,$$

where ϱ^* is a function of fluid sources caused by the difference between two fields of velocities. In other words, the function $\varrho^* = \varrho^*[(\underset{\sim}{w}-\underset{\sim}{v}),t]$ and satisfies the following evolution equation:

$$(3.15)\quad \frac{\partial \varrho^*}{\partial t} + [\varrho^*(w_k - v_k)]_{,k} = 0,$$

which seems to be natural one.

Introducing (3.14) into (3.13) we arrive to the following equations of motion:

$$(3.16)\quad \begin{aligned} &\sigma_{ij,j} + (\varrho - \bar{\varrho})X_i = (\varrho - \bar{\varrho})\dot{v}_i + \frac{1}{2}\varrho^*(w_i - v_i) + F_i ,\\ &\sigma_{,i} + \bar{\varrho}X_i = \bar{\varrho}\dot{w}_i + \frac{1}{2}\varrho^*(w_i - v_i) - F_i , \end{aligned}$$

where $\underset{\sim}{F}(F_i)$ is the force of interaction between two components ot the medium in consequence of their relative motion.

When introducing (3.14) and (3.16) into (3.9), we arrive at the local form of the first law of thermodynamics

$$(3.17)\quad \dot{U}_1 = (\sigma_{ij}\dot{\varepsilon}_{ij} + \sigma\dot{\Theta}) - [q_{k,k} - \bar{\varrho}c_{\varrho f}(w_k - v_k)\vartheta_{,k}] + F_k(w_k - v_k).$$

The equation (3.17) informs us that changes of internal energy are caused by the work done by the external forces (the first bracket), by the heat conducted and transported (the second bracket), and by the internal friction due to the relative motion between the skeleton and fluid (the third bracket).

4. SECOND LAW OF THERMODYNAMICS. DARCY'S LAW. HEAT CONDUCTION EQUATION.

For the time being we omit the souces of entropy (caused by the internal friction), which is marked by the asterisk, and we write the part of the entropy corresponding to the flow of heat $\dot{Q}$, i.e.

$$(4.1)\quad \dot{S}^* = -\int_{(\Omega)} \frac{q_{k,k}}{T}\,d\Omega + \int_{(\Omega)} \frac{\bar{\varrho}c_{\varrho f}(w_k - v_k)\vartheta_{,k}}{T}\,d\Omega .$$

The sources manifest themselves by their influence on the temperature. The entropy exchange with the surronding of the volume Ω by the way of heat is equal to the integral

$$(4.2)\quad \dot{S}_e = -\int_{(A)} \frac{q_k - \bar{\varrho}c_{\varrho f}(w_k - v_k)\vartheta}{T}\,n_k dA$$

and the irreversible part:

$$(4.3)\quad \dot{S}_i^* = -\int_{(\Omega)} \frac{[q_k - \bar{\varrho}c_{\varrho f}(w_k - v_k)\vartheta]\,\vartheta_{,k}}{T^2}\, d\Omega .$$

The changes of entropy (4.3) must locally satisfy the Causius-Duhem inequality, i.e.

$$-[q_k - \bar{\varrho}c_{\varrho f}(w_k - v_k)\vartheta]\,\vartheta_{,k} \geq 0,$$

which implies the heat conduction law

$$(4.4)\quad q_i = -\lambda\vartheta_{,i} + \bar{\varrho}c_{\varrho f}(w_i - v_i)\vartheta; \qquad \lambda > 0.$$

Just obtained law includes the influence of the heat transport caused by the relative motion of the free fluid.

The total rate of changes of the entropy has to satisfy the relation

$$\dot{S}_i = \dot{S} - \dot{S}_e \geq 0,$$

which implies the local inequality

$$(4.5)\quad \dot{s}_{i1} = \dot{s}_1 + \left[\frac{q_k - \bar{\varrho}c_{\varrho f}(w_k - v_k)\vartheta}{T}\right]_{,k} \geq 0.$$

Now, we make use of the first law of thermodynamics (3.17) and rewrite our inequality as follows:

$$(4.6)\quad T\dot{s}_1 - \dot{U}_1 + (\sigma_{ij}\dot{\mathcal{E}}_{ij} + \sigma\dot{\Theta}) + F_k(w_k - v_k) + \\ - \frac{[q_k - \bar{\varrho}c_{\varrho f}(w_k - v_k)\vartheta]\,\vartheta_{,k}}{T} \geq 0.$$

We introduce the Helmholtz free energy $F_1 = F_1(\mathcal{E}_{ij}, \Theta, T)$ making use of Legendre transformation

$$(4.7)\quad F_1 = U_1 - Ts_1,$$

which transforms our inequality in the form depending on the free energy. This last form permits to constat that it is always satisfied, if it holds for the following equalities:

$$(4.8)\quad \sigma_{ij} = \frac{\partial F_1}{\partial \mathcal{E}_{ij}}, \qquad \sigma = \frac{\partial F_1}{\partial \Theta}, \qquad s_1 = -\frac{\partial F_1}{\partial T},$$

and for one inequality:

$$(4.9)\quad F_k(w_k - v_k) - \frac{q_k - \bar{\varrho}c_{\varrho f}(w_k - v_k)\vartheta}{T}\,\vartheta_{,k} \geq 0.$$

Its first term is related to the internal heat sources caused by the friction between the fluid and skeleton due to the relative motion. The second therm was already discussed and it implies the heat conduction law (4.4). In order to simplify our considerations we suppose that evry term has to satisfy the inequality independently. In the oposite case one might be obtained some secondary effects as, for example, a generalization of so-called Dufour or Soret effects. Taking into account our simplification we write

(4.10) $F_k(w_k - v_k) \geq 0.$

Consequently we can write

$$F_i = \sum_{n=1,3,.}^{N} b_n(w_i - v_i)^n ; \qquad b_n \geq 0.$$

Therefore, the linear relation has the form

(4.11) $F_i = b(w_i - v_i).$

When omitting body and inertia forces, as well as, the thermal effects in the second of equations (3.16), we arrive at the relation

(4.12) $F_i = \sigma_{,i} = b(w_i - v_i),$

known as the Darcy's law. The coefficient b corresponds here to the flow resistance. It is obvious that the validity of the relation (4.12) is limited by suitable Reynolds number.

The free enrgy is a scalar function; so it must be a function of the invariants of strain state and temperature. Our main aim is to present a theory as simple as possible. Therefore, we take into account only two first invariants of the strain state and write:

(4.13) $F_1 = F_1(I_1, D_2, \Theta, T); \qquad I_1 = \mathcal{E}_{kk} = \mathcal{E}, \quad D_2 = I_2 - \frac{1}{3} I_1, \quad I_2 = \mathcal{E}_{ij}\,\mathcal{E}_{ij}.$

We expend the free energy (4.13) into a Taylor series in the vincinity of the natural state with respect to the kinematic variables neglecting terms of higher order than squares. Next, taking into account the formulae (4.8) we write the relations linking the stresses and strains in the medium:

(4.14) $\sigma_{ij} = 2N\mathcal{E}_{ij} + \left(K\mathcal{E} + L\Theta + \frac{\partial F_1}{\partial I_1}\right)\delta_{ij}, \qquad \sigma = M\Theta + L\mathcal{E} + \frac{\partial F_1}{\partial \Theta},$

where

(4.15)

$K = \dfrac{\partial^2 F_1(0,0,0,T)}{\partial I_1^2}$ – the moduli of the skeleton volumetric deformation;

$L = \dfrac{\partial^2 F_1(0,0,0,T)}{\partial I_1 \partial \Theta}$ – the coupling coefficient of the skeleton and fluid volumetric deformations;

$N = \dfrac{\partial F_1(0,0,0,T)}{\partial D_2}$ – the shear moduli of the skeleton;

$M = \dfrac{\partial^2 F_1(0,0,0,T)}{\partial \Theta^2}$ – the moduli of the fluid volumetric deformation.

Let us suppose that the aggregate has the possibility of free thermal deformations and is submitted to the action of a temperature field only. In such a case $\sigma_{ij} = 0$, $\sigma = 0$, and the relations (4.14), with known thermal dilatations:

$$\varepsilon_T = 3\alpha_s \vartheta, \qquad \Theta_T = 3\alpha_f \vartheta,$$

where α_s and α_f are the coefficients of the linear thermal expansions of the skeleton and fluid, respectively, permit to write:

$$(4.16)\qquad \begin{aligned} -\frac{\partial F_1}{\partial I_1} &= [(2N + 3K)\alpha_s + 3L\alpha_f]\vartheta = \gamma_s \vartheta , \\ -\frac{\partial F_1}{\partial \Theta} &= 3(L\alpha_s + M\alpha_f)\vartheta = \gamma_f \vartheta . \end{aligned}$$

The first term of the series, i.e. $F_1(0,0,0,T)$ is determined from the relation known in thermodynamics:

$$c_\Omega = c_{\varepsilon,\Theta} = T\left(\frac{\partial S_1}{\partial T}\right)_{\varepsilon,\Theta} = - T\left(\frac{\partial^2 F_1}{\partial T^2}\right)_{\varepsilon,\Theta} ,$$

where $c_\Omega = c_{\varepsilon s} + c_{\Theta f}$ is the sum of the specific heats of the components of the medium per unit volume. The integration of above relation with respect to temperature leads to the evaluation of $F_1(0,0,0,T)$:

$$(4.17)\qquad F_1(0,0,0,T) = - \int_{T_0}^{T} dT \int_{T_0}^{T} \frac{c_\Omega}{T}\, dT.$$

We have already interpreted all coefficients in the function of free energy and now we can write all introduced functions of state in the evident form. Following the regular practice we arrive to the last sought equation, namely, to the equation of heat conduction:

$$(4.18)\qquad \lambda \nabla^2 \vartheta + W_0 = c_\Omega \dot{\vartheta} + \frac{c_{\sigma s} - c_{\varepsilon s}}{3\alpha_s} \dot{\varepsilon} + \frac{c_{\sigma f} - c_{\Theta f}}{3\alpha_f} \dot{\Theta} ,$$

where

$$(4.19)\qquad W_0 = b(w_k - v_k)(w_k - v_k)$$

is the source function.

REFERENCES

1. Biot M.A., J. Acoust. Soc. Am., 28, 108 (1956).
2. Biot M.A., J. Acoust. Soc. Am., 34, 1254 (1963).
3. Deresiewicz H., Pecker C., Acta Mech., 16, 45 (1973).

4. Derski W., Bull. Acad. Pol. Sci., cl.IV, 11, 26 (1978).
5. Derski W., Rock and Soil Mechanics, Elsevier (1988).
6. Green A.E., Craine R.E., Naghdi P.M., Q. J. Mech. Appl. Math., 23, 171 (1970).
7. Keller J.U., Non-Equilib. Thermod., 1, 67 (1976).
8. Scheidegger A.E., The Physics of Flow through Porous Media, McMillan (1960).
9. Zolotarev P.P., Inzh. Zb., 5, (1965).
10. Zwikker C., Kosten C.W., Sound Absorbing Materials, Elsevier (1949).